Power Systems

More information about this series at http://www.springer.com/series/4622

Siddhartha Kumar Khaitan · James D. McCalley
Chen-Ching Liu
Editors

Cyber Physical Systems Approach to Smart Electric Power Grid

Springer

Editors
Siddhartha Kumar Khaitan
Electrical and Computer Engineering
Iowa State University
Ames Iowa
USA

James D. McCalley
Electrical and Computer Engineering
Iowa State University
Ames Iowa
USA

Chen-Ching Liu
Energy Systems and Innovation Center (ESIC)
Washington State University
Pullman
Washington
USA

ISSN 1612-1287 ISSN 1860-4676 (electronic)
Power Systems
ISBN 978-3-662-45927-0 ISBN 978-3-662-45928-7 (eBook)
DOI 10.1007/978-3-662-45928-7

Library of Congress Control Number: 2014958754

Springer Heidelberg New York Dordrecht London

Printed on acid-free paper

Springer-Verlag GmbH Berlin Heidelberg is part of Springer Science+Business Media
(www.springer.com)

Preface

Cyber-physical systems (CPSs) are the class of systems that offer close interaction between cyber and physical components. Due to deep intertwining of physical and software components, CPSs feature high degree of automation, integration at multiple temporal and spatial scales and reconfiguring dynamics. CPSs are expected to play a major role in the development of next-generation systems ranging from handheld devices to warehouse-scale systems. For this reason, CPSs have been actively researched in the recent years. This book brings together the recent advancements, implementation and future directions in the field of cyber physical systems. The book covers several aspects related to CPSs, such as modeling and simulation, resiliency, security, design of network infrastructure and the application of cyber-physical design paradigm in domains of practical interest, such as power system.

Following is a brief introduction to the chapters in the book.

The chapter entitled, "Modeling and Simulation of Network Aspects for Distributed Cyber-Physical Energy Systems" presents an aspect-oriented approach to modeling smart grid communication, fault modeling and timing concerns on a distributed state estimation application study. The prototyping of the approach is performed using the heterogeneous modeling environment Ptolemy II. The simulation results highlight the benefits of aspect-oriented models for network and middleware scalability in cyber-physical energy system models.

The chapter entitled, "A Service-Oriented, Cyber-Physical Reference Model for Smart Grid" presents a cyber-physical reference model for smart grid. The proposed reference model is based on service-oriented computing paradigm and is also capable of handling the hard real-time aspects of smart grid. The chapter also presents the development steps of a smart grid application according to the proposed reference model.

The chapter entitled, "Real Time Modeling and Simulation of Cyber-Power System" discusses the challenges in the development of cyber-power testbed. It also presents realization of a cyber-power testbed by integration of Real Time Digital Simulator (RTDS) and Network Simulator 3 (NS3). The studies conducted on the testbed help in understanding the complex relationship between cyber and physical domains, and the impacts of an attack on the cyber-physical power system.

The chapter entitled, "Cyber physical Approach to HVDC grid control" presents a cyber-physical approach to design of HVDC (high voltage direct current) control system architectures. It further describes the communication system architectures needed for centralized and distributed operation and control of HVDC grids. It also describes two applications for distributed control of DC grids which utilize the concepts presented in the chapter.

The chapter entitled, "Smart Buildings in the Smart Grid: Contract-Based Design of an Integrated Energy Management System", presents a compositional framework for implementing an optimal supply-following strategy in a cyber-physical power grid supplying services for an HVAC (heating, ventilation and air conditioning) system for a building. It uses the concept of assume-guarantee contracts to formalize the requirements of the grid and the building subsystem. Using this, an optimal control scheme for both the grid and the building can be formulated.

The chapter entitled, "Decision-Support Tools for Renewables-Rich Power Systems: A Stochastic Futures Approach" presents a framework and algorithm set for day-ahead generation scheduling which utilizes the coupling between cyber- and physical- resources in the power system. It also presents a method for unit scheduling for the day-ahead market based on the probabilistic renewable energy generation futures.

The chapter entitled, "Cyber security of smart grid communications: risk analysis and experimental testing" presents a cyber-security evaluation of active distribution grids which feature a high level penetration of distributed renewable sources. The chapter focuses on voltage control in medium voltage grids connecting distributed energy resources with the aim of analyzing the most relevant security scenarios of a system which implements such control application.

The chapter entitled, "Reliable and Scalable Communication for the Power Grid" presents methods for developing network topologies of smart devices for enabling multi-route discovery in a power grid. Since multi-route pathways ensure timely delivery of critical data, the methods presented in the chapter enable the design of reliable and scalable cyber network structure in a power grid.

The chapter entitled, "Biologically Inspired Hierarchical Cyber-Physical Multi-Agent Distributed Control Framework For Sustainable Smart Grids" presents a hierarchical framework based on flocking theory for improving stability of a cyber-physical smart grid. It also presents strategies for leveraging physical coupling of the system for designing a control scheme with the aim of identifying and mitigating data corruption and attacks.

The chapter entitled, "Cyber-Physical Security of Substations in a Power Grid" presents an approach to enhance the cyber security of substations in a cyber-physical power system. Using a cyber-physical substation testbed, it evaluates the resilience of the substations towards various vulnerabilities and intrusion scenarios.

The chapter entitled, "Cyber-Attacks in the Automatic Generation Control" analyzes the impact of a cyber-attack on the AGC (automatic generation control) signal. It uses reachability and optimal control theory to study the the existence of an attack pattern that can disturb the power system. Dynamic simulations on a IEEE-118 bus network are performed to show the performance of their methodology.

The chapter entitled, “Intrusion Detection for CPS Real-Time Controllers” presents mechanisms for time-based intrusion detection which work by detecting the execution of unauthorized instructions in real-time CPS environments. The chapter also develops techniques to detect intrusions in a self-checking manner by the application and through the operating system scheduler.

The chapter entitled, “Against Data Attacks on Smart Grid Operations: Attack Mechanisms and Security Measures” discusses data attack models and techniques for enabling fast detection of such attacks. It also presents several cost-effective countermeasures using data encryption, access authentication and meter protection.

Students, researchers, system-designers and professionals interested in cyber-physical system approach to power grids will find this book very useful and timely. We hope that this book will inspire even greater efforts in the area of design and management of cyber-physical systems.

Fall 2014 (USA)

Siddhartha Kumar Khaitan
Ames, IA, USA
James D. McCalley
Ames, IA, USA
Chen-Ching Liu
Ames, IA, USA

List of Contributors

Akkaya, Ilge
Electrical Engineering and Computer Science Department
University of California, Berkeley
California, USA

Andersson, Goran
Power Systems Laboratory
ETH Zurich
Zurich, Switzerland

Babazadeh, Davood
Department of Industrial Information and Control Systems
KTH - Royal Institute of Technology
Stockholm, Sweden

Bhat, Balasubramany
Department of Computer Science
North Carolina State University
North Carolina, USA

Biswas, Saugata
School of Electrical Engineering and Computer Science
Washington State University
Washington, USA

Chen, Ying
Department of Electrical Engineering
Tsinghua University
Beijing, China

Donde, Vaibhav
Pacific Gas and Electric
California, USA

Dondossola, Giovanna
Ricerca sul Sistema Energetico - RSE Spa.
Dpt T&D Technologies
Milan, Italy

Esfahani, Peyman Mohajerin
Department of Information Technology and Electrical Engineering
ETH Zurich
Zurich, Switzerland

Govindarasu, Manimaran
Department of Electrical and Computer Engineering
Iowa State University
Iowa, USA

Grijalva, Santiago
School of Electrical and Computer Engineering
Georgia Institute of Technology
Georgia, USA

Jiang, Jiayi
School of Electrical Engineering and Computer Science
Washington State University
Washington, USA

Hong, Junho
School of Electrical Engineering and Computer Science
Washington State University
Washington, USA

Kim, Jinsub
Oregon State University
Kundur, Deepa
Department of Electrical and Computer Engineering

Department of Electrical and Computer Engineering
University of Toronto
Ontario, Canada

Lee, Edward A.
Electrical Engineering and Computer Science Department
University of California, Berkeley
California, USA

Liu, Chen-Ching
School of Electrical Engineering and Computer Science
Washington State University
Washington, USA

Liu, Juhua
ABB US Corporate Research
North Carolina, USA

Liu, Ren
School of Electrical Engineering and Computer Science
Washington State University
Washington, USA

Liu, Yan
Electrical Engineering and Computer Science Department
University of California, Berkeley
California, USA

Lygeros, John
Automatic Control Laboratory
ETH Zurich
Zurich, Switzerland

Maasoumy, Mehdi
Department of Electrical Engineering and Computer Sciences
University of California, Berkeley
California, USA

Margellos, Kostas
Department of Industrial Engineering and Operations Research
University of California, Berkeley
California, USA

Mohan, Sibin
Information Trust Institute
University of Illinois at Urbana-Champaign
Illinois, USA

Mueller, Frank
Department of Computer Science
North Carolina State University
North Carolina, USA

Nordstrom, Lars
Department of Industrial Information and Control Systems
KTH - Royal Institute of Technology
Stockholm, Sweden

Nuzzo, Pierluigi
Department of Electrical Engineering and Computer Sciences
University of California, Berkeley
California, USA

Roy, Sandip
School of Electrical Engineering and Computer Science
Washington State University
Washington, USA

Sangiovanni-Vincentelli, Alberto
Department of Electrical Engineering and Computer Sciences
University of California, Berkeley
California, USA

Srivastava, Anurag
School of Electrical Engineering and Computer Science
Washington State University
Washington, USA

Tariq, Muhammad Umer
School of Electrical and Computer Engineering
Georgia Institute of Technology
Georgia, USA

Terruggia, Roberta
Ricerca sul Sistema Energetico - RSE Spa.
Dpt T&D Technologies
Milan, Italy

Tong, Lang
School of Electrical and Computer Engineering
Cornell University
New York, USA

Vellaithurai, Ceeman B
Schweitzer Engineering Laboratories, Inc.
Washington, USA

Vrakopoulou, Maria
Power Systems Laboratory
ETH Zurich
Zurich, Switzerland

Wei, Jin
Department of Electrical and Computer Engineering
The University of Akron
Ohio, USA

Wolf, Marilyn
School of Electrical and Computer Engineering
Georgia Institute of Technology
Georgia, USA

Zimmer, Christopher
Department of Computer Science
North Carolina State University
North Carolina, USA

Contents

Modeling and Simulation of Network Aspects for Distributed Cyber-Physical Energy Systems

Ilge Akkaya, Yan Liu, and Edward A. Lee

Abstract. Electric power grids are presently being integrated with sensors that provide measurements at high rates and resolution. The abundance of sensor measurements, as well as the added complexity of applications trigger a demand for cyber-physical system (CPS) modeling and simulation for evaluating the characteristics of appropriate network fabrics, timing profiles and distributed application workflow of power applications. Although simulation aids in the pre-deployment decision making process, system models for complex CPS can quickly become impractical for the purposes of specialized evaluation of design aspects. Existing modeling techniques are inadequate for capturing the heterogeneous nature of CPS and tend to inherently couple orthogonal design concerns. To address this issue, we present an aspect-oriented modeling and simulation paradigm. The aspect-oriented approach provides a separation between functional models and cross-cutting modeling concerns such as network topology, latency profiles, security aspects, and quality of service (QoS) requirements. As a case study, we consider a three-area smart grid topology and demonstrate the aspect-oriented approach to modeling network and middleware behavior for a distributed state estimation application. We also explore how aspects leverage scalable co-simulation, fault modeling, and middleware-in-the loop simulation for complex smart grid models.

1 Introduction

Emerging cyber-physical energy systems (CPES) broadly depend on high-throughput sensor measurements to execute distributed control tasks. Integration of *smart* sensors

Ilge Akkaya · Edward A. Lee
University of California, Berkeley
e-mail: {ilgea,eal}@eecs.berkeley.edu

Yan Liu
Concordia University, Montreal, Canada
e-mail: yan.liu@concordia.ca

S.K. Khaitan, J.D. McCalley, and C.-C. Liu (eds.), *Cyber Physical Systems Approach to Smart Electric Power Grid*, Power Systems, DOI: 10.1007/978-3-662-45928-7_1

into the grid enables high-fidelity and trustworthy data to become available at monitoring centers in real-time, which has the potential benefit of dramatically improving the accuracy of existing contingency and state estimation applications, and enabling novel grid control techniques.

Traditionally, the primary concern of power engineers has been to provide correct and efficient design of algorithms that run on power grid hardware. Given the unprecedented volumes of streaming data available on the smart grid, this requirement alone is no longer sufficient. Data volume combined with real-time aggregation and processing requirements of applications bring about a unique challenge for existing communication and data management tools. Deployment of decentralized computation resources and coordination among these become key to realizing the next generation data intensive real-time tasks on the grid [9].

The communication infrastructure for distributed application management encompasses middleware, communication networks, and software platforms. Responsiveness and scalability play an essential role within this framework not only for efficiency, but also for the correctness of the distributed applications. Such applications include distributed state estimation, contingency analysis and grid stability monitoring, which rely on real-time sensor measurements produced at geographically distributed synchrophasor devices.

The qualitative evaluation of this ubiquitous application domain requires systematic modeling formalisms that enable abstraction of system dynamics. The modeling challenges are threefold: (i) defining modeling formalisms, which provide multi-view models that enable a separation of cross-cutting concerns, such as communication fabrics, and implementation-specific middleware characterizations; (ii) composing multiple models in one heterogeneous simulation environment or equivalently, enabling determinate composition of computation models; (iii) correctly representing application-specific timing characteristics of the distributed applications by the aid of a common notion of time among distributed model components.

This paper studies a network modeling approach for CPES infrastructure, which leverages separation of cross-cutting concerns in the system, and consequently enables development of scalable and easy-to-analyze CPES models. We demonstrate the benefit of aspect-oriented modeling (AOM) that presents a separation of power domain applications and implementation-specific aspects such as network communication and middleware for fine-grained data coordination. We prototype the AOM methodology using the actor-oriented modeling and simulation environment Ptolemy II (4.1). Ptolemy II is a framework that specializes in addressing intrinsic CPS challenges such as timing, concurrency and heterogeneous composition and has been extensively used for industrial CPS design [11, 25].

To demonstrate the vast capabilities of the aspect-oriented CPES models, we consider a distributed state estimation (DSE) application, which operates on time-synchronized phasor data collected from the transmission grid. We perform simulation studies on a three-area distributed grid architecture, composed with a prototype network and middleware infrastructure, and explore how AOM facilitates design and analysis of network and middleware requirements for time-critical distributed applications. We present analysis and results that demonstrate a vast set of potential benefits of AOM for network and middleware design for distributed CPES applications.

The rest of the chapter is organized as follows: We survey recent developments on CPES modeling and simulation tools and methodologies in Section II, followed by the detailed analysis of application requirements for CPES applications in Section III. In Section IV, we introduce the main modeling modeling approach and explain the aspect-oriented modeling methodology for the energy systems domain. Section V presents simulation results with emphasis on temporal characterization of model execution. We finally discuss possible extensions in Section VI and provide concluding remarks in Section VII.

2 Related Work

The use of computer models for power system analysis have been an actively investigated research topic for several decades. Early models and software for transmission and distribution systems emerged when computers started to become popular for business purposes. Following the rapid development in distributed and parallel computing, distributed power system modeling and analysis has become a widely studied research topic of the last decade [10, 30, 21, 20].

Existing electric power simulators provide well-proven point tools for transmission networks (e.g., Siemens PSS/E) and distribution networks (e.g., GridLab-D [8], general optimal power flow [32]). Recent work has also focused on enabling power system and network co-simulation. GridSpice [https://code.google.com/p/gridspice/], which composes several projects (such as GridLab-D and MatPower) in a single simulation package, provides a simulation environment that allows modeling the interactions between all parts of the electrical network including generation, transmission, distribution, storage and load models [3]. GridSpice builds upon a cloud-based architecture, in which the client user interface is accessed through the Google App Engine and the simulation package is deployed on a public cloud such as Amazon EC2. This cloud based architecture is able to handle the increasing computational demand of power system simulation even for large-scale smart grid scenarios. The network modeling aspect has not yet been part of simulation.

Co-simulation has also been addressed by several research studies recently, which have integrated continuous system simulators (e.g., OpenDSS, PSLF) with a Discrete-Event network simulator (e.g., NS2, OPNET) to simulate cyber-physical smart-grid behavior [15, 26].

Several approaches utilize the Common Information Model (CIM) as a model to combine several traditional power system analyses [4, 23]. These approaches focus on the interoperability of different energy management subsystems. CIM contains standard-based entities and attributes to present the semantics of power system entities and their organization. [23, 18].

The aspect-oriented modeling paradigm in Ptolemy II builds upon concepts that have first been introduced in the Metropolis project [5], in which the use of *Quantity*

Managers have introduced a mechanism to annotate and synchronize different model resources, leading to the intrinsic separation of concerns (SoC) within the model. AOM has also been studied in the context of the Unified Modeling Language (UML) [13, 24], however, the concept of an *aspect* in Ptolemy II provides a fundamentally different approach to the AOM paradigm. The Ptolemy implementation of AOM offers an actor-oriented mechanism to associate aspects with executable models that are defined by deterministic concurrent semantics tied to one or more of many supported models of computation (MoC) [29]. The details of AOM in Ptolemy II will be discussed further in Sections 4.1-4.2.

3 Application Requirements and Modeling Challenges

In recent years, power grids have undergone dramatic changes in the fields of grid monitoring and control due to the increasing number of phasor measurement units (PMUs) deployed to produce real-time synchrophasor data that capture the power system dynamics. The swarm of synchrophasors that actively fetches real-time data, which is then utilized for supervisory control in the distributed power grid, has become the root cause of the need to reevaluate the entire data flow design of the grid. PMUs generate precisely time stamped measurements at rates that typically range between 10-60 samples per second. This high data rate enables power applications to operate at significantly higher frequencies compared to traditional Supervisory Control And Data Acquisition (SCADA) based applications.

Wide Area Monitoring and Control (WAMC) systems benefit from this dramatic increase in the rate and redundancy of grid measurements. One such WAMC application is state estimation. State estimation collects field measurements and solves the system-wide nonlinear state equations based on redundant PMU measurements to estimate the system state variables at each iteration. The results are estimates of state variables in the grid, e.g., voltage magnitude, power injection, power flow, and power factors. These estimates are critical inputs for related power system operational tools, such as contingency analysis, optimal power flow analysis, economic dispatch and automatic generation control.

Combining high-rate phasor measurements with low-rate SCADA data for distributed state estimation has been an emerging research interest of power engineers [16, 19, 22]. The volumes of available PMU data has been a major improvement to the quality of WAMC applications, however, at the expense of increased load at data centers, network fabric, and computation nodes.

In addition to the real-time requirements on data processing, phasor data has also imposed constraints on historian components. Given a control center with hundreds of PMUs installed, archived sensor data can accumulate to the order of terabytes, within only a 30-day period [14]. In effect, it becomes extremely inefficient to utilize a single centralized coordinator to collect and archive all the available data from corresponding balancing areas. One approach to alleviating the burden on computing resources has been to distribute the computation across sub-areas.

At run time, the power system dynamics require data flow to be coordinated for the implementation of algorithms that utilize the data. For instance, varying sensor data rates and imperfect clocks at the sensor end requires data from different sensors to be time-synchronized at the *middleware* layer. Hence, the underlying infrastructure (including middleware and network fabric) plays a key role in satisfying both the functional and non-functional requirements of the power application. We use the term *functional model* to describe the designed behavioral model of a system, whereas the remaining model may be interpreted as the implementation-specific configuration details. The functional requirements include coordination of the data flow to distributed state estimators by data synchronization, aggregation and coordination of multiple data source-destination pairs. The non-functional requirements include the following *aspects*:

- *Scalability*. The middleware needs to support a large number of sensor devices and their intercommunications since they become a significant factor in determining the temporal and functional properties of distributed applications that consume data streams.
- *Low latency and time-predictability*. Optimizing the worst-case latency is particularly important to meet deadlines of time-sensitive applications. Previous work has demonstrated that heavy-tailed latency behavior in middleware is directly observable in the end-to-end completion times of distributed algorithms that require synchronization of an extensive number of sensor streams [1, 2]. Even a simple rule of coordination based on enumerating the expected number of data streams can cause up to 45% overhead in the middleware layer.
- *Reconfigurability*. Power engineers are often required to revise algorithms to improve the accuracy of WAMC applications. The data flow through the power grid consequently needs to be revised to work with the reconfigurations. The distributed programming software should allow adding new computation nodes, revising current components and integrating additional sensor streams into the distributed system. It is required for the modeling methodology to take into consideration this reconfigurable behavior of distributed power applications.

Given the application requirements, a monolithic model representation of distributed applications has an immediate disadvantage of quickly leading to a complex model that blends functional models with implementation details of middleware and communication infrastructure. The monolithic model would intrinsically couple the sensor-to-node data path with the communication infrastructure, which is a poor design choice as it entangles independent design aspects and tremendously increases model complexity. Moreover, such models develop to be neither scalable nor reconfigurable, especially when the distributed model topology is to be altered.

The modeling approach presented in Section 4.3 demonstrates an aspect-oriented CPES design methodology that separates functional application entities and communication infrastructure. Each model can further be customized with application-specific timing characteristics and an appropriate computation model in an actor-oriented environment.

4 Modeling Approach

Power system dynamics and the communication layer mutually affect one another in a closed-loop distributed sensing and control environment, constraining these two systems to be designed as a combined CPS. Nonetheless, a unified approach that encapsulates continuous dynamics, discrete-event communication models, network layer requirements, possible fault tolerance, and QoS requirements in a single model would be too complex to be useful in any form for system designers.

In the light of these requirements, we present an *aspect-oriented modeling* methodology, which enables the overall design to provide a clear separation of cross-cutting concerns from component interactions in the model. We prototype the AOM approach in Ptolemy II.

Ptolemy II is a heterogeneous actor-oriented design environment that supports hierarchical composition of computation models, which enables continuous dynamics to be integrated with the discrete-event AOM framework. Also, there exists extensive work on enabling Functional Mock-up Units (FMUs) for determinate co-simulation and various other applications that are supported by the Ptolemy framework [6, 31]. The rest of this chapter will study a promising approach to modeling complex network and middleware aspects of complex energy systems. Since the focus of this chapter is mainly the aspect-oriented composition of cross-cutting modeling concerns, integration of continuous system dynamics to the framework will not be discussed in depth, however, it will be noted that this capability has already been studied in the context of Ptolemy II environment [12, 27].

4.1 Ptolemy II

Ptolemy II is an actor-oriented modeling and simulation tool for heterogeneous system design [29]. Actors in Ptolemy II are concurrently executed components that communicate via messages called *events* sent and received through actor *ports*.

An *actor* in Ptolemy is annotated with a set of parameters, input and output ports along with an inner state representation which itself can be a graphical submodel. Actor execution at a particular level of the model hierarchy is governed by a component named a *director*, which implements a desired MoC. Figure 1a demonstrates an example model that represents two sensor clusters that transfer data streams through the communication network. The streams are later processed at high performance computing (HPC) nodes. A middleware component connects these two nodes and facilitates intermediate data exchange. In Figure 1, the Discrete-Event (DE) Director is used in the top-level model, enforcing DE semantics on events produced by the PMUCluster actors.

Ptolemy also provides entities called *decorator*s, which are objects that can be associated with other objects as parameters and provide services to those objects they are associated with. *Aspects* in Ptolemy, which are discussed in the following section, rely on the *decorator* mechanism to provide services to the actors they are associated with.

4.2 Aspect-Oriented Modeling in Ptolemy II

Ptolemy is an actor-oriented framework. Modeling cross-cutting concerns in an actor-oriented environment entails an unambiguous syntax for associating an aspect with an actor object. In the context of CPES models, we concentrate on *communication aspects* that provide cross-cutting refinements to the network resource contention and inter-component behavior modeling. In a Ptolemy model, a *communication aspect* is associated with other model components via parameters tied to actor ports [7].

As an example, in Figure 1b, a subsystem network model that features two network aspects have been presented. The input ports of `Node1` and `Node2` actors are associated with the `NetworkModel` aspect, as annotated on the incoming links to the regarding ports. This association implies that any event destined to these ports would first be forwarded to the `NetworkModel` for processing, before they are routed back to the original destination. In Figure 1b, any event sent from `PMUCluster1` to `Node1` are routed to port `A1` of the `NetworkModel`. Upon being merged with events from port `A2` and being processed by a `Server` actor in time stamp order, the events are routed to their original destinations in the functional model, i.e, the `Node1` actor. The `NetworkModel` aspect aims at modeling the resource contention of the single server that is processing messages from two different data sources. Note that a communication aspect itself is an actor with input and output ports, and may have an internal composite actor representation, as in this example. It is common for a communication aspect to alter time stamps and values of incoming events before relaying these to their respective destination ports. In this example, the `Server` actor, which is part of the `NetworkModel` definition, applies a constant processing delay to incoming events, while the payloads of the events remain unaffected.

As seen in the dialogue box in Figure 1b, the association of the `Node` actors with the `NetworkModel` aspect is realized by an `enable` parameter of the input port. The process for associating the `Middleware` aspect with `Node` actors follows the same pattern.

Following the introduction of the AOM semantics, it would provide insight to compare the models given in Figures 1a and 1b. Note that these two models consist of the same set of actors, which are connected in a different topology in the two realizations. The `NetworkModel` and `Middleware` actors appear as regular actors that belong to the functional model in the first case, whereas they have been implemented as communication aspects in the second variant. Note that the internal functional representation of these two actor blocks remain identical across 1a-1b. The former approach, illustrated in in Figure 1a serially incorporates network and middleware behavior into the functional model. Specifically, Figure 1a is ambiguous as in whether `PMUCluster1` and `PMUCluster2` are both communicating with `Node1` and `Node2`, or the `NetworkModel` is providing a peer-to-peer (P2P) channel between pairs of actors. This is not a feasible design choice due to two reasons: (i) network and middleware components are typically models of helper infrastructure that are highly dependent on design choices and are usually subject to frequent alterations due to dynamic hardware requirements and availability [9], therefore it is not desirable to couple these with the core functionality of a system (ii) having a single-view model with serially connected components introduces an intrinsic impedance to the scalability of the model.

4.3 Domain-Specific Models

We represent key entities in distributed power applications including sensors, data concentrators, computing clusters and middleware in the framework of aspect-oriented modeling. In the context of the smart electric grid, it is natural to represent network topologies and middleware components as communication aspects. The top level model for a three-area distributed application topology is depicted in Figure 2. The section describes each entity of the model. The execution details of this model will be discussed further in section 5.

4.3.1 Sensor Clusters

Sensors are the main sources of data that enable applications that run on distributed power systems. The sensors considered for smart-grid applications are PMUs, also known as *synchrophasors*, due to providing synchronous phasor data at rates varying in the range of 10-60 samples per second. Due to the overwhelming number of sensors in the power grid, it is not feasible to model each sensor as an individual component. For the interest of distributed applications, one practical abstraction is to model clusters of sensors that belong to the same area in a single component that is parameterized by the number of PMUs that it encapsulates. The `PMUCluster` actor given in Figure 2 follows this methodology and generates samples of multiple data streams at each iteration, where the number of streams corresponds to the number of PMUs in this area.

4.3.2 Data Concentrators

Data concentrators are intermediate historian components that are commonly utilized in distributed power systems architectures. In the considered model (see Figure 2), Phasor Data Concentrators (PDCs) are actors representing the data concentration units that receive data streams from PMU Clusters in a FIFO fashion and perform relaying prior to sending data to computation nodes and the middleware.

4.3.3 Computation Clusters

The computation clusters are abstract components that may correspond to hardware that process sensor data for various purposes (e.g., GPUs, HPC Clusters). In the common sense of a distributed application, the computation units cannot perform autonomously and are capable of producing local estimates of algorithm outcomes. The model for these computation clusters are named `Area I-III` in Figure 2.

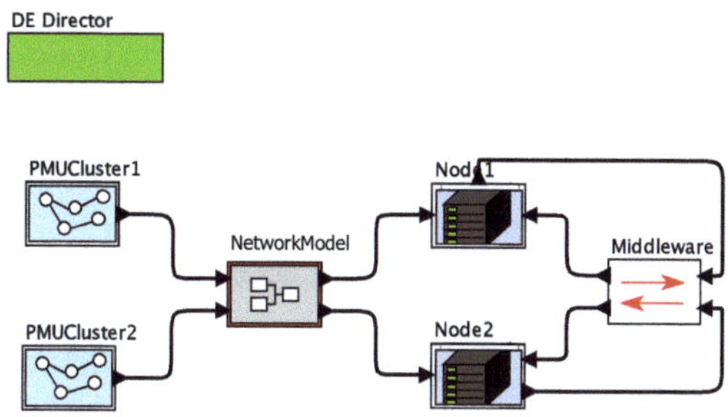

(a) A simple communication model in Ptolemy II

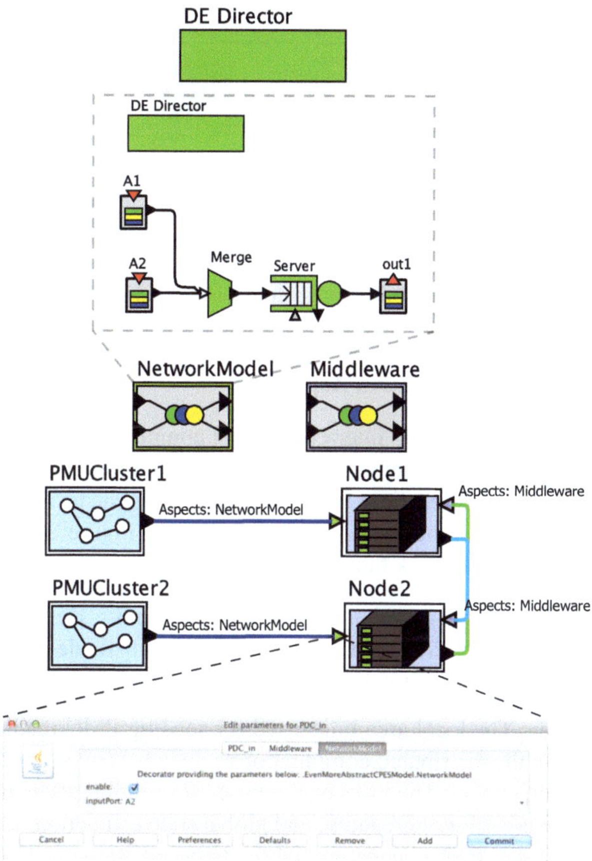

(b) Model refinement using Ptolemy aspects

Fig. 1 Comparison of Monolithic and Aspect-Oriented Modeling Approaches

Each `Area` establishes communication with the neighboring computation clusters and exchanges local estimates, until global consensus is established.

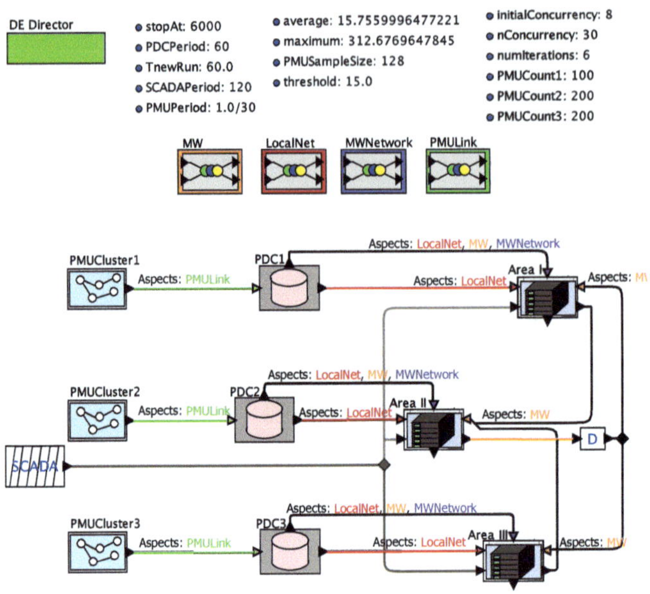

Fig. 2 Aspect-Oriented communication model for a distributed grid application

4.3.4 Modeling Communication Aspects

The use of aspects for modeling inter-component communication enables different network topologies to be implemented in separate composite models in a cross-cutting way, such that multiple links between components can be mapped to a single communication aspect. In the top-level model given in Figure 2, four aspects are used to model different network fabrics:

- `LocalNet`: Network aspect that models inter-area links. Depicted in Figure 3, this aspect models each link as a server with probabilistic delay characteristics. Three types of communication links are defined in the scope:
  - PDC to Area: The local area links used for sending PMU readings to `Areas` to be used in the distributed computations
  - PDC to Middleware : The links that are used for sending PMU readings to the middleware layer for aggregation and global state estimation
  - Area to Area : P2P connection fabric between `Areas`. The connections implicitly determine the topological layout of the neighboring areas in the power grid. In the studied topology, pairwise communications are established between Areas I-II and II-III.
- `MWNetwork`: This intermediate component models the network that connects the historians (PDCs) of each area to the middleware (`MW`). In Figure 4, this network layer models a single channel that carries the PMU data coming from all three areas into the middleware.

- `PMULink`: In Figure 5, this aspect is an aggregate of parallel dedicated links that connect each Phasor Measurement Unit (PMU) to the local PDC.

In all of the listed communication models, probabilistic component delays are modeled as a function of physical system characteristics including physical link length, propagation speed of light in fiber, network packet length, link capacity and a constant queuing delay. These variables characterize the smart-grid communication latency based on the NASPINet specification [17].

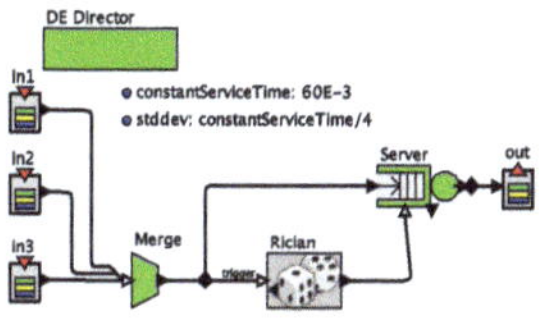

Fig. 3 Sub-Model for MWNetwork aspect

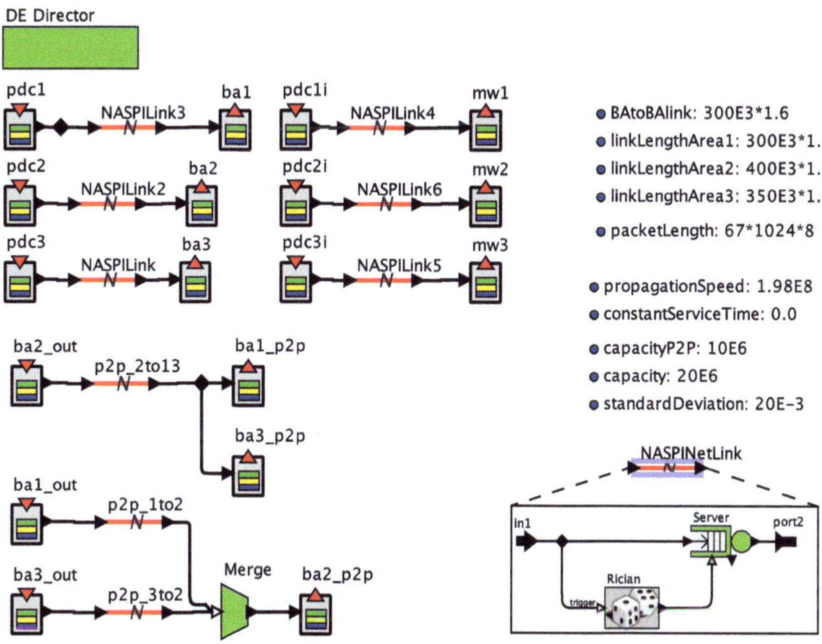

Fig. 4 Sub-Model for LocalNet aspect

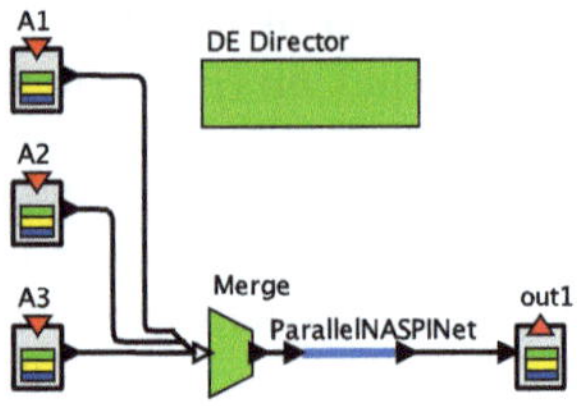

Fig. 5 Sub-Model for PMULink aspect

4.3.5 Serial Composition of Aspects

In a general setting, links between components may be associated with more than one aspect at a time. In such condition, the processing order of events as they are pipelined through aspects is determined by the order in which aspects are associated with the input port. In Figure 2, the pairwise data communication between actors {`PDC1, PDC2, PDC3`} and {`Area I, Area II, Area III`} respectively, are mediated through three communication models, specified as three composite communication aspects named `LocalNet, MW`, and `MWNetwork`, whose respective models are given in Figures 3, 7 and 4. Events generated by the PDCs are processed by the `LocalNet`, then are handed over to `MW` and finally to `MWNetwork` before eventually being delivered to the north input port of the computation nodes (`Area I`). Figure 6 demonstrates the communication traversal from `PDC1` to `Area I`. The time stamp of the original event produced by `PDC1` is being modified by the three communication actors before the event is handed over to the final destination, `Area I`. Note that in the case that communication aspect also

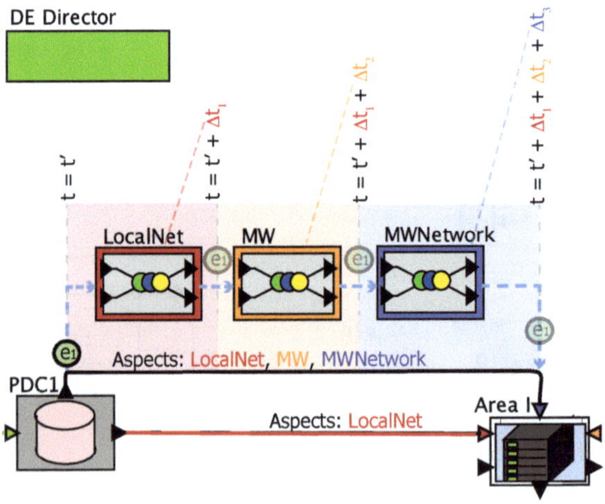

Fig. 6 Sequential event handover between multiple aspects associated with the same link

addresses fault conditions, as in a packet erasure channel, the event may be *dropped* by one of the aspects and never be delivered to the destination port.

4.3.6 Modeling Middleware Structure

The middleware configuration is studied separately from the network layer. The reasoning is the additional role of performing aggregation and time-alignment of packets of the middleware layer. These roles can be elaborated as follows:

- *Aggregation*. The initial task of the middleware is to time-align and aggregate the individual sensor data into a single file that only contains information of faulty or missing data from all areas of sensors. As modeled in Figure 7, lower branch, it combines extracted faulty information into a globally broadcast file. local runs eliminate data with these time stamps in order to maintain data consistency among clusters.
- *Control of global convergence*. The second task of the middleware is to receive local estimates from distributed areas and to declare global convergence of state. As a result of this requirement, in general, P2P communication between computation nodes of different areas must be coordinated by the middleware. The model is presented in Figure 7.

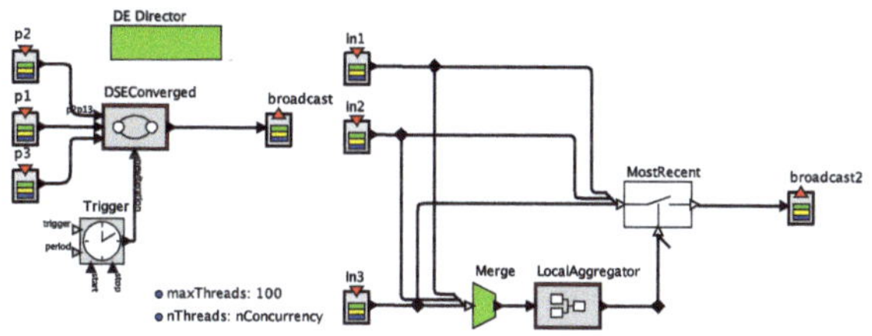

Fig. 7 Sub-Model for MW aspect: [Left branch] Algorithm convergence control [Right branch] Aggregation and processing simulation at middleware level

5 Case Study: Distributed State Estimation

We consider the Distributed State Estimation (DSE) application to evaluate the use cases of aspect-oriented CPES models. Improving the accuracy and robustness of DSE has been an actively investigated research topic [19]. The algorithms are commonly exercised on centralized computation nodes for verification, nonetheless, the actual deployment performance and correctness under a distributed setting remains unexplored [10]. There is a shortfall of distributed system testbeds on which DSE can be deployed and evaluated, so that the characteristics of the run time behavior in the presence of communication, processing and middleware delays can be accounted for. In practice, the volume of data communication between state estimators until convergence of estimators is a function of underlying power system dynamics and data quality. Understanding the characteristics of communication latency in a model-based design environment leverages better design of DSE algorithms that scale in size of the system and data volume [9].

DSE is a comprehensive candidate application that demonstrates the middleware requirements for time-centric CPES applications due to its data intensive nature that forces multi-area communication and coordination. Moreover, several different network fabrics are involved in DSE architectures, for which aspect-oriented modeling proves efficient. From a distributed system perspective, the challenge is to autonomously coordinate the data flow and access within the distributed model. The core of this system architecture is a middleware that mediates data exchange between predetermined grid partitions. Local results from each area are transferred to the middleware to be aggregated and time-aligned. The middleware also coordinates the faulty PMU readings of the entire system, and broadcasts this information to all remote state estimators for global situational awareness. An additional decision component, which is also part of the middleware, receives intermediate state estimation results and notifies each area when global convergence for the estimation model has been achieved. Additional P2P communication between subsystems occur without the need of a coordinator.

We consider the aspect-oriented network model in the context of DSE using aspects and actor components defined in Section 4.3. The simulation enables *what-if analysis* for different scenarios of network delays and middleware configurations. The simulation results provide insight on temporal properties of large a class of distributed power applications at the design level.

5.1 Overview of the Top Level Model

The DSE application is triggered by two main sources of data, which are collected from (i) PMU and (ii) SCADA sensors. Since SCADA data is available at much lower frequencies compared to PMU data, we consider PMU data to be the main trigger for the application data flow.

As shown in Figure 2, the functional model consists of PMU clusters that generate sensor readings at a rate of 30 samples per second, delivered to PDCs for relaying, which are then sent to `Areas` for local executions of the DSE algorithm. In parallel to the computation task, the data is additionally sent to the middleware (`MW`) layer, where it is aggregated to identify faulty or missing data into an aggregate index file, which is then broadcast to Areas for global situational awareness.

For a single iteration of the DSE application, it is expected that the areas exchange multiple estimates of local state with the neighboring computation nodes until global convergence of state has been attained. A convergence message issued by the middleware declares the finalization of the active iteration of the DSE algorithm.

5.2 *End-to-end Simulation*

The use cases of the aforementioned functional model annotated with network and middleware aspects can be multiple. Initially, such functional models are essential for evaluating architectures under test for achievability of end-to-end latency and performance goals. Decoupling the functional model from the implementation-dependent network and middleware components triggers a convenient experimentation process for implementation decisions to be made over wide-area smart-grid communication design. For instance, means of communications and protocols which are feasible to be deployed on the power grid communications, remains an active topic of debate. In this formalism, it is possible to replace each network component with a candidate communication model (e.g., GSM, PLC, IEEE 802.11) and evaluate the end-to-end performance in the presence of the candidate model.

5.3 *What-If Analysis*

To explore satisfiability of requirements discussed in Section 3 by the proposed model, we carry out a *what-if* analysis under corner cases of distributed system behavior. Since the middleware design has flexible choices of architectural options, such concerns to be studied include whether a certain middleware architecture can accommodate application deadlines or whether middleware scales as smart grid components increase in size and connectivity over time. The following subsections address the temporal requirement schemes for (i) network congestion, (ii) network faults, (iii) meeting strict application deadlines, and (iv) maintaining a desired mean end-to-end run time for applications.

5.3.1 Network Congestion Analysis

It is often desirable to test an end-to-end power application under worst-case scenarios in terms of latency. The complex network that interconnects distributed components in the topology is the main source of timing uncertainty in such applications. A reduced channel capacity or a burst of sensor data packets may dominate link capacity to cause local or global deadlines to be missed. We carry out what-if analysis on the model by defining custom stress tests on network components. As an example, we consider the `LocalNet` component, which is modeled to have a link capacity of 10 Mbps per area. As a stress test, we assume a 60% capacity loss on the link that connects Area I to the middleware, on the topology presented in Figure 2. Figure 8 illustrates the distributions of end-to-end run times under stress and under normal configurations. The simulation results demonstrate the effect of the local link capacity loss on the distribution of end-to-end execution times. Under the network congestion, the majority of execution times exceed the worse case run time observed in the no congestion scenario.

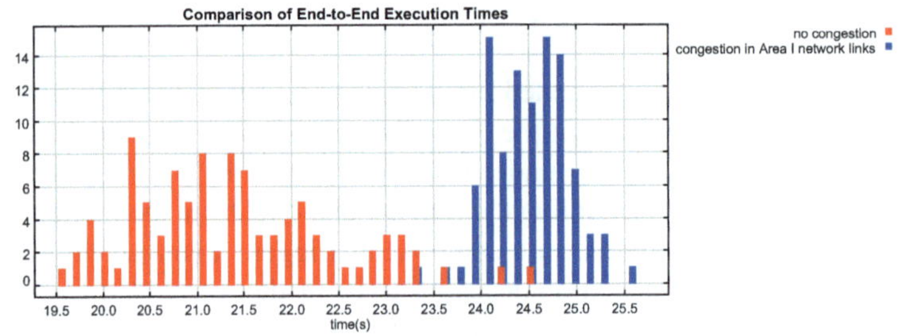

Fig. 8 Effect of local link capacity loss on end-to-end DSE run times

5.3.2 Network Fault Modeling

It is a common scenario to consider network failures and probabilistic faults during packet transmission in designing a network process. When accounting for end-to-end performance, delay and failure characteristics may have a large impact on the satisfiability of application requirements given a platform of choice. Since faults of this nature are usually considered to be *external* artifacts to system behavior than being a part of the intrinsic system, aspects facilitate modeling fault behavior as cross-cutting injections to the model.

We consider a packet erasure channel scenario, where the probability of each packet being dropped at a channel is given by a Bernoulli random variable, parameterized with a packet drop probability p. In Ptolemy, a *modal model* actor defines an extended finite state automaton, in which each state may implement an arbitrary MoC internally. Modal models also support probabilistic transitions, where a probabilistic guard expression of a transition is defined by an expression `probability(`p`)`, which evaluates to *true* with probability $p \in [0,1]$. For visualization purposes, we consider a packet erasure channel with 10% loss. With this probabilistic fault model implemented as an aspect, a PMU-to-PDC link with packet loss can be realized. Figure 9 illustrates a portion of the top level model that now includes a new aspect, `PacketLossFault`, that is associated with the input ports of the PDC components. In Figure 10, a subsequence of the input and output packets are plotted in the `PacketLossFault` aspect. Note that the packet drop behavior will process the events that have already been handled by the `PMULink` aspect. This is why variable inter-arrival times are observed in the input packet stream. It can be seen that some packets that are present in the input are not relayed to the output. The empirical probability of packet loss follows the loss probability that is a parameter of the `PacketLossFault` aspect.

5.3.3 Middleware Scalability for Distributed Applications with Fixed Deadlines

We consider a time-critical application scenario, in which a distributed application has a hard deadline to satisfy. One common class of applications with such deadline characteristics arise from integration of the middleware layer with physical grid dynamics. While the deterministic CPS integration between physical plants and the

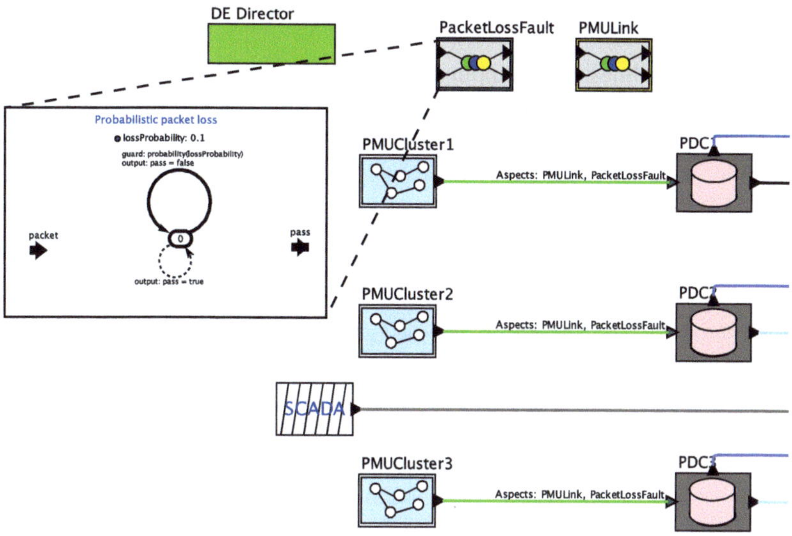

Fig. 9 Sub-model with packet loss fault injection at the PMU-to-PDC link

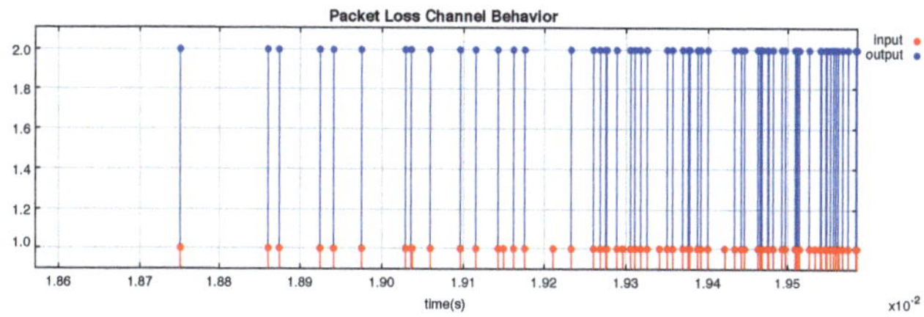

Fig. 10 Packet erasure channel behavior with 10% loss probability

cyber layer is a major research interest, we point the reader to [28] for further details and focus on the networking aspect in this section.

The end-to-end model starts execution with a baseline middleware resource allocation scenario and requests an increase to the allocated resources in the middleware layer each time a deadline is missed, subject to a maximum middleware thread pool capacity. 500 PMU streams are considered to be available in the electrical power grid, distributed to three areas as $nPMU = \{100, 200, 200\}$, where nPMU_i denotes the number of PMUs at Area i. The assumption of the application scale already exceeds the scenario in previous research that provide commercial distributed software solutions [18]. Figure 11 depicts an an application scenario that assumes a deadline of $\tau = 20$ s. The simulation model assumes an initial concurrency level of 8 (simulated by 8 parallel middleware processing queues), that is dynamically incremented upon a missed deadline τ during execution. The simulation trace reveals that this dynamic scaling policy results in a steady-state middleware configuration, under which missed deadlines are avoided for the most part.

Likewise, deadlines can be applied to sub-models too, for instance, the designer could set a latency deadline on the `PMULink` perform adaptive design variations on this link.

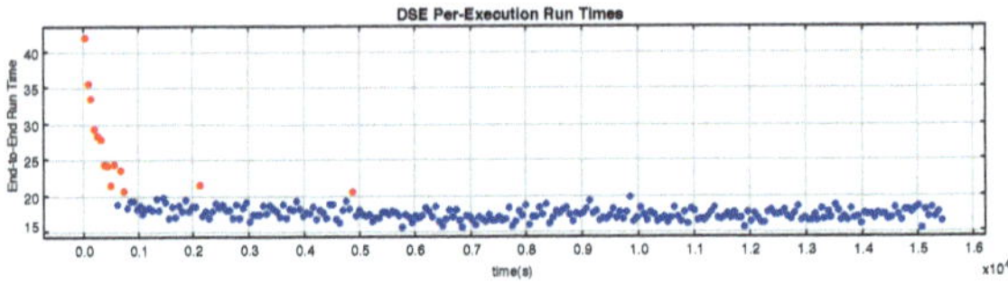

(a) Run-Times of middleware adaptation. [Blue] Executions within deadline, [Red] Executions that missed the deadline

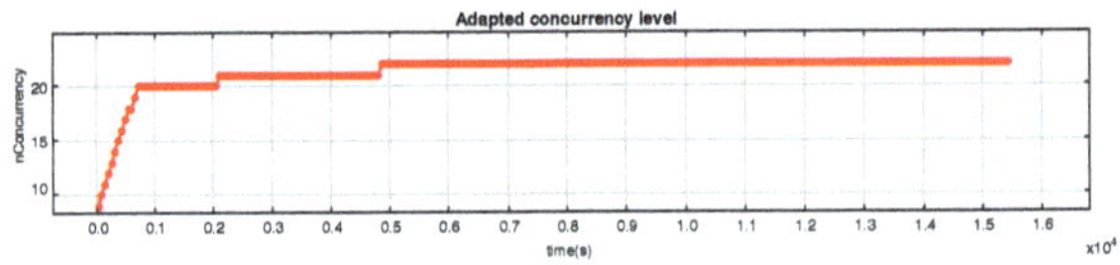

(b) Middleware concurrency level adaptation

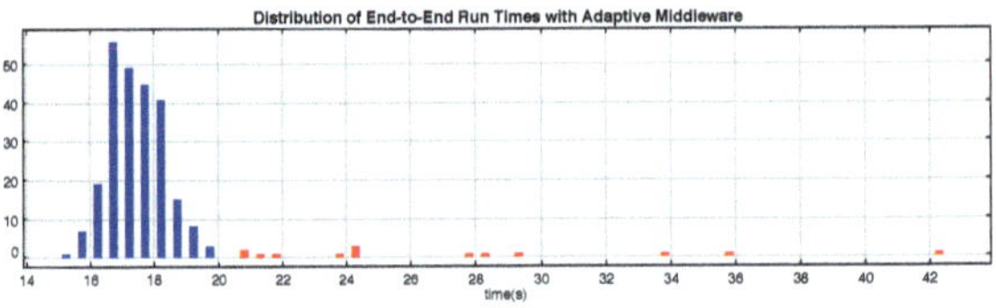

(c) End-to-end run time histogram of overall execution with adaptive middleware capacity

Fig. 11 Middleware adaptation for fixed-deadline distributed applications

5.3.4 Middleware Scalability for Distributed Applications with Variable Deadline

A variant of the above scenario is desired to evaluate middleware scalability for applications that don't have a fixed deadline, but require maintaining a desired average end-to-end run time. The resources in the model (network latency, middleware concurrency level) are modeled in a stochastic sense, which in turn may suggest dynamic analysis of application deadlines. To maintain a desired average run time subject to probabilistic model delay, a manually-tuned proportional-integral (PI) controller is implemented. Th PI controller is used to control the middleware concurrency level, which is modeled as a thread-pool with variable concurrency. The error signal provided to the PI controller is the difference between the the desired and current end-to-end run time. The PI output is quantized to the nearest integer to be used as the correction signal to the middleware concurrency level. Figure 12 demonstrates simulation results for a sample three-area application with a desired end-to-end run time of 20 s. As shown in Figures 12a-12c, with the PI controller in the loop, execution time is eventually stabilized around the desired run time of 20 s, with a steady-state concurrency level of 19.

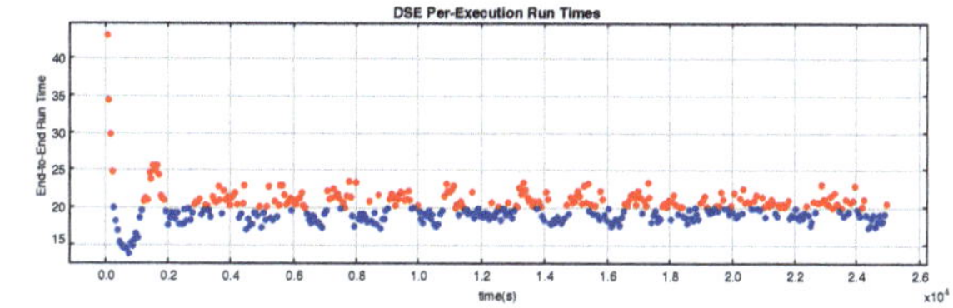

(a) Per-execution DSE run time [Blue] Below desired, [Red] Above desired

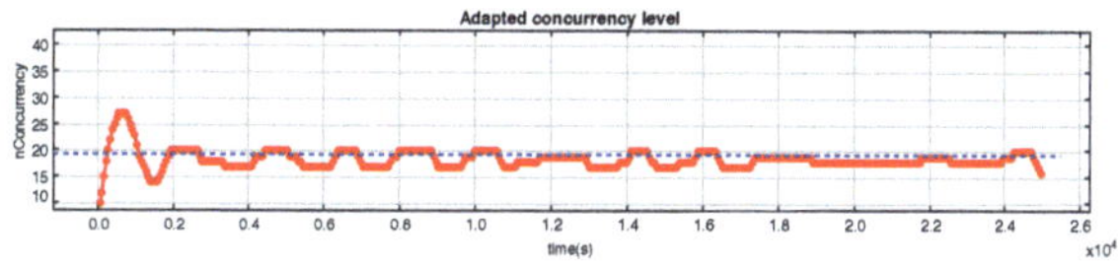

(b) Middleware concurrency level adaptation

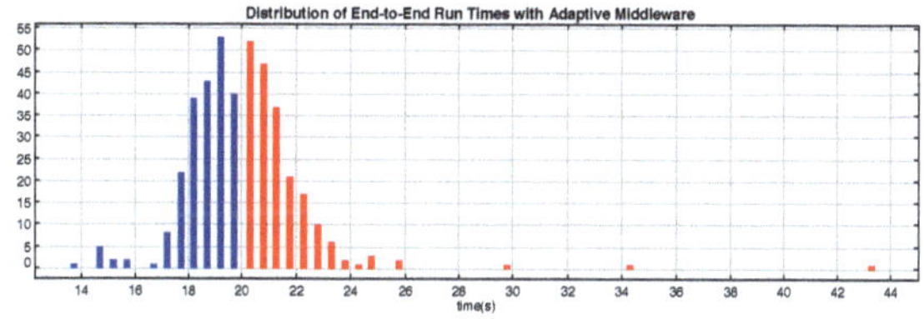

(c) End-to-end run time histogram of overall execution with desired run time

Fig. 12 Middleware adaptation for desired average run time of 20s

6 Discussions

6.1 Hardware-in-the-Loop Simulation

A common method for evaluation of complex real-time embedded systems is hardware-in-the-loop (HIL) simulation. Aspect-oriented simulation models for network and middleware topologies can be used to evaluate real-time control algorithms using smart grid components such as relays and PMUs. Aspects can act as an interface in the simulation-hardware boundary, while integrating physical system components under evaluation into the control simulation. An example to such HIL study would be to replace the PMU clusters in Figure 2 with actual PMU hardware.

6.2 Middleware-in-the-Loop Simulation

The single-server network models assumed so far may fall short in demonstrating the actual complex network and middleware fabric required for grid deployment. AOM can also facilitate the software integration process for evaluating actual middleware architectures into the model, enabling middleware-in-the loop simulation. Figure 13 demonstrates an alternative implementation of the `Middleware` aspect with Java connectors (`MIFTransmitter` and `MIFReceiver`) to a proprietary middleware implementation. An application that uses this middleware has been introduced in [1], where an actual middleware is invoked within the simulation loop. The middleware is implemented using a JMS (Java Message Service) interface that connects to ActiveMQ [http://activemq.apache.org/], an open-source Apache messaging server. Hence, the run times of a three-area system performing DSE are collected as the per-packet real-time latency introduced by ActiveMQ.

In Figure 13, the Discrete-Events delivered to the `MIFTransmitter` are internally converted to JMS messages, processed through the ActiveMQ, received at the `MIFReceiver` and eventually issued as discrete-events sent to the rest of the simulation flow. To incorporate the timing characteristics of the ActiveMQ fabric, the real-time processing latency is computed by the Java connector actors and reflected to model time by altering time stamps of the outgoing events. The ActiveMQ implementation does not follow an adaptively scalable structure as assumed in Figures 11 and 12. Moreover, parallel queues in this framework are configured statically upon initialization of the ActiveMQ server. Based on simulation results, the extension to ActiveMQ should include a monitoring mechanism of collecting run times and a trigger to create new queues for missed deadlines. The conclusions suggest that this modeling approach, which enables integrating actual middleware with smart grid network models would guide the actual middleware development process.

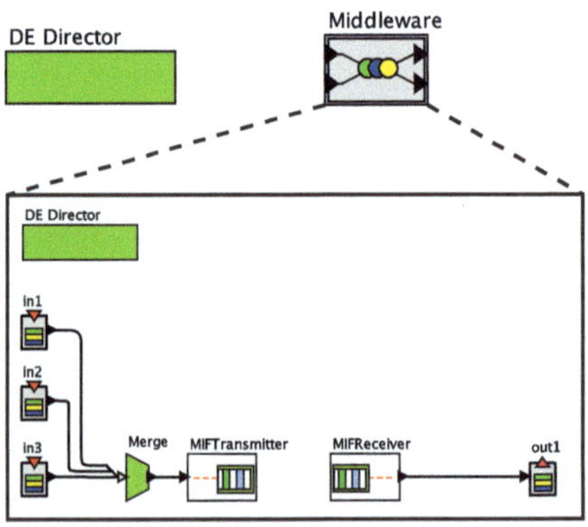

Fig. 13 Middleware-in-the-Loop Simulation

7 Conclusion

In this chapter, we have introduced a novel design methodology for data intensive distributed CPES applications. We studied an aspect-oriented design paradigm, which focused on decoupling cross-cutting aspects in functional CPES models. A general distributed CPES architecture has been presented and was populated with network and middleware models that can be altered without any modifications to the functional model.

Following the discussed modeling paradigm, we performed an extensive simulation study using the example of the DSE application, which demonstrated how end-to-end simulation can be performed to yield analysis of algorithm execution times and to evaluate resource requirements for desired application timing profiles. We also demonstrated how AOM enables performance comparisons among different network topologies that are integrated with a fixed functional grid application. The discussions were presented on an abstract high-level application scenario to highlight that they seamlessly apply to a wide family of WAMC applications.

Further improvements of the modeling paradigm will include integration of fault and attack models and anomaly detection techniques to simulate data and network quality in the AOM setting. This will contribute to developing robustness and reliability features as part of distributed CPES models. It will also be interesting integrate the AOM paradigm followed for network modeling with a co-simulation of physical grid dynamics and demonstrate the benefit of the modeling abstractions for this integration.

Acknowledgements. The authors would like to thank Shuai Lu, Senior Engineer at the Pacific Northwest National Laboratory (PNNL) for his insightful guidance on the Power Engineering background aspects of the chapter.

References

1. Akkaya, I., Liu, Y., Gorton, I.: Modeling and analysis of middleware design for streaming power grid applications. In: Proceedings of the Industrial Track of the 13th ACM/IFIP/USENIX International Middleware Conference, MIDDLEWARE 2012, pp. 1:1–1:6. ACM, New York (2012), `http://doi.acm.org/10.1145/2405146.2405147`, doi:10.1145/2405146.2405147
2. Akkaya, I., Liu, Y., Lee, E.A., Gorton, I.: Modeling uncertainty for middleware-based streaming power grid applications. In: Proceedings of the 8th Workshop on Middleware for Next Generation Internet Computing, MW4NextGen 2013, pp. 4:1–4:6. ACM, New York (2013), `http://doi.acm.org/10.1145/2541608.2541612`, doi:10.1145/2541608.2541612
3. Anderson, K., Du, J., Narayan, A., Gamal, A.E.: Gridspice: A distributed simulation platform for the smart grid. In: 2013 Workshop on Modeling and Simulation of Cyber-Physical Energy Systems (MSCPES), pp. 1–5 (2013), doi:10.1109/MSCPES.2013.6623311

4. Andrén, F., Stifter, M., Strasser, T.: Towards a semantic driven framework for smart grid applications: Model-driven development using CIM, IEC 61850 and IEC 61499. Informatik-Spektrum 36(1), 58–68 (2013), `http://dx.doi.org/10.1007/s00287-012-0663-y`, doi:10.1007/s00287-012-0663-y
5. Balarin, F., Watanabe, Y., Hsieh, H., Lavagno, L., Passerone, C., Sangiovanni-Vincentelli, A.: Metropolis: An integrated electronic system design environment. Computer 36(4), 45–52 (2003)
6. Broman, D., Brooks, C., Greenberg, L., Lee, E.A., Masin, M., Tripakis, S., Wetter, M.: Determinate composition of fmus for co-simulation. In: Proceedings of the Eleventh ACM International Conference on Embedded Software, p. 2. IEEE Press (2013)
7. Cardoso, J., Derler, P., Eidson, J.C., Lee, E.A., Matic, S., Zhao, Y., Zou, J.: Modeling timed systems. In: Ptolemaeus, C. (ed.) System Design, Modeling, and Simulation using Ptolemy II, pp. 355–393. Ptolemy.org, Berkeley (2014), `http://ptolemy.org/books/Systems`
8. Chassin, D., Schneider, K., Gerkensmeyer, C.: GridLAB-D: An open-source power systems modeling and simulation environment. In: IEEE/PES Transmission and Distribution Conference and Exposition, pp. 1–5 (2008), doi:10.1109/TDC.2008.4517260
9. Chen, Y., Huang, Z., Liu, Y., Rice, M., Jin, S.: Computational challenges for power system operation. In: 2012 45th Hawaii International Conference on System Science (HICSS), pp. 2141–2150 (2012), doi:10.1109/HICSS.2012.171
10. Chen, Y., Huang, Z., Liu, Y., Rice, M.J., Jin, S.: Computational challenges for power system operation. In: Proceedings of the 2012 45th Hawaii International Conference on System Sciences, pp. 2141–2150. IEEE Computer Society (2012)
11. Derler, P., Lee, E.A., Vincentelli, A.S.: Modeling cyber-physical systems. Proceedings of the IEEE 100(1), 13–28 (2012), doi:10.1109/JPROC.2011.2160929
12. Eker, J., Janneck, J.W., Lee, E.A., Liu, J., Liu, X., Ludvig, J., Neuendorffer, S., Sachs, S., Xiong, Y.: Taming heterogeneity-the Ptolemy approach. Proceedings of the IEEE 91(1), 127–144 (2003)
13. Elrad, T., Aldawud, O., Bader, A.: Aspect-oriented modeling: Bridging the gap between implementation and design. In: Batory, D., Blum, A., Taha, W. (eds.) GPCE 2002. LNCS, vol. 2487, pp. 189–201. Springer, Heidelberg (2002)
14. Gibson, T., Kulkarni, A., Kleese-van-Dam, K., Critchlow, T.: The feasibility of moving pmu data in the future power grid. In: CIGRE Canada Conference on Power Systems: Promoting Better Interconnected Power Systems, Hallifax, NS, Canada (2011)
15. Godfrey, T., Mullen, S., Dugan, R.C., Rodine, C., Griffith, D.W., Golmie, N.: Modeling smart grid applications with co-simulation. In: 2010 First IEEE International Conference on Smart Grid Communications (SmartGridComm), pp. 291–296. IEEE (2010)
16. Gomez-Exposito, A., Abur, A., de la Villa Jaen, A., Gómez-Quiles, C.: A multilevel state estimation paradigm for smart grids. Proceedings of the IEEE 99(6), 952–976 (2011)
17. Hasan, R., Bobba, R., Khurana, H.: Analyzing naspinet data flows. In: IEEE/PES Power Systems Conference and Exposition, PSCE 2009, pp. 1–6. IEEE (2009)
18. Hazra, J., Das, K., Seetharam, D.P., Singhee, A.: Stream computing based synchrophasor application for power grids. In: Proceedings of the First International Workshop on High Performance Computing, Networking and Analytics for the Power Grid, HiPCNA-PG 2011, pp. 43–50. ACM, New York (2011), `http://doi.acm.org/10.1145/2096123.2096134`, doi:10.1145/2096123.2096134

19. Jiang, W., Vittal, V., Heydt, G.: A Distributed State Estimator Utilizing Synchronized Phasor Measurements. IEEE Transactions on Power Systems 22(2), 563–571 (2007), doi:10.1109/TPWRS.2007.894859
20. Khaitan, S.K., McCalley, J.D.: Cyber physical system approach for design of power grids: A survey. In: 2013 IEEE Power and Energy Society General Meeting (PES), pp. 1–5 (2013)
21. Khaitan, S.K., McCalley, J.D.: Design techniques and applications of cyberphysical systems: A survey. IEEE Systems Journal PP(99), 1–16 (2014), doi:10.1109/JSYST.2014.2322503
22. Korres, G.: A distributed multiarea state estimation. IEEE Transactions on Power Systems 26(1), 73–84 (2011)
23. Koziolek, A., Happe, L., Avritzer, A., Suresh, S.: A common analysis framework for smart distribution networks applied to survivability analysis of distribution automation. In: 2012 International Workshop on Software Engineering for the Smart Grid (SE4SG), pp. 23–29 (2012), doi:10.1109/SE4SG.2012.6225713
24. Krechetov, I., Tekinerdogan, B., Garcia, A., Chavez, C., Kulesza, U.: Towards an integrated aspect-oriented modeling approach for software architecture design. In: 8th Workshop on Aspect-Oriented Modelling (AOM 2006), AOSD, vol. 6. Citeseer (2006)
25. Lee, E.A.: Cyber physical systems: Design challenges. In: International Symposium on Object/Component/Service-Oriented Real-Time Distributed Computing (ISORC), pp. 363–369. IEEE, Orlando (2008), http://dx.doi.org/10.1109/ISORC.2008.25
26. Lin, H., Sambamoorthy, S., Shukla, S., Thorp, J., Mili, L.: Power system and communication network co-simulation for smart grid applications. In: 2011 IEEE PES Innovative Smart Grid Technologies (ISGT), pp. 1–6 (2011), doi:10.1109/ISGT.2011.5759166
27. Liu, J., Liu, X., Lee, E.A.: Modeling distributed hybrid systems in Ptolemy II. In: Proceedings of the 2001 American Control Conference, pp. 4984–4985. IEEE (2001)
28. Matic, S., Akkaya, I., Zimmer, M., Eidson, J.C., Lee, E.A.: PTIDES model on a distributed testbed emulating smart grid real-time applications. In: Innovative Smart Grid Technologies (ISGT-EUROPE). IEEE, Manchester (2011), http://chess.eecs.berkeley.edu/pubs/857.html
29. Ptolemaeus, C. (ed.): System Design, Modeling and Simulation using Ptolemy II (2014), http://ptolemy.org/books/Systems
30. Tomsovic, K., Bakken, D.E., Venkatasubramanian, V., Bose, A.: Designing the next generation of real-time control, communication, and computations for large power systems. Proceedings of the IEEE 93(5), 965–979 (2005)
31. Wetter, M.: Co-simulation of building energy and control systems with the building controls virtual test bed. Journal of Building Performance Simulation 4(3), 185–203 (2011)
32. Zimmerman, R., Murillo-Sánchez, C., Thomas, R.: Matpower: Steady-state operations, planning, and analysis tools for power systems research and education. IEEE Transactions on Power Systems 26(1), 12–19 (2011), doi:10.1109/TPWRS.2010.2051168

A Service-Oriented, Cyber-Physical Reference Model for Smart Grid

Muhammad Umer Tariq, Santiago Grijalva, and Marilyn Wolf

Abstract. This chapter presents a cyber-physical reference model for smart grid. Most of the early smart grid applications have been developed in an ad-hoc manner, without any underlying framework. The proposed reference model addresses this issue and enables the design of smart grid as a robust system that is extensible to the future. The proposed reference model is based on service-oriented computing paradigm and is compatible with the existing service-oriented technologies, used in enterprise computing, such as Web Services. However, it also extends these technologies for handling the hard real-time aspects of smart grid by introducing resource-aware service deployment and quality-of-service (QoS)-aware service monitoring phases. According to the proposed reference model, each smart grid scenario is characterized by three elements: (1) an *application model* that describes the smart grid applications to be supported by the system as a set of resource- and QoS-aware service descriptions, (2) a *platform model* that describes the smart grid platform as a set of computing nodes, communication links, sensors, actuators, and power system entities, and (3) a set of algorithms that enable resource-aware service deployment, QoS-aware service discovery, and QoS-aware service monitoring. This chapter also presents typical development steps of a smart grid application according to the proposed reference model. Moreover, this chapter identifies a number of technological requirements that can enable the development of smart grid applications according to the proposed reference model. Although the development of these required technologies is a topic of ongoing research, this chapter identifies some potential solution approaches, based on state-of-the-art techniques from real-time systems literature. The case study of a demand response application has been employed to explain the various aspects of the proposed smart grid reference model.

Muhammad Umer Tariq · Santiago Grijalva · Marilyn Wolf
Georgia Institute of Technology, Atlanta
e-mail: {m.umer.tariq,sgrijalva,marilyn.wolf}@gatech.edu

S.K. Khaitan, J.D. McCalley, and C.-C. Liu (eds.), *Cyber Physical Systems Approach to Smart Electric Power Grid*, Power Systems, DOI: 10.1007/978-3-662-45928-7_2

1 Introduction

Traditional electric power grid is not capable of reliably handling the imminent deployment of large amounts of renewable energy sources because of their intermittent nature [10][18]. This deficiency has resulted in a worldwide effort towards realizing the vision of a smarter electric power grid [16]. This vision of a *smart grid* proposes to overlay the electric grid with a more extensive computation and communication infrastructure. Unfortunately, most of the early smart grid applications have been developed in an ad-hoc manner, without any underlying framework. This lack of an underlying framework has resulted in a set of isolated smart grid technologies, standards, and applications that are difficult to integrate and extend for the future [13].

In the past, various other complex engineering domains have faced similar problems in their early days. Research communities for those engineering domains overcame this problem by developing a *reference model* for the domain that could enable clear communication among different stakeholders and inform the development of an integrated set of technologies and standards for that domain [19] [14]. In this chapter, we leverage this idea of a reference model and propose a cyber-physical reference model for the domain of smart grid. The proposed reference model for smart grid will enable the development of smart grid technologies, standards, and applications in a robust, integrated, and flexible manner.

The proposed reference model for smart grid is based on service-oriented computing (SOC) paradigm, as this paradigm is uniquely capable of handling the large scale, open nature, and long lifecycle of smart grid scenarios. However, the traditional SOC paradigm, used in enterprise computing domain through popular technologies such as Web Services, cannot be directly applied to smart grid, because this traditional paradigm is not capable of handling the hard real-time aspects of smart grid [4]. Therefore, the proposed smart grid reference model extends the traditional SOC paradigm by introducing resource-aware service deployment and QoS-aware service monitoring phases. According to the proposed service-oriented reference model for smart gird, each smart grid scenario is characterized by three elements:

1. An *application model* that describes the smart grid applications to be supported by the system as a set of resource- and QoS-aware service descriptions.
2. A *platform model* that describes the smart grid platform as a set of computing nodes, communication links, sensors, actuators, and power system entities.
3. A set of algorithms that achieve resource-aware service deployment, QoS-aware service discovery, and QoS-aware service monitoring.

The proposed reference model is capable of describing relevant characteristics of a wide set of smart grid scenarios as it has a sufficiently rich set of features. Moreover, the proposed reference model is generic enough to be reconciled with the existing smart grid standards and technologies, but still provides valuable guidance for the evolution of these standards and technologies into an integrated and consistent set of future smart grid standards and technologies.

In this chapter, we explain various elements of the proposed reference model and present typical development steps of a smart grid application according to the proposed reference model. We also identify some technologies that must be developed before the proposed smart grid reference model could be applied in practice. Although the development of these technologies is a work-in-progress, we present promising solution approaches, based on some state-of-the-art techniques from real-time systems literature. In this chapter, we have used the case study of a demand response scenario in order to explain various aspects of the proposed reference model.

The rest of the chapter is organized as follows. Section 2 explains what is a reference model and how it can be employed successfully for the domain of smart grid. Section 3 presents the details of the proposed reference model for smart grid. This section also presents a development methodology for smart grid applications according to the proposed reference model. Section 4 identifies some technological requirements of the proposed reference model and presents some promising approaches for meeting those requirements. Section 5 presents the case study of a demand response scenario. Section 6 presents the conclusion.

2 What Is a Reference Model?

A reference model for a domain is an abstract conceptual framework, consisting of a small number of interlinked and unifying concepts for that domain. A reference model is designed to enable clear communication about the domain among various stakeholders. A reference model is not a standard or implementation technology in itself. However, it does "inform" the development of a set of compatible standards and technologies for a certain domain [14] [2].

In the past, the concept of a reference model has been successfully employed in various domains to enable the development of a coherent set of technologies and standards for that domain. Following are some examples of reference models developed for various domains:

- Open Systems Interconnection (OSI) Reference Model for communication systems [19].
- Agent Systems Reference Model (ASRM) for multi-agent systems [15].
- National Institute of Standards and Technology (NIST) Reference Model for software engineering environments [2].
- National Institute of Standards and Technology (NIST) Reference Model for project support environments [1].
- Task-based Reference Model for real-time computer systems [12].

Similarly, the development of an appropriate reference model for smart grid can not only ensure clear communication among different stakeholders, but also help in the process of developing a coherent and consistent set of standards and technologies for smart grid. However, any reference model for smart grid must be based on concepts that are generic enough to be reconciled with existing standards (such as various NIST and IEC standards related to smart grid [13]) and technologies (such as real-time operating system and middleware [7] [11]), but still provide valuable guidance for the evolution of existing standards and technologies into a consistent and coherent set of future standards and technologies. In this chapter, we propose a reference model for smart grid that relies on the unification of concepts from the domains of service-oriented computing [4] and cyber-physical systems [17] [9] [8].

3 Reference Model for Smart Grid

This section presents the details of the proposed reference model for smart grid. The proposed reference model is based on service-oriented computing paradigm. Although service-oriented computing paradigm is currently being used in enterprise computing through Web Services technology, it cannot be directly applied to the domain of smart grid because of the hard real-time aspects of smart grid applications. On the other hand, the task-based reference model used in typical distributed, real-time systems, such as automotive and avionics, cannot be directly applied to the domain of smart grid as this reference model is not capable of handling the large scale, open nature, and long lifecycle of smart grid scenarios [12]. Our proposed reference model essentially extends the traditional service-oriented computing paradigm by introducing resource-aware service deployment and QoS-aware service monitoring phases.

According to the proposed reference model for smart gird, each smart grid scenario is characterized by three elements:

1. An *application model* that describes the smart grid applications to be supported by the system as a set of resource- and QoS-aware service descriptions.
2. A *platform model* that describes the smart grid platform as a set of computing nodes, communication links, sensors, actuators, and power system entities.
3. A *set of algorithms* that achieve resource-aware service deployment, QoS-aware service discovery, and QoS-aware service monitoring.

Figure 1 shows all the three elements of the proposed service-oriented reference model for smart grid. Figure 2 shows the major steps involved in the development

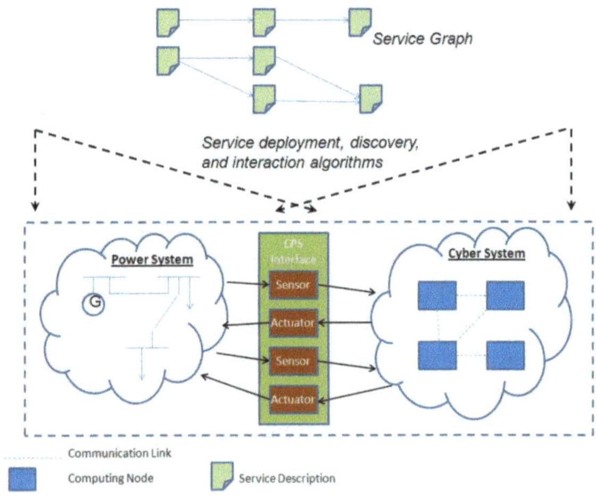

Fig. 1 Reference model for smart grid

of smart grid application according to the proposed reference model. In the *platform porting* step, a generic, service-based computing platform is ported to all the heterogeneous computing nodes involved in a smart grid scenario. In the *service modeling* step, the smart grid application is modeled as a set of services that interact with each other as well as with physical entities through sensing and actuation. In the *service implementation* phase, the implementation code for the services is developed. Both *service modeling* and *service implementation* steps contribute to the development of service descriptions. The service description of a service not only defines the messages that a service exchanges with other services, but it also defines sensing and control actions that the service takes on the co-located physical entities. Moreover, a service description identifies the quality-of-service (QoS) constraints on message exchanges with other services and platform resource requirements of a service. A service description also identifies various modes of operation of a service for various QoS fault scenarios.

In the *service deployment* phase, all the services are deployed on their associated computing nodes. This leads to the *service discovery* step, where all the services involved discover their peer services. This step could be performed online or offline depending on the nature of the smart grid application. In the *service interaction* step, services involved in a smart grid application interact by sending messages to each other. During the *service interaction* step, services switch between different modes of operation if QoS faults occur. Finally, through a *service update* phase, this smart grid reference model supports system maintenance and system updates. In the *service update* phase, services involved in the smart grid application are updated. These services again pass through *service implementation* and *service deployment* steps. Again, the *service update* step can be designed to work online or offline depending on the nature of smart grid application.

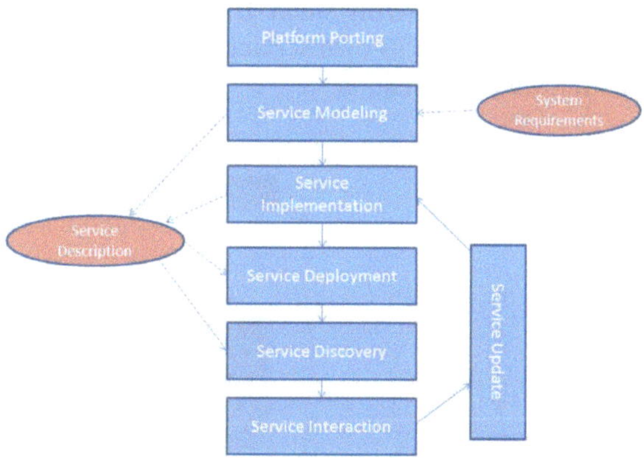

Fig. 2 Development steps of a smart grid application according to the proposed reference model for smart grid

4 Technological Implications of a Service-Oriented Reference Model for Smart Grid

As noted earlier in Section 2, the reference model for a domain helps in the process of developing a consistent set of standards and technologies for that domain. In this section, we present some technological implications of the proposed reference model for smart grid. In particular, we identify some technological requirements for enabling the development of smart grid applications according to the reference model, proposed in Section 3. We also present some potential solutions that can meet these requirements and are based on some state-of-the-art techniques from the domains of real-time systems and embedded control systems.

4.1 Technological Requirements

In order to enable the development of smart grid applications according to the proposed reference model, three major technological requirements are the following:

- A service description language.
- A service-based computing platform for smart grid computing nodes with support for resource-aware service deployment and QoS-aware service interaction.
- A service compiler

Figure 3 shows the role played by these technologies to enable the smart grid application development according to the proposed service-oriented reference model.

4.1.1 Service Description Language

According to the proposed reference model for smart grid, a service description plays a central role. Any smart grid application is modeled as a set of interacting services, each with its own service description. These service descriptions contain the following information:

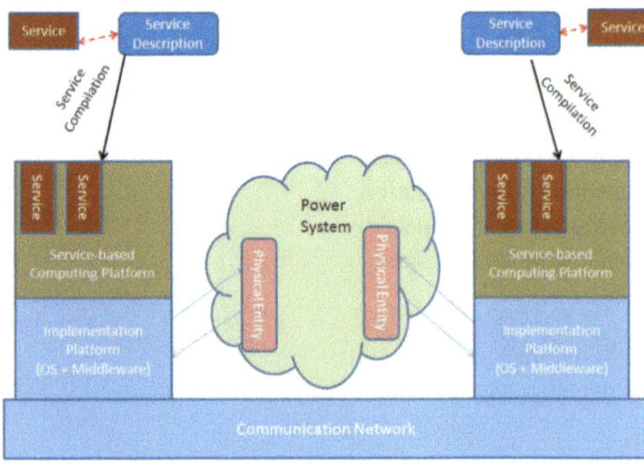

Fig. 3 Technological requirements of proposed smart grid reference model

Service Interface

The *service interface* section of a service description describes the messages that the service exchanges with other services and sensing and control actions that a service takes on the co-located physical entities. This section also identifies the QoS constraints on these messages and sensing and control actions.

Service Resources

The *service resources* section of a service description describes platform resource requirements of a service in order to satisfy the QoS constraints identified in the *service interface* section.

Service Modes

Unlike automotive and avionics systems, smart grid is a wide-area system. As a result, QoS constraints on message exchange among computing nodes of smart grid cannot be guaranteed by the communication subsystem. Therefore, service description for a service must contain a section which defines different modes of operation of the service for different QoS-fault scenarios.

In order to develop service descriptions that contain the above mentioned information (*service interface*, *service resources*, and *service modes*), an appropriate service description language (SDL) is required.

4.1.2 Service-Based Computing Platform for Smart Grid Computing Nodes

To enable the development of smart grid applications according to the proposed service-oriented reference model, each smart grid computing node must have an appropriate service-based computing platform that can support resource-aware service deployment and QoS-aware service interaction. A typical smart grid scenario involves a heterogeneous set of computing nodes with different processors, operating systems, and middleware technologies. Therefore, the required service-based computing platform must be capable of being ported onto these heterogeneous computing nodes.

4.1.3 Service Compiler

In order to properly deploy a service onto the service-based computing platform, an appropriate service compiler is required. This service compiler must be able to read the service description (specified using an appropriate service description language) and decide whether a certain computing node has enough resources to successfully deploy this service such that the service can meet its QoS constraints.

4.2 Potential Solution Approaches

In Section 4.1, a number of technological requirements have been identified that must be met before the proposed reference model can be utilized for the development of smart grid applications. Although the development of technologies that meet the identified requirements is a topic of ongoing research, in this section, we identify some potential solutions for each of these technological requirements. The proposed solutions are grounded in some state-of-the-art techniques, reported in the literature of embedded control systems and real-time systems.

Our proposed solutions are influenced heavily by the research on Giotto [6], a programming language for embedded control systems, and E Machine [5], a virtual machine that serves as the target for compilation of Giotto programs. Figure 4 shows the Giotto and E Machine configuration for a typical distributed real-time system. This configuration has been applied to local-area, distributed, real-time systems (such as automotive and avionics systems). We propose to extend this research for wide-area, distributed, real-time systems (such as smart grid), where QoS constraints on the message exchange cannot be guaranteed by the communication subsystem. In particular, we propose to extend Giotto programming language into the required *service description language*, E Machine into the required *service-based computing platform*, and Giotto compiler into the required *service compiler*. Figure 5 shows the technological requirements (originally shown in Figure 3) with the proposed solution approaches.

4.2.1 Service Description Language

A Short Review of Giotto Programming Language

The typical development process for an embedded control system can be divided into two steps: *control design* and *software implementation*. During the *control design* phase, a control engineer models the plant behavior and disturbances, derives the feedback control laws, and validates the performance of plant under the

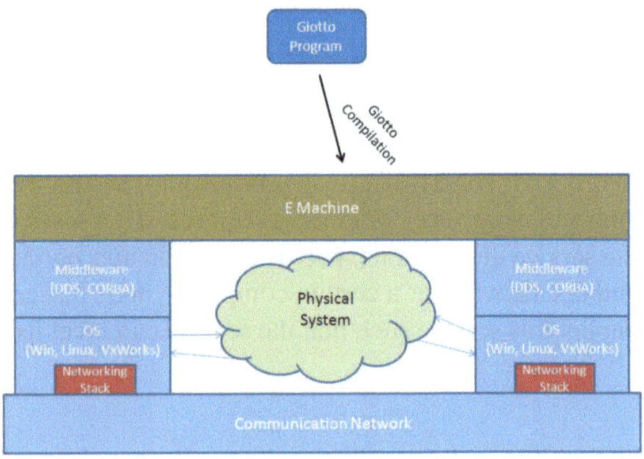

Fig. 4 Typical configuration of Giotto and E Machine for embedded control systems

influence of feedback controller through mathematical analysis and simulations. During the *software implementation* phase, a software engineer breaks down the feedback controller's computational activities into tasks and associated timing constraints on the completion of these tasks. Then, the software engineer develops code for these tasks in a traditional programming language (such as C) and assigns priorities to these tasks so that the tasks could meet their timing constraints while being scheduled on a processor by the scheduler of a real-time operating system (RTOS).

Giotto programming language aims to bridge the communication gap between control engineer and software engineer by providing an intermediate level of abstraction between control design and software implementation [6]. Giotto language syntax can be used by a Giotto program to specify time-triggered sensor readings, actuator updates, task invocations, and mode transitions. Then, a Giotto compiler must be used to compile (an entirely platform independent) Giotto program onto a specific computing platform. The compiler must preserve the functionality as well as the timing behavior specified by the Giotto program.

Figure 6 shows the major elements of Giotto syntax: *task*, *mode*, *driver*, *port*, and *guard*. *Task* is the basic functional unit of Giotto language and represents a periodically executable piece of code. Giotto *tasks* communicate with each other as well as with sensors and actuators. However, in Giotto, all data communication occurs through *ports*. In a Giotto program, there are mutually disjoint sets of task ports, sensor ports, and actuator ports. Task ports are further divided into task input ports, task output ports, and task private ports. Each *task* also has an associated function f (implemented in any sequential programming language) from its input ports and private ports to its output ports and private ports. According to Giotto semantics, sensor ports are updated by the environment while task ports and actuator ports are updated by the Giotto program.

Driver represents a piece of code that transports values between two *ports*. A *driver* can also have an associated *guard*, which is some boolean-valued function on the current values of certain *ports*. The code associated with the *driver* only executes if the *guard* of the *driver* evaluates to *true*. According to Giotto semantics, a *task* is an application-level code that consumes non-negligible amount of CPU

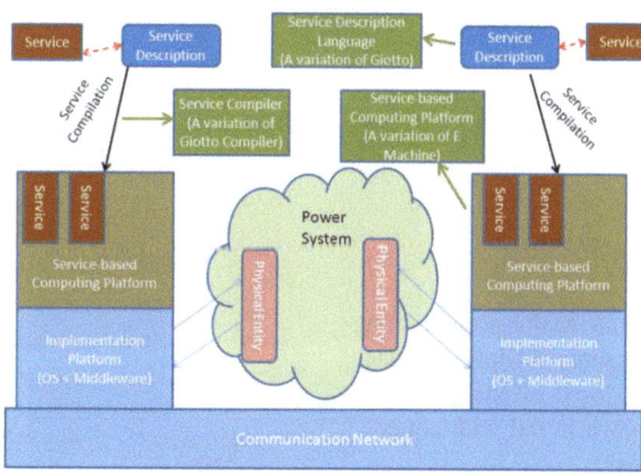

Fig. 5 Technological requirements with potential solution approaches

time, while *driver* is a system-level code that can be executed instantaneously before the environment changes its state.

At the highest level of abstraction, a Giotto program is essentially a set of *modes*. At a certain instant of time, Giotto program can only be in one of its *modes*. However, during its execution, a Giotto program transitions from one *mode* to another based on the values of different *ports*. These possible *mode* transitions are specified in Giotto syntax through *mode swithces*. A *mode switch* specifies a target *mode*, switch frequency, and a guarded *driver*. Formally, a Giotto *mode* is made up of several concurrent *tasks*, a set of *mode switches*, a set of mode *ports*, a set of actuator updates, and a period. Each *task* of a *mode* specifies its frequency of execution per *mode* period. While Giotto program is in a certain *mode*, it repeats the same pattern of *task* executions for each *mode* period.

Figure 6 shows a Giotto program with two *modes*, m1 and m2. *Mode* m1 has two *tasks*, t1 and t2, while *mode* m2 has only one *task*, t3. *Mode* m1 has a period of 10ms, while *mode* m2 has a period of 20ms. *Task* t1 has a frequency of 2, while *task* t2 has a frequency of 1. This means that as long as Giotto program is in *mode* m1, *task* t1 executes every 5ms while *task* t2 executes every 10ms. Moreover, in this example, there is a *mode switch* from *mode* m1 to *mode* m2 with a switch frequency of 2. This implies that the *mode switch* condition (provided by the *guard* of *driver* d5) is tested every 5ms.

Extension of Giotto as a Service Description Language

Although Giotto was originally proposed as a programming language for embedded control systems, it can also be used as the service description language, required by the proposed smart grid reference model, with certain extensions. As outlined in Section 4.1.1, a service description must specify *service interface*, *service resources*, and *service modes*. The current Giotto syntax is capable of specifying all these requirements, except for the input and output messages of a service and QoS

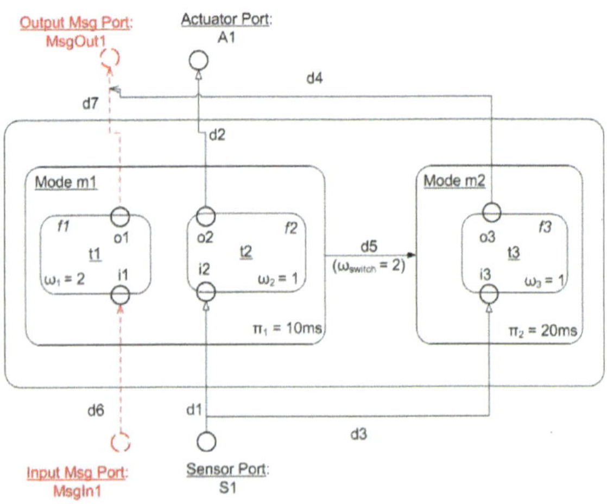

Fig. 6 Major programming elements of Giotto language. Proposed extensions are shown in red with dotted lines.

constraints associated with these messages. In order to overcome this deficiency, we propose to extend Giotto syntax with two new types of *ports*: *input message port* and *output message port*. We also propose to attach the following attributes with these new ports: *TimeSinceLastUpdate* and *DelayInLastUpdate*. These attributes could be used in the guard conditions, present in mode swithces. As a result, Giotto can be used to specify mode switches based on the violation of QoS constraints associated with message exchanges among services. Figure 6 also shows these proposed extensions to Giotto syntax.

4.2.2 Service-Based Computing Platform for Smart Grid Computing Nodes

A Short Review of E Machine

Earlier in this chapter, we have described Giotto, a platform-independent programming language for embedded control systems. In real-time systems literature, development of Giotto compilers for various computing platforms has been reported [6]. However, while developing these Giotto compilers, researchers have found it useful to have an intermediate language, which does not support the high-level concepts of Giotto but still provides a lower level platform-independent semantics for mediating between physical environment and software tasks [5]. The concept of such an intermediate language has evolved into *E code*. Moreover, in the literature, the term *Embedded Machine* or *E Machine* has been used for a virtual machine that interprets the *E code* [5].

The proposed *E code* essentially has the following three instructions:

1. *Call driver*
2. *Release task*
3. *Future E code*

In the *E Code* terminology, a *task* is a piece of application-level code, whose execution takes non-zero time. When invoked with its parameters, a *task* implements a computational activity and writes the results to *task* ports. On the other hand, a *driver* is a piece of system-level code that typically enables a communication activity. For example, a *driver* can provide sensor readings as arguments to a *task* or load *task* results from its ports to an actuator. It is assumed that the execution of a *driver* takes logically zero time.

Call driver instruction starts the execution of a *driver*. As the *driver* is supposed to execute in logically zero time, the *E Machine* waits until the driver completes execution before interpreting the next instruction of E code. *Release task* instruction hands off a task to the operating system. Typically, the task is put into the ready queue of the operating system. Scheduler of the operating system is not under the control of the *E Machine*. The scheduler may or may not be able to satisfy the real-time constraints of the *E code*. However, a compiler (which takes into account the platform resources) checks the time safety of *E code*, generated from a higher level language, such as Giotto. Such a compiler attempts to rule out any timing violations by knowing the worst-case execution time (WCET) of all the tasks and by applying the schedulability results available in the real-time systems literature.

Future E code instruction marks a block of *E code* for execution at some future time. This instruction has two parameters: a trigger and the address of the block of *E code*. The trigger is evaluated with every input event (such as clock, sensor, or task output) and the block of *E code* is executed as soon as the trigger evaluates to true.

E Machine as the Foundation of Service-Based Computing Platform

The smart grid reference model, proposed in this chapter, requires each smart grid computing node to support a computing platform that can support resource-aware service deployment and QoS-aware service monitoring. Since *E Machine*, summarized in the last section, supports resource-aware deployment and QoS-aware execution of Giotto programs, and we have already proposed a Giotto-based service description language in Section 4.2.1, it is natural to leverage *E Machine* as the foundation of required service-based computing platform. However, as noted in the last section, *E code* must be generated by an appropriate compiler to ensure time safety. Therefore, the required service-based computing platform must combine the *E Machine* with an appropriate service compiler that ensures resource-aware service deployment on *E Machine*. However, the service compiler code itself is not hard real-time in nature. Therefore, we propose a design of the service-based computing platform that combines two virtual machines: a hard real-time Embedded Machine (E Machine) and a soft real-time Compiler Machine (C Machine). E Machine executes the hard real-time service code and C Machine executes the soft real-time code for service compiler. The resulting service-based computing platform is shown in Figure 7.

4.2.3 Service Compiler

Giotto compilers, reported in the literature, typically work in two phases: *platform independent* phase and *platform dependent* phase. *Platform independent* phase

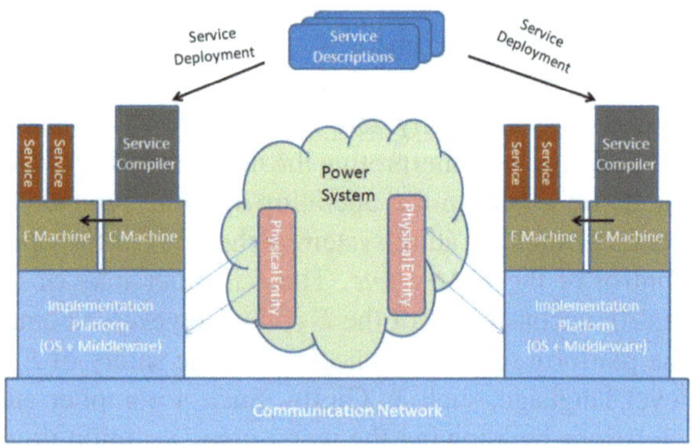

Fig. 7 Potential solution approach for the requirement of a service-based computing platform

generates *E code* from Giotto program; while *platform dependent* phase checks the time-safety of generated *E code* for a particular platform with known worst-case execution times and scheduling schemes. In the last two sections, we have proposed a service description language based on Giotto programming language and service-based computing platform based on *E Machine*. Therefore, it is possible to leverage the existing Giotto compilers and extend them into appropriate service compilers that can ensure resource-aware service deployment of services onto heterogeneous smart grid computing nodes.

5 Case Study: Demand Response

In this section, we present a case study of the application of the proposed reference model to a canonical smart grid application, demand response [3]. Figure 8 shows the demand response scenario under consideration. The power system topology for

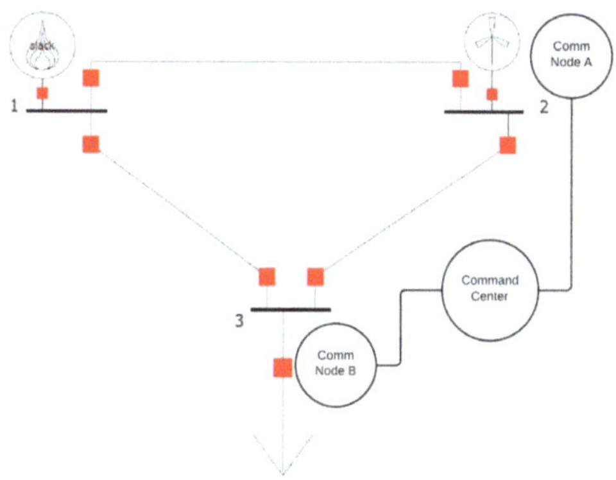

Fig. 8 System topology for the demand response scenario

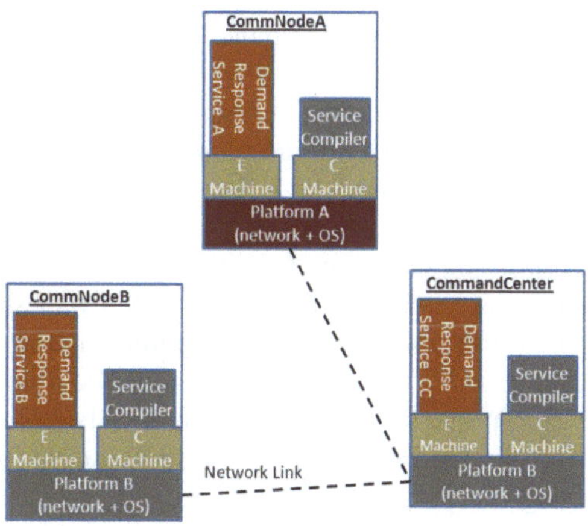

Fig. 9 Demand response case study (with proposed solution approach)

the demand response scenario consists of a wind generator at Bus2, an elastic load at Bus3 that tries to follow the intermittent wind generation at Bus2, and a gas generator at Bus1 that provides the slack. The communication system topology for this demand response scenario consists of three computing nodes: CommNodeA (co-located with wind generator), Command Center, and CommNodeB (co-located with elastic load).

According to the smart grid reference model, proposed in this chapter and shown in Figure 1, this demand response scenario can be described by an *application model*, a *platform model*, and a set of algorithms that facilitate application deployment on the platform. The *application model* for this demand response scenario consists of three services: *DemandResponseServiceA*, *DemandResponseServiceB*, and *DemandResponseServiceCC*. The *platform model* for this demand response scenario consists of three computing nodes (CommNodeA, CommNodeB, ControlCenter), one sensor (co-located with CommNodeA), one actuator (co-located with CommNodeB), three buses, three branches, two generators, and one load. Moreover, the application is installed on this platform by deploying *DemandResponseServiceA* on CommNodeA, *DemandResponseServiceB* on CommNodeB, and *DemandResponseServiceCC* on CommandCenter.

Figure 2 had shown the development steps of a smart grid application according to the proposed smart grid reference model. For the demand resposnse case study, presented in this section, the *platform porting* step consists of installing an appropriate service-based computing platform on the three computing nodes involved: CommNodeA, CommNodeB, and CommandCenter. The *service modeling* step consists of decomposing the demand response application into three services: *DemandResponseServiceA*, *DemandResponseServiceB*, and *DemandResponseServiceCC*. The *service implementation* step consists of developing service descriptions for these three services in an appropriate service description language. The *service deployment* step consists of deploying the three services, *DemandResponseServiceA*, *DemandResponseServiceB*, and *DemandResponseServiceCC* on three computing nodes, CommNodeA, CommNodeB, and CommandCenter, respectively. The *service discovery* step consists of these three services establishing initial contacts (for example, by setting up the lower level transport sockets). During the *service interaction* phase, *DemandResponseServiceA* reads from the co-located sensor and sends periodic messages to *DemandResponseServiceCC*, which in turn sends periodic messages to *DemandResponseServiceB*. The *service update* step will be required if we want to re-configure this demand response application (for example, by adding a new elastic load).

The rest of this section presents the application of proposed technological solutions to this case study. Figure 9 shows the cyber subsystem of demand response case study, where a service-based computing platform (consisting of a combination of E Machine and C Machine) has been ported onto each of the computing nodes and the appropriate service has been deployed on that computing platform through a service compiler. Table 1 shows the service description of *DemandResponseServiceB* using the proposed Giotto-based service description language, while Figure 10 shows the same service description graphically. *DemandResponseServiceB*

Table 1 Service Description of *DemandResponseServiceB* using the Proposed Giotto-based Service Description Language

```
Sensor Ports
    port customerOverride type binary
Actuator Ports
    port genPower type double
Input Message Ports
    port reqPower type double
Output Message Ports
    port status type binary
Task Input Ports
    port i1 type double
    port i2 type binary
Task Output Ports
    port o1 type double
    port o2 type binary
Task Private Ports

Tasks
    task t1 input i1 output o1 o2 function f1
    task t2 input i2 output o2 function f2

Drivers
    driver d1 source reqPower customerOverride
              guard g1 destination i1 i2 function h1
    driver d2 source o1 guard g2 destination genPower
              function h2
    driver d3 source o2 guard g3 destination status
              function h3
    driver d4 source customerOverride guard g4
              destination i2 o2 function h4
    driver d5 source customerOverride guard g5
              destination i2 function h5
    driver d6 source customerOverride guard g6
              destination i1 i2 o1 o2 function h6

Modes
    // Normal operating mode
    mode m1 period 10000ms ports i1 i2 o1 o2
      frequency 1 invoke task t1 driver d1
      frequency 1 update d2
      frequency 1 update d3
      frequency 1 switch m2 driver d4

    // User override mode
    mode m2 period 1000ms ports i2 o2
      frequency 1 invoke task t2 driver d5
      frequency 1 update d3
      frequency 2 switch m1 driver d6

Start m1
```

```
function f1( ) {
    o1 = i1;
    o2 = true;
}
function f2( ) {
    o2 = false;
}
function h1( ) {
    i1 = reqPower;
    i2 = customerOverride;
}
function h2( ) {
    genPower = o1;
}
...
...

binary guard g1( ) {
    return true;
}
...
...
binary guard g4( ) {
    return customerOverride;
}
binary guard g5( ) {
    return true;
}
binary guard g6( ) {
    return !customerOverride;
}
```

[a] Some guard and driver functions have been omitted to avoid unnecessary details.

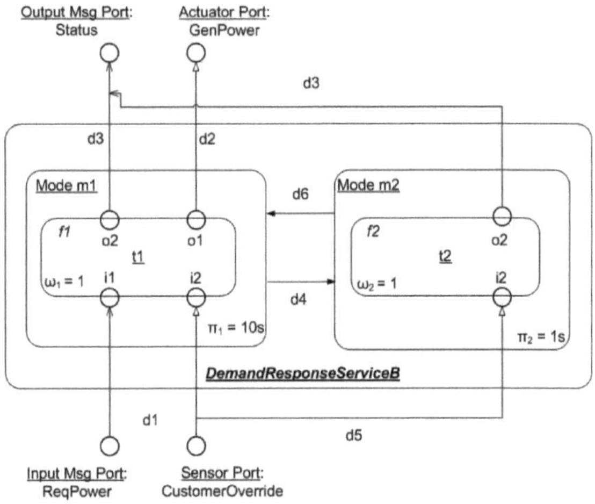

Fig. 10 Graphical representation of service descriptions for *DemandResponseServiceB*

consists of two *modes*: m1 (representing the normal operating mode) and m2 (representing the operating mode when the customer overrides the operation of demand response application). *Driver* d4 and *guard* g4 combine to describe the mode switch condition from m1 to m2, while *driver* d6 and *guard* g6 describe the mode switch condition from m2 to m1. Mode transitions between m1 and m2 occur based on the value of sensor port *customerOverride*, which represents the binary status of an application override user interface mechanism available to the customer. According to the service description, shown in Table 1, *mode* m1 has a period of 10000ms and it has a *mode switch* with the target *mode* of m2 and a frequency of 1, indicating that the mode switch condition is tested once every mode period. Therefore, mode switch condition from m1 (normal mode) to m2 (user override mode) is tested every 10 seconds.

6 Conclusion

Early smart grid applications have been developed in an ad-hoc manner without any underlying framework, resulting in a set of incompatible and inflexible smart grid technologies and standards. However, in the past, various engineering domains have successfully employed the concept of a reference model to enable clear communication among stakeholders and to serve as the underlying framework for the development of a consistent set of standards and technologies for that domain. In this chapter, we have presented an underlying reference model for smart grid that can enable the development of a set of compatible smart grid standards and technologies that are extensible to the future. The proposed reference model is based on some suitable extensions to the traditional service-oriented computing paradigm as this paradigm is uniquely suitable to handle the large scale, open nature, and long lifecycle of smart grid applications.

In this chapter, we have also identified some technological requirements (such as service description language, service-based computing platform, and service compiler) for enabling smart grid application development according to the proposed reference model. Although development of suitable technologies, which can meet these requirements, is a topic of ongoing research, we have presented potential solution approaches based on state-of-the-art techniques from real-time systems literature.

References

1. Brown, A., Carney, D., Feiler, P., et al.: A project support environment reference model. In: Proceedings of the ACM Conference on TRI-Ada, pp. 82–89 (1993)
2. Brown, A.W., Earl, A.N., McDermid, J.: Software engineering environments: automated support for software engineering. McGraw-Hill, New York (1992)
3. Conejo, A., Morales, J., Baringo, L.: Real-time demand response model. IEEE Transactions on Smart Grid 1(3), 236–242 (2010)
4. Erl, T.: Service-oriented architecture: concepts, technology, and design. Prentice Hall, New Jersey (2005)
5. Henzinger, T.A., Kirsch, C.M.: The embedded machine: predictable, portable real-time code. ACM Transactions on Programming Languages and Systems (TOPLAS) 29(6), 33 (2007)
6. Henzinger, T.A., Horowitz, B., Kirsch, C.M.: Giotto: A time-triggered language for embedded programming. Proceedings of the IEEE 91(1), 84–99 (2003)
7. Kopetz, H.: Real-time systems: design principles for distributed embedded applications. Springer, New York (2011)
8. Khaitan, S.K., McCalley, J.D.: Cyber physical system approach for design of power grids: a survey. In: Proceedings of the IEEE Power and Energy Society General Meeting, pp. 21–25 (2013)
9. Khaitan, S.K., McCalley, J.D.: Design techniques and applications of cyber physical systems: a survey. IEEE Systems Journal PP(99), 1–16 (2014)
10. Kundur, P.: Power system stability and control. McGraw-Hill, New York (1994)
11. Laplante, P.A., Ovaska, S.J.: Real-time systems design and analysis: tools for the practitioner. IEEE Press, New York (2012)
12. Liu, J.W.S.: Real-time systems. Prentice Hall, New Jersey (2000)
13. NIST Special Publication 1108R2, NIST Framework and Roadmap for Smart Grid Interoperability Standards, Release 2.0. (2012), `http://www.nist.gov/smartgrid/upload/NIST_Framework_Release_2-0_corr.pdf` (cited July 27, 2014)
14. OASIS Standard, OASIS Reference Model for Service Oriented Architecture (1999), `http://docs.oasis-open.org/soa-rm/v1.0/soa-rm.pdf` (cited May 21, 2014)
15. Regli, W.C., Mayk, I., Dugan, C.J., et al.: Development and specification of a reference model for agent-based systems. IEEE Transactions on Systems, Man, and Cybernetics, Part C: Applications and Reviews 39(5), 572–596 (2009)
16. World Economic Forum Report, Accelerating Successful Smart Grid Pilots (2010), `http://www.weforum.org/reports/accelerating-successful-smart-grid-pilots` (cited July 27, 2014)

17. Wolf, W.: Cyber-physical system. Computer 42(3), 88–89 (2009)
18. Wan, Y.: A Primer on Wind Power for Utility Applications. Technical Report: National Renewable Energy Laboratory (2005), http://www.nrel.gov/docs/fy06osti/36230.pdf (cited July 27, 2014)
19. Zimmermann, H.: OSI reference model–The ISO model of architecture for open systems interconnection. IEEE Transactions on Communications 28(4), 425–432 (1980)

Real Time Modeling and Simulation of Cyber-Power System

Ceeman B. Vellaithurai, Saugata S. Biswas, Ren Liu, and Anurag Srivastava

Abstract. Ongoing smart grid activities have resulted in proliferation of intelligent devices and associated Information and Communication Technologies (ICT) to enable enhanced system monitoring and control. Integration of ICT has led to an increase in the number of cyber assets and requires cyber-physical study for system analysis. In order to realize the vision of a smarter grid, it is necessary to understand the complex relationship between cyber and physical domains, and potential impacts on the power grid due to successful cyber-physical attacks. In order to understand this coupling, cyber physical test bed can help to model and simulate the smart grid with sufficient level of detail. In this chapter, an introduction to the smart electric grid and the challenges associated with the development of cyber-power test bed is presented. The integration of Real Time Digital Simulator (RTDS) and Network Simulator 3 (NS3) to realize a real time cyber-power test bed is discussed with the implementation of an example application.

Keywords: Application testing, Cyber-power system, Cyber security, Device testing, Network simulator 3, Real time, Smart grid.

1 Introduction

1.1 Electric Power Grid

The primary aim of the Electric Power Grid (EPG) is to reliably deliver power to load centers with high level of service continuity and minimal cost, while minimizing the

Ceeman B. Vellaithurai
Schweitzer Engineering Laboratories, Inc. (SEL), Pullman, WA, USA
e-mail: ceeman_vellaithurai@selinc.com

Saugata S. Biswas · Ren Liu · Anurag Srivastava
Washington State University, Pullman, WA, USA
e-mail: {saugatasbiswas,liuren248}@gmail.com,
asrivast@eecs.wsu.edu

S.K. Khaitan, J.D. McCalley, and C.-C. Liu (eds.), *Cyber Physical Systems Approach to Smart Electric Power Grid*, Power Systems, DOI: 10.1007/978-3-662-45928-7_3

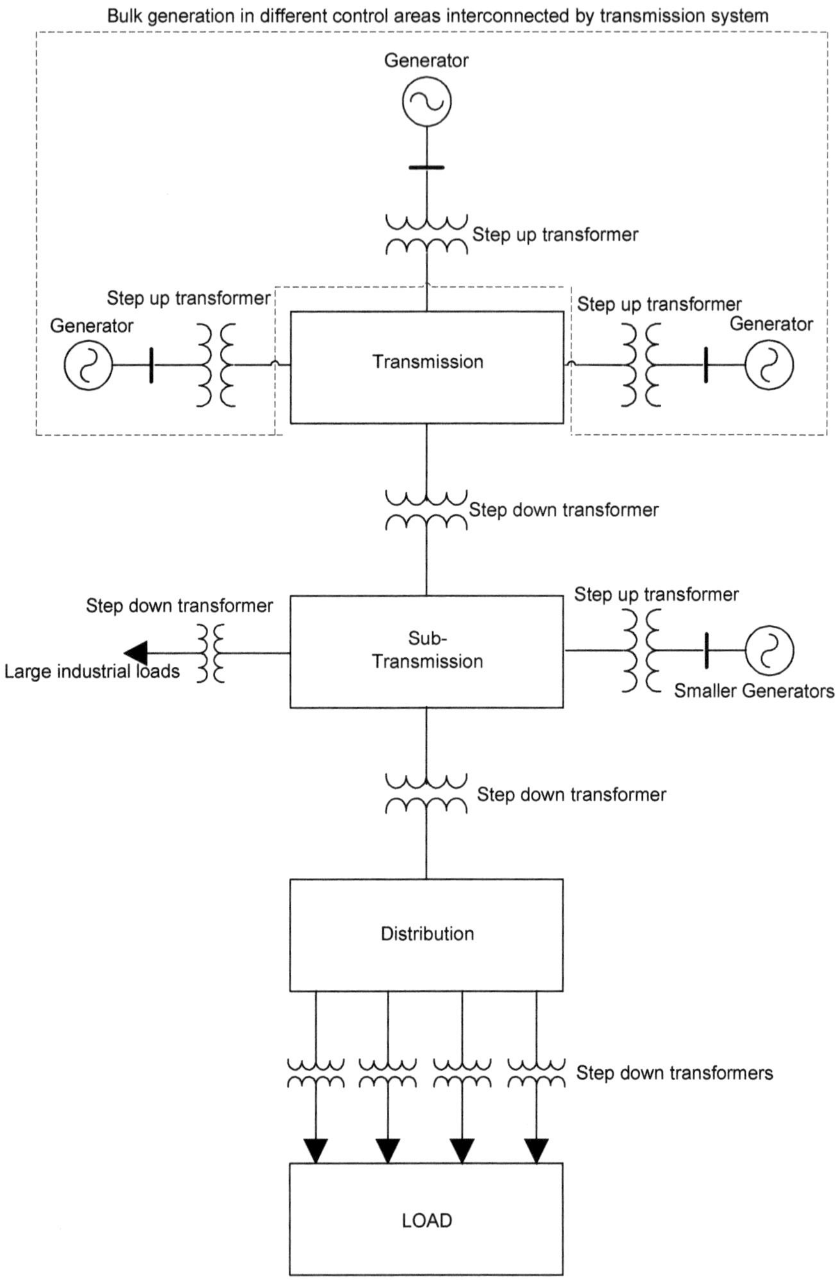

Fig. 1 Basic structure of electric power grid

impact of component failures. The physical EPG consists of four major domains: generation, transmission, distribution, and load.

Fig 1 describes the basic structure of EPG [1]. The generation system consists of several large generators generally located away from load centers. Distributed generation may feed into the grid at the subtransmission or distribution level. The power generated by these generators is delivered to the load centers through the transmission and distribution systems. To minimize power loss in transmission of electric power, the voltage level at generating substations is stepped up to high voltage levels. Transmission level systems have a meshed topology, while distribution systems are usually radial.

The power system has to maintain a constant balance between the electric power generated and consumed. This criterion needs to be satisfied in order for the power system to be stable and operate in synchronism within a specified band around rated frequency. In the past few decades, with the advent of power electronic devices, High Voltage Direct Current (HVDC) links [2] are also being used for bulk power transfer.

1.2 Power System Monitoring

The EPG is a complex network and dynamic system of systems. It is therefore necessary to monitor the system continuously and take appropriate control actions as required. Various monitoring and control methodologies have been used over time, usually driven by the needs and available technologies of that time. The earliest method of data acquisition for monitoring the EPG involved scanning of remote terminal units (RTU) in a sequential order to obtain measurements. This process usually took several minutes due to vast number of devices to scan in addition to communication network constraints. With the use of Supervisory Control and Data Acquisition (SCADA) systems, and upgrade to the advanced communication networks, the time taken to acquire data from RTUs was reduced from several minutes to a few seconds. The measurement devices are polled every few seconds to collect data in a routinely. SCADA systems are widely used throughout the EPG and industrial control systems for monitoring and control purposes. The data acquired from the devices need to be processed through a State Estimator to get more accurate system state estimate and to remove bad data. Power flow studies and other stability studies rely heavily on the collected data to determine control and mitigation strategies for contingencies in the system.

1.3 Power System Control

The control systems employed in the EPG can be broadly classified into local and wide area controls. Due to data availability constraints, most of the control systems that have been implemented so far make use of local information to take control decisions and actions. A brief survey of the control methodologies employed in the power grid are discussed here [3].

1.3.1 Local Controls

Local control systems typically make use of the data available within a single substation. These control systems do not take into account the state of the system in other locations to take control decisions and actions. The control action may be opening/closing of circuit breakers to reroute power, changing transformer taps in response to terminal voltage, switching in capacitors or reactors to alter power flow, or changing generator mechanical input to control output power.

Power System Protection: The EPG is prone to faults such as a tree branch falling on conductor leading to a short circuit. These faults are typically characterized by high current flows resulting in heating or burning of equipment. In order to protect the devices in the power system, several protection techniques are applied as needed. These systems usually have redundancy to protect against equipment failures.

Voltage Control: Voltage is generally maintained at the required level either through changing the taps of a tap changing transformer or by use of switched capacitor or reactor banks to provide/absorb reactive power. Power electronic devices have augmented these capabilities in recent years.

Generator Control: Generators may use a combination of local controls such as governor control, excitation control and power system stabilizers for controlling power input, voltage output and damping oscillations respectively.

Power Flow Control: Similar to voltage control, this type of control typically involves the use of switched capacitors or reactors in the transmission lines to either reduce or increase the effective line impedance or angle. Traditional methods involved the use of slow response controls such as AC phase shifting transformers. Power electronic devices provide faster control.

Note that, several of these above local control can also be coordinated and need not to be always based on only local measurements.

1.3.2 Wide Area Controls

Controls that require information from not just the local devices but also remote devices through use of communication channels are classified as wide area controls. The scope of wide area controls may vary involving just two substations to multiple substations. A few typical wide area control methods are described in this section.

Coordinated Frequency Control: The balance between power generation and load has to be maintained at all times. If there is any imbalance, the generators begin to speed up or slow down depending on whether generation is higher or lower. Governors provide a local and fast control to regulate speed by changing the mechanical power input. A second level of control called Automatic Generation Control (AGC) utilizes the generator output data from the generators and sends raise or lower commands to the governor control to maintain inter-tie schedules and frequency. This control is generally slow as it involves collection of data from the different generators and running algorithms to determine appropriate set points for the generators to minimize control area error. In the North American power grid, this control may take place every few seconds as the frequency requirements are strict.

Coordinated Voltage Control: In addition to local voltage control, wide area voltage control applications have been employed on a limited scale to achieve coordinated voltage regulation across the system through actuation of local devices.

Remedial Action Scheme(RAS): These schemes involve elaborate systems that may trip generators, loads, or transmission lines in response to a contingency. A RAS typically require extensive data for offline simulations and studies to determine control actions for particular contingencies.

1.4 Evolving Smart Electric Grid

The EPG has remained largely unchanged over the last few decades. According to the U.S. Department of Energy report, the average demand for electricity in the past two decades has been increasing at the rate of 2.5 percent annually [4]. At the same time electric grid is going through several changes including generation mix, load types, electricity markets, difficulty in building new transmission lines, and environmental constraints. A long list of blackouts in the past has pointed to the need for continued improvements. Energy storage needs, need for better visibility and situation awareness, automated control, ,and sustainable energy are some of the key factors which have generated the push towards to the development of a smarter grid. Analysis of past blackouts in the North American grid have shown that the lack of visibility of the grid and unavailability of high resolution information to make critical decisions were the main cause of the blackouts. Operators in control centers are trained to make informed decisions based on their knowledge of the system. However, the response times during critical periods are too short for operator intervention highlighting the need for automated systems. Once the system enters the cascading stage of a blackout, an operator can hardly take any corrective control actions. The state of the grid needs to be monitored continuously and appropriate action need to be taken to prevent a blackout condition. This may involve islanding the grid into several sub systems, shedding loads, or a combination of both.

The smart grid is a major upgrade to the electric grid infrastructure for improved efficiency, reliability and safety, with smooth integration of renewable and alternate energy sources, through use of automated control and modern communication technologies. The technology needed to realize a smarter grid such as processing power to handle large amount of data, remote access for monitoring and control, and automation to enable self healing capabilities need to be integrated with the EPG.

The U.S. Energy Independence and Security Act of 2007 directed the National Institute of Standards and Technology (NIST) to lead the related research work of smart grid. According to the NIST report [5], the smart grid has the following key characteristics:

1. Enables informed participation by customers
2. Accommodates all generation and storage options
3. Enables new products, markets and services
4. Provides the required power quality for a range of needs
5. Optimizes asset utilization and operates efficiently
6. Operates resiliently to disturbances, attacks and natural disasters.

Fig 2 shows the framework for a smart grid as defined by the NIST report [5]. In addition to the power delivery domains, the smart grid also includes the following domains: electricity markets, system operation and service providers. Each domain and their respective sub-domains have a group of actors and applications. Information flows within each domain as well as between domains. Actors may not be restricted to just one domain. For instance, distribution service providers may have actors not only in the distribution domain, but in the operations and markets domain as well.

1.4.1 Advanced Power Grid Communication Networks

In the existing power system, the various communication requirements of the grid are supported by independent and often dedicated networks. For example, data delivery between substations and control centers is a dedicated independent network in most cases. In a typical communication network used in the grid at present, fiber optic cables are generally used between critical substations and control center. Fiber optic cables can be laid along the transmission lines in the power system for data transmission. All Dielectric Self Supporting (ADSS) fiber optic cables are installed along the transmission lines using the same tower support infrastructure. Redundancy is provided to cover for failure of one or two links in the system. If it is not feasible to lay a fiber optic line, private WiMAX networks are used. For distribution network communications such as Advanced Metering Infrastructure (AMI) and Distributed Automation (DA), low speed networks with bandwidth in the range of 200 kbps are used. In some cases, public communication lines may be leased. Multiprotocol Label Switching (MPLS) is used for managing the Internet Protocol (IP) network traffic. Some of the different service segregation used to differentiate traffic is telemetry protection, AMI, SCADA and enterprise access. With the replacement of old bandwidth restricted networks with high capacity fiber optic links, the first step towards realization of fast inter domain information flow has been taken. However, in order to fully realize the goals of a smart grid, it is necessary to expand the information network inter-connectivity so that information may flow securely between the different domains in the smart grid.

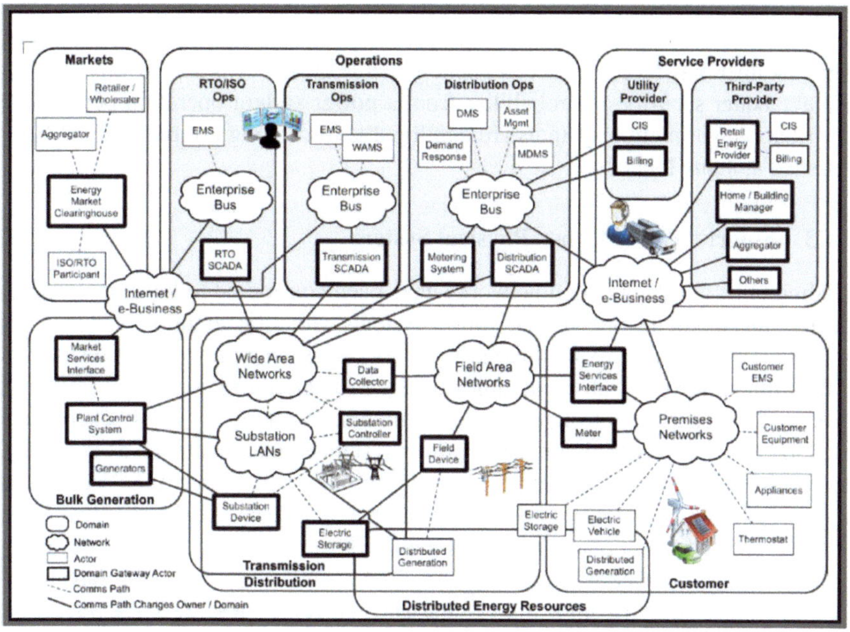

Fig. 2 Structure of a smart grid

1.4.2 Advancements in Monitoring and Control Systems

Recent developments in the field of measurements have led to the development of Phasor Measurement Units (PMU), which can provide data at a rate greater than thirty samples per second. The major advantage of this device is the availability of phasors values taken with reference to a single time source provided usually by a Global Positioning System (GPS) clock. This essentially means that all the measurements are taken on a common reference and a linear state estimator can be used to filter bad data in presence of PMU at all buses. The filtered data is used for real time monitoring and control purposes. The wide area controls described in Sect 1.1.3.2 can be greatly improved with the use of PMU data. Frequency control is inherently restricted by the speed, at which governor controls respond to changes. However, the availability of data at such high resolutions is of great value to RAS and voltage control algorithms to implement real time control. Operators in control centers have a better situational awareness of the grid. Adaptive control algorithms, contingency mitigation strategies, and self healing capability are some of the goals that can be attained through the use of real time data. A real time voltage stability monitoring algorithm using PMU data is described in Sect 1.4.

The availability of data and automated control in the distribution system has been very limited. This is fast changing with the implementation of advanced metering infrastructure as part of the smart grid initiative. With the installation of these smart meters, the degree of resolution of distribution systems will be improved greatly.

Additionally, through customer participation, it will be possible to raise or lower load levels at varying times of the day benefiting both the utility and the customer. The customer may receive monetary benefit, while the utility is able to vary load and achieve better security and reliability from a power system operation perspective. Distribution automation systems are gaining traction among utilities and are being installed at many locations.

1.4.3 Smart Grid: A Cyber-Physical System

Traditionally, communication networks have been considered to be support infrastructure that aid in the operation of power systems with little attention paid to cyber-security. The vulnerability of smart grid cannot be assessed as two separate metrics: cyber vulnerability and physical vulnerability. In a smart grid, the compromise of a cyber-asset such as a control, protection, or monitoring device by an attacker maybe used to cause damage to the physical power system components such as generators and transformers. Depending on the severity of the attack, it may take a long time to replace/bring these devices back to the service. In February 2014, the Wall Street Journal reported that a planned attack on a California substation involving sniping of transformers resulted in repairs that required twenty seven days to complete [6]. This illustrates the difficulty in servicing and replacing these devices. Successful cyber-attacks typically make use of vulnerability in the communication protocol, routing, or authentication of a cyber-asset to install malware, deny legitimate services, or directly intrude into an information system [7]. The level of physical consequences due to a cyber-attack is dependent on the nature and depth of the attack. Thus, the smart grid should be treated to its true nature of being a cyber-physical system (CPS) and security of the grid should be assessed with this view. CPS security systems must be able to differentiate between physical and cyber-attacks and respond accordingly. It is important to guard against coordinated cyber-physical attacks as the potential consequences may be severe.

1.4.4 Need for Cyber-Physical Security Analysis

Over the past few years, there have been several reports on industrial control systems vulnerability and victims of cyber-attacks. In March 2007, Idaho National Laboratory conducted an experiment in which physical damage was caused to a diesel generator through the exploitation of a security flaw in its control system by disabling the sync check element in the protective relay [8]. In April 2009, the Wall Street Journal reported that cyber spies had penetrated the U.S. electrical grid and left behind software programs that could be used to disrupt the system. The most significant of cyber-attacks on industrial control systems was Stuxnet, which happened in 2010. Stuxnet, a large complex piece of malware with many different components and functionalities, targeted Siemens industrial control systems and exploited four zero day vulnerabilities running Windows operating systems [9]. Increasing connectivity within the smart grid, interfacing with legacy devices, proliferation of access points, use of internet for remote access, common operating systems and platforms contribute to the increase in risk factor. From Fig 2, it is clear that there is a proliferation of access points and routers scattered across the domains. This makes it possible for intruders to gain access to other domains. For example, attackers can start intruding into the system from a home network and work their way gradually into the enterprise networks and gain privileged information. This may then be used in arbitrage for illegal economic benefits or for causing unwanted operations in the power system. This necessitates the development and maintaining of authentication procedures and relevant best practices for cyber-physical security.

In general, the cyber security requirements of a system deployed in response to cyber threats include three main properties: confidentiality, integrity and availability [10]. These three properties are designed around the cyber paradigm and are not directly applicable for cyber-physical system security. However, these properties help in establishing basic security requirements. Confidentiality prevents an unauthorized user from obtaining secret or private information. Integrity prevents an unauthorized user/attacker from modifying the information. Availability ensures that a resource is available to the legitimate user when needed.

2 Modeling and Simulation of Cyber-Power System

A survey of the design methodologies adopted for cyber-physical system design, modeling and simulation; and the underlying issues involved are discussed in [11] [12]. In order to conduct cyber-physical analysis of the smart grid, it is necessary to develop modeling and simulation methodologies with sufficient detail. Fig 2 shows the different domains that exist in the smart grid. Of particular interest from a power system operation perspective are the generation, transmission, distribution ,and system operation domains. There are a number of tools available for modeling and simulation of these domains; traditionally the focus of power system modeling. Integration of the following simulators/devices is required to realize a tightly coupled cyber-power simulator: power system simulator, communication network simulator/emulator, data measurement and collection simulator, and end user application simulator. Digital power system simulators are usually discrete time based and communication system simulators are usually discrete event based. Data measurement and collection devices need to be simulated to measure and export power system data to end user applications through the communication simulator/emulator.

A modular approach to the development and integration of these simulators is preferable as it may not be necessary to have all the domains integrated together at all times. For example, in order to test an application running in a control center, it is enough to model the power system with data measurement and collection devices, communication network and control center. If any specific inter-dependency needs to be modeled, modules for that particular domain can be added. The simulation environments used for studying the coupling of the systems maybe broadly classified as centralized simulation environments and co-simulation environments.

2.1 Centralized Simulation Environments

These environments involve the development of a single simulator for the purpose of modeling and simulating both the power and communication networks. Such an implementation would help in alleviating the problem of time synchronization and coupling between the different components in the simulation. However, significant effort needs to be directed towards the development and validation of detailed models incorporating both static and dynamic behaviors of the system components. The major challenge associated with this methodology is the need for implementation of comprehensive validated models for both the power and communication system. Additionally, since these are generally implemented using software packages, it may be possible to test applications but not the physical devices. Power System Computer Aided Design (PSCAD) is a power system simulation tool. Implementation of a synchrophasor device in the simulator is discussed in [13]. Through the implementation of communication network components as discussed in [14], the simulator can be used for cyber-physical simulation. However, these models need to be tested and validated extensively.

2.2 Co-simulation Environments

A more practical and feasible approach is to keep the simulation of power and communication systems in different simulators and integrate them through a common framework to work together. The common framework is used to realize the required time synchronization and data flow interface between the two simulators. The main advantage is that industry grade commercial tools can be used for creating a cyber-physical simulation environment.

The Electric Power and Communication Synchronizing Simulator (EPOCHS) was the first one of this kind of simulation environments. Two different commercial power system simulators PSCAD/Electromagnetic Transients including DC (EMTDC) and Power System Load Flow (PSLF) were integrated with Network Simulator 2 (NS2). The connection between the simulators is realized through the implementation of a Run Time Infrastructure responsible for maintaining the same time scale on all the simulators. This is required due to the different time scales of the two simulators. The Global Event driven Co-simulation framework (GECO) combines the PSLF and NS2 to provide a co-simulation framework. The main goal here is the modeling and simulation of wide area monitoring, protection and control schemes [17]. The test bed developed at University of Arizona called Test Bed for Analyzing Security of SCADA Control System (TASSCS) is used for SCADA analysis [18]. It uses OPNET system-in-the-loop (SITL) emulation along with PowerWorld simulator. It is primarily used for research activities related to intrusion detection. SCADA Cyber Security Testbed [19] is another platform that is similar to TASSCS except that Real-Time Immersive Network Simulation Environment (RINSE) is used to simulate the cyber system. The Virtual Control Systems Environment (VCSE) developed by Sandia National Laboratory uses OPNET and PowerWorld simulator [20]. The Virtual Power System Test bed (VPST) developed at the University of Illinois at Urbana-Champaign utilizes RINSE and PowerWorld simulator [21]. The major advantage of co-simulation is the ability to integrate simulators, emulators and physical hardware devices for integrated simulation.

2.3 *Guidelines for Modeling and Simulation of Cyber-Power System*

This chapter focuses on the modeling of power system, data measurement services, data collection services, and communication networks used to deliver data to the control center. The models used in different power system simulators might vary slightly in the level of detail but the underlying core generally remains the same. The basic models used in power system simulators are already available. For data measurement and collection models, it is preferable to have these models implemented inside the power simulator. In some simulators, the data measurement and collection devices are not modeled explicitly. For simulation of communication system networks in the power system, generalizations and acceptable assumptions need to be made to design a feasible system. The layout of communication networks is affected by the structure of power system as the communication links are usually laid along transmission lines with substations representing nodes in the network. Guidelines that can be used for modeling of communication systems for the power grid are presented in the next section.

2.3.1 Communication Network

Two types of communication network topologies are possible: point to point star topology and mesh topology. The star and mesh topologies are shown in Fig 3. Star topologies are essentially single hop networks where a packet originating at the source makes just one hop to reach the destination. Such networks are very costly to build for large systems, since it involves building dedicated lines for each substation and may not be economically feasible. In the distribution system where few localized control points maybe connected to devices in the network through dedicated communication networks over short distances, star topology may be used. Mesh topologies on the other hand involve the use of a single communication link by multiple nodes thereby increasing link utilization. This leads to a more efficient use of the available infrastructure. These networks are multi-hop networks where a packet might need to be routed through multiple nodes before reaching its destination.

The communication network for a given power system is generally derived using a top down approach.
1) The first step is to reduce the power system network topology into substations/nodes to obtain the nodal representation of the network. Usually a bus represents a substation in a power system layout. However, in the event that a transformer is present between two buses, it is reasonable to assume that the transformer and the associated buses are located in the same substation. For distribution systems, control

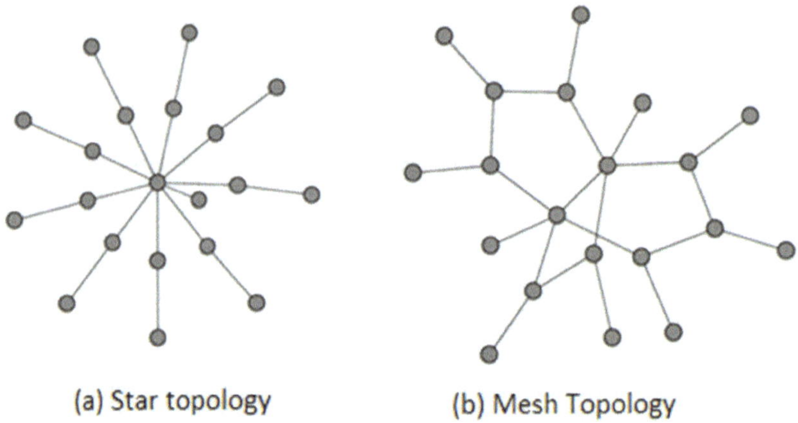

Fig. 3 Communication network topologies

or monitoring points which collect and/ transmit data can be represented as nodes. Fig 4 shows the IEEE 14 bus power system one line diagram and the reduced communication network nodal diagram.

2) Fiber optic cables can be laid along the transmission lines in the power system. It is reasonable to deduce the length of these data transfer lines from the transmission line length. The length of the transmission lines are derived using [15] for appro-

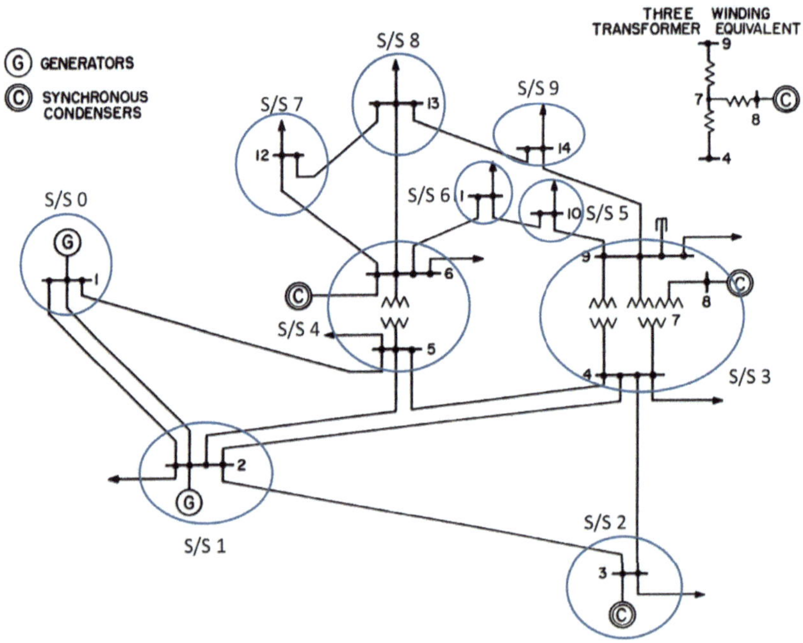

Fig. 4 Communication network node assignments

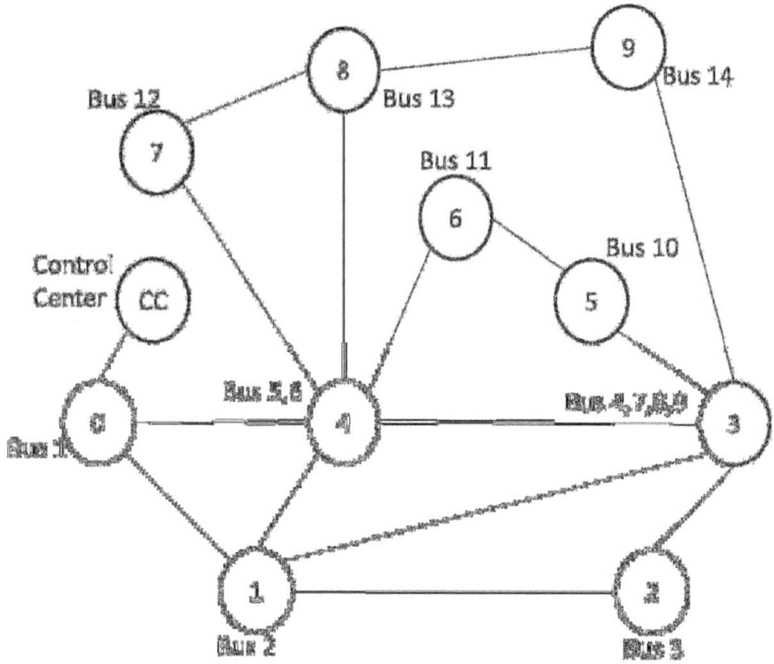

Fig. 5 Communication model for the IEEE 14 bus system

priate voltage levels. For distribution level voltages, it is reasonable to assume a per mile reactance of 0.5 ohm. Fig 5 shows the communication network topology derived from the one line diagram of power system. In this figure, the location of control centers, special protection scheme centers and other regional control points are also added as nodes. This forms the top level view of the communication network.

3) For transmission level systems, multiple control centers maybe assumed to be present depending on the size of the system. Control centers are usually located near the reference bus or near strategic locations. For distribution systems, certain nodes may have control devices associated with them.

4) The next level of the communication network view is the intra node view. Intra-node communications are local area network interactions and can be modeled accordingly. In a power system substation, it can be assumed that a single or multiple server(s) provide access to the IEDs and relays at that substation through a dedicated gateway. Wired or wireless network or a combination of these can be used. At the transmission level, it is reasonable to assume that each node will have a gateway associated with it. Distribution network nodes may not always be equipped with gateways as the information may pass through a radial network with very few devices. A generalized intra node view is depicted in Fig 6. The intra-node view depicts the different devices that could be part of a node.

One of the objectives of a smart grid is to achieve interoperability between multiple vendors, and compatibility with legacy systems. Legacy systems refer to the devices, which do not possess advanced technology and are in use in the power system. It is worth noting that a lot of electromechanical relays are still in use and have not been replaced by digital relays completely. For a transmission level system node, it is possible to have all the devices shown in Fig 6. For a distribution system level node, it is likely that an automation controller and few metering devices will be present. Substation computers may act as a means for providing authentication to engineering access used to modify the settings in the relays remotely. It can be used as an asset management device. Additionally, it may be used to execute local monitoring and control tools.

2.3.2 Energy Management System Modeling

In the Energy Management System (EMS) domain, of particular interest, is the control center and regional transmission operator centers where end use power system applications are implemented and state of the system is constantly monitored on a Human Machine Interface (HMI). The application software packages used are typically from multiple vendors. A common model as defined by the IntelliGrid Architecture (IGA) [16] depicted in Fig 7 can be used to model these control centers to achieve inter-operability. The major advantage of following this model is that the interfacing between the data collection devices and the application will be universal. The applications can be developed to support integration with this common

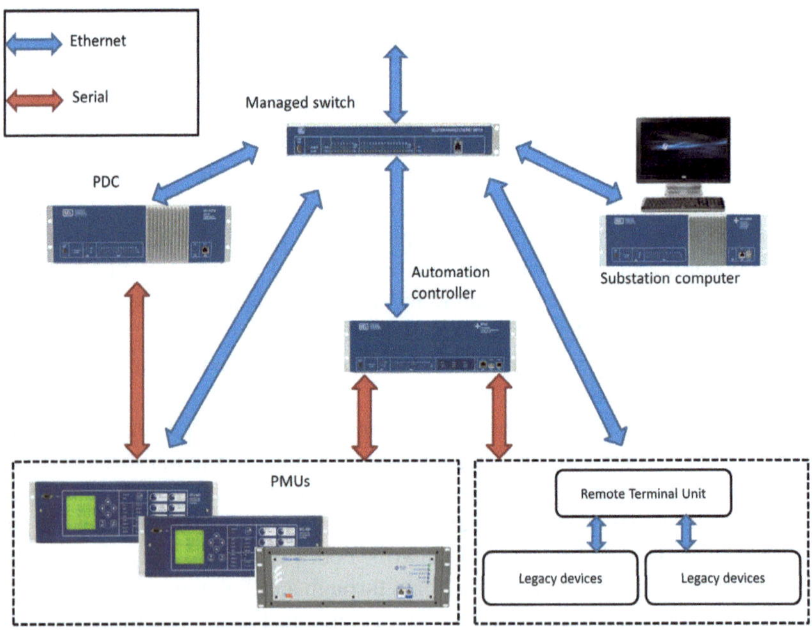

Fig. 6 Generalized intra node view

interface. Object Linking and Embedding (OLE) for Process Control Standard (OPC) is used for implementing this data transfer interface in this work. The OPC specification defines a set of objects, interfaces and methods for use in process control and manufacturing automation applications to facilitate interoperability. The OPC Data Access (OPC DA) specification is used to read and write data in real time. OPC Historical Data Access (OPC HDA) is used for access and retrieval of archived data. A OPC DA server needs to be implemented for the data aggregation device, and a OPC DA client needs to be implemented for interface with the application to be used. For example, if Citect SCADA software implements OPC DA server and MATLAB supports OPC DA client. This allows for Citect SCADA to be used for aggregation of Distributed Network Protocol (DNP) 3.0 data and MATLAB for data processing. Through implementation of OPC DA server for other data aggregation devices and software it will be possible to facilitate interoperability seamlessly.

In the event that OPC DA implementations are not possible, there are alternate options to transfer data from data aggregation device to the application. If SEL 5073 software Phasor Data Concentrator (PDC) is to be used, then a python script can be written to directly access the storage database and retrieve data. In case OpenPDC is used, then additional options exist. Data can be exported from OpenPDC into a SQL database or to a Comma Separated Value (CSV) file and data may be fed into the applications as necessary.

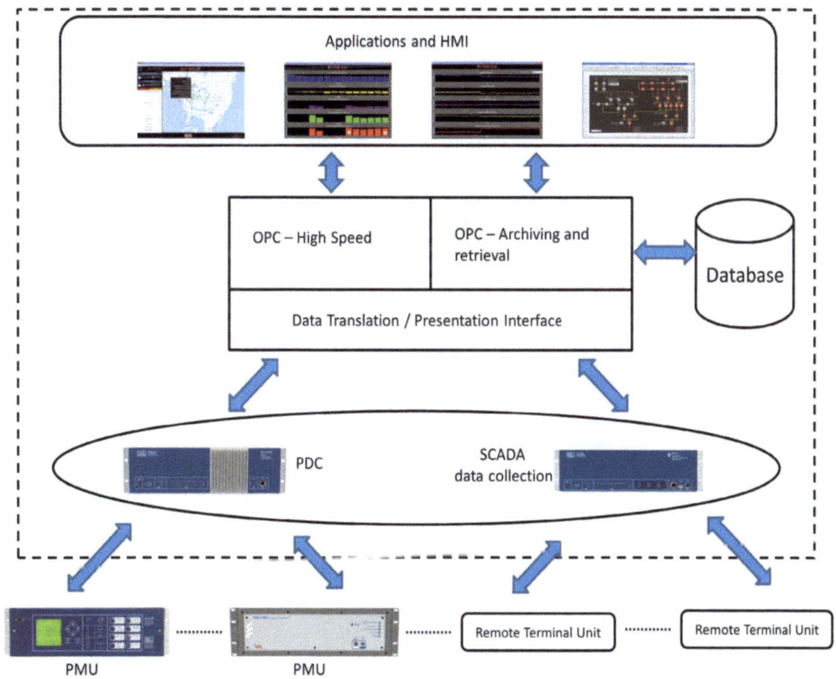

Fig. 7 EMS layer architecture

3 A Real Time Cyber-Power Test Bed

The development and implementation of a comprehensive cyber-power test bed consisting of simulation, emulation and real devices is discussed in this section. Real Time Digital Simulator (RTDS) is used to simulate the power system while network simulator -3 (NS3) is used to simulate the communication system. The data measurement and collection system comprises of commercial hardware, software and simulated devices. The major advantage of the test bed lies in the modular approach. The test bed can be segregated into four separate layers; physical power system layer, monitoring systems layer, communication layer, and energy management layer. The developed test bed is explained through the implementation and simulation of a electric power transmission level system.

3.1 Test Bed Components

Fig 8 is a self explanatory picture showing the different devices and the typical interconnections between them. This figure shows the base test bed without the communication simulator integrated into it. RTDS is used for power system and

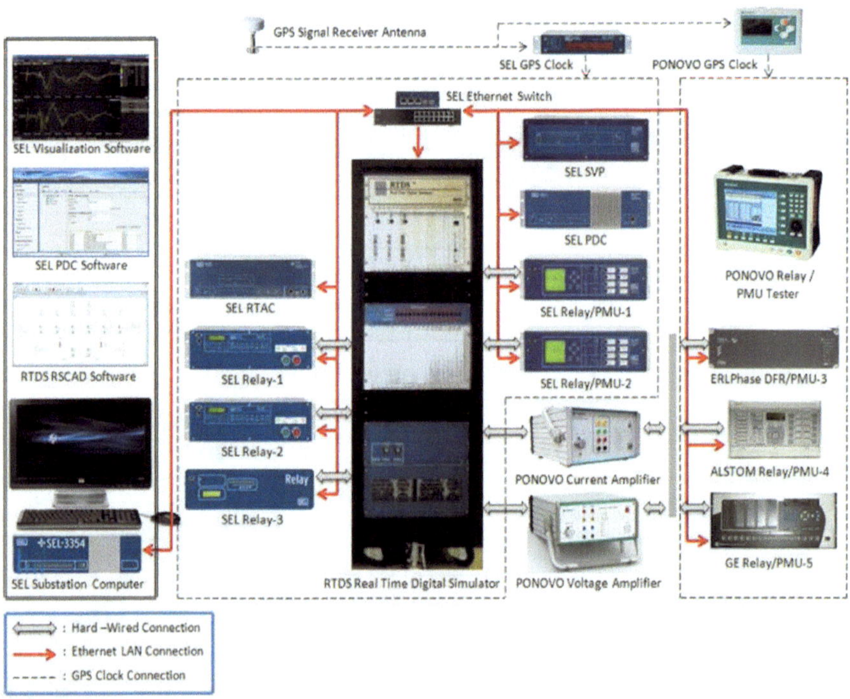

Fig. 8 Smart grid research test bed at WSU

sensors simulation. Intelligent Electronic Devices (IED) from multiple vendors such as SEL, GE, ERLPhase, Alstom are used in the test bed. These IEDs may combine functionality of a relay and phasor measurement unit (PMU) can interface with the RTDS to receive measurement signals. For IEDs that have a low level input interface for measurement signals, hardwired interface is provided with the RTDS. For devices without this option, amplifiers are used as shown in the figure. Automation controllers such as the Synchrophasor Vector Processor (SVP), and Real Time Automation Controller (RTAC) serve to implement and test local or centralized algorithms. These devices receive measurement information from the IEDs through the Ethernet Local Area Network (LAN) connection as shown in the figure. Software and hardware PDC from SEL are used for data aggregation at substations. OpenPDC is an open source PDC software maintained by the Grid Protection Alliance (GPA) is also used at the EMS layer. All the devices in the test bed are time synchronized by a single GPS clock which provides IRIG-B signals for this purpose. The test bed is designed for implementation and testing of smart grid devices and algorithms.

3.2 *Physical Power System Layer*

Advancements in the field of power system simulators have facilitated the near real time simulation of the power system. The RTDS is a virtual power system simulator designed for real time simulation with a typical time step of fifty microseconds, if no power electronic devices are modeled. This means that the state of the power grid is updated every fifty microseconds. Even though the simulation is discrete time based, due to the number of points computed within a given power system cycle, the simulation approximates the continuous time power system appropriately. The RTDS draft software module includes accurate power system component models required to represent the complex elements of the physical power system. The network solution technique employed in the RTDS is based on nodal analysis. The underlying solution algorithms are those introduced in [27] known as Dommel's algorithm. Dommels solution algorithm is used in most digital simulation tools designed for the study of electromagnetic transients.

The RTDS can be interfaced with external devices through dedicated analog and digital signal interface cards. Support for use of DNP3 protocol, software PMUs compliant with the IEEE C37.118.1 standard, GOOSE messaging and IEC 61850-9-2 sampled value messaging for power system voltages and current are also available [28]. RTDS is a commercial tool and is used extensively by academia, research organizations, service providers, and utilities for real time simulation. The basic model library provided by RTDS can be extended through implementation of user defined models. The use of RTDS enables hardware in the loop simulation through signal interface devices. Analog and digital signals can be exported and imported into the RTDS simulation environment through these dedicated devices. Hence, both monitoring and control environments are supported inherently.

For simplicity, a relatively small IEEE 14 bus test case is presented in the following sections The steady state values obtained during normal system conditions has been verified to validate the test case model.

3.3 Communication Layer

NS3 is used for the purpose of emulating the communication system for the simulated power system. NS3 has a modular implementation and contain a core library which takes care of the generic aspects of the simulator and a library dealing with specifying simulation time objects, schedulers and events. Protocol entities are written to be closer to the real world implementation. Packet implementations are based on the real data packets in order to enable communication between simulated agents and external world. This makes it suitable for emulation purposes. NS3 is running in a dedicated server to emulate the communication network in real time. It is to be noted here that the real time implementation of NS3 uses the system time to schedule events. This time is synchronized to a high precision GPS clock input which is used to time synchronize the devices in the test bed. NS3 is an open source simulator offering great flexibility in developing modules. This is of particular interest as a separate module such as security module can be implemented easily as long as the user has an understanding of NS3 development. The emulated communication network is protocol independent and can be used to transfer any data packet from source to destination.

The derivation of the communication network using the top down approach has already been presented in Sect 1.2.3.1. The reduced network for the IEEE 14 bus system has been shown in Fig 9. In NS3, each node represents a gateway to which the devices belonging to a particular private LAN can communicate. For communication with devices belonging to another private network, the gateway routes data to the receiving private network gateway.

Fig 9 depicts the communication between two private networks. In order to understand information flow, it is important to distinguish the layers involved. The power system simulation layer and EMS layer are not shown here for sake of simplicity. The communication network is represented by the nodes in the layer, interconnected by lines which represents the transmission media between the LANs. Assume that each node represents the gateway of that particular LAN number. The monitoring network is represented by the PMU at LAN 9, and PDC at LAN 0. Consider that the PMU at LAN 9 is going to send data to a PDC at LAN 0. The device in LAN 9 will view the gateway inside NS3 as being on its local physical link. It forwards the data packet to be sent to the PDC at LAN 0 to the gateway. In this way, the packet to be sent from node 9 to node 0 enters the NS3 simulation where the communication network is being simulated. The gateway node inside NS3 decides the shortest path to send the packet to the destination and forwards the packet to the next hop based on the routing table. In this example, it is node 3. In this way, the packet follows the red line and reaches node 0 gateway. The gateway then forwards the packet to the intended device. The delays that are experienced by the data packets are described as follows:

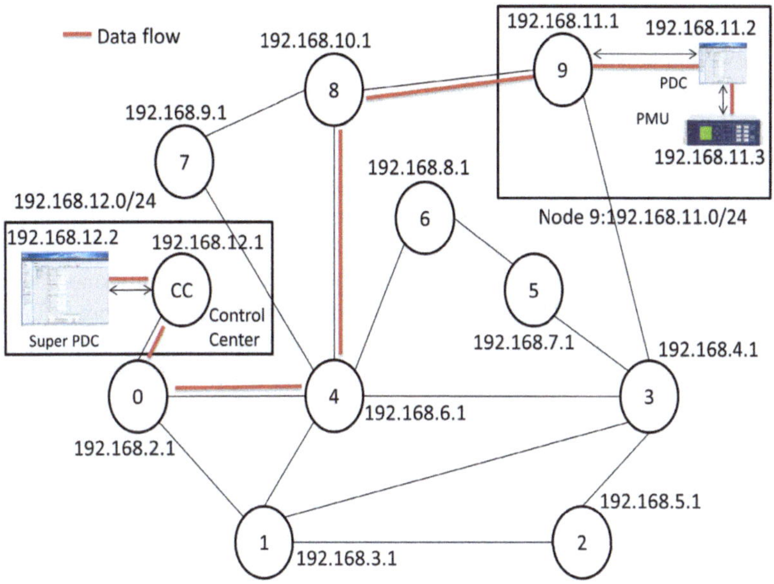

Fig. 9 Communication between two private networks

1) Network Processing Delay: These delays are incurred when the network gateways, firewalls and servers decide what needs to be done with an incoming packet. The delay depends on the network equipment technology and the specific processing function.

2) Signal Propagation Delay: It is the time taken for a signal to travel in the physical propagation medium, and depends on the medium itself and the distance. The propagation speed of the signal through fiber optic medium is about seventy percent of the speed of light in vacuum.

3) Transmission Delay: There is a definitive time delay for a packet to be completely pushed on to the physical link layer. This delay is called the transmission delay and is dependent on the bandwidth of the link and packet size.

4)Queuing Delay: This kind of delay occurs when multiple packets from an ingress port need to be routed to the same egress port. One packet is transmitted at a time, and a queue is maintained to hold the remaining packets. The queuing delay experienced by a packet is the time that a packets waits in a queue before being processed by a node. The total end network latency is the sum of all these delays.

Hence, a packet that is transmitted from a device at LAN 9 to a device at LAN 0, experiences all these delays as a result of communication network emulation by NS3.

3.4 Power Systems Monitoring Layer

The monitoring systems layer of the test bed is represented by the intra node substation view. It comprises of the sensor devices such as potential transformers (PT) and current transformers (CT), and the IEDs that use the signals from the sensor devices. A substation might have several devices interconnected and interacting with each other through IEC 61850 compliant protocols or other standard protocols. The interaction between the devices is flexible and configurable according to user requirements. In this example, it is assumed that the transmission level system is fully PMU enabled and that each bus has at least one PMU.

The substation view for such a system is shown in Fig 10. Only one PMU is shown here as an example. Depending on the node of interest, the number of PMUs might be higher with possible interaction between the PMUs. The PMU output data stream is concentrated in a PDC for transmission to super PDCs. Each substation is considered to be on its own private network with access to other private networks through its gateway. Engineering access is used to gain access to the private network to change settings and configuration files in the relays from the control center. A substation computer may or may not be present locally to make use of the data archived to run any algorithms and serve as an asset management device.

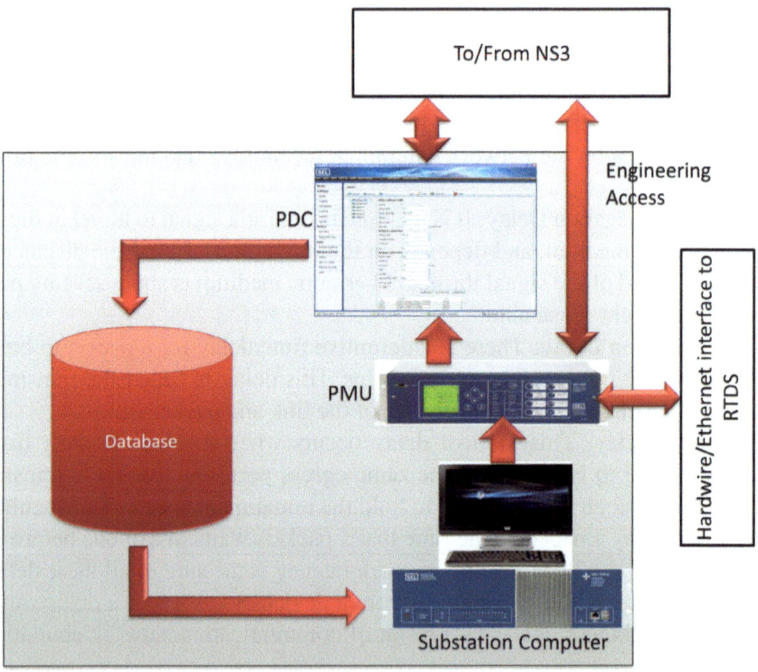

Fig. 10 Substation view for transmission system

3.5 Energy Management System Layer

The EMS layer of the test bed is represented by the control center. It mainly consists of a super PDC for collection of data from all the other nodes for data archiving. This data stored in the database maybe used by the real time and non real time applications.

Fig 11 shows the control center view for the PMU enabled transmission system. Here the control center has a super PDC, which aggregates data from all the PDCs in the system. Human Machine Interface (HMI) and visualization tools may be used depending on the application or algorithm to be implemented in the test bed. In this configuration, it is assumed that no other super PDCs are available except at the control center.

For the IEEE 14 bus system, with a PMU at each bus, fourteen PMUs are required to monitor the entire system. The communication topology is made up of ten nodes, eleven if the control center is included. Each node is considered to be a substation and houses one software or hardware PDC. This brings the total number of PDCs used to eleven including the super PDC at the control center. The test bed can support up to six hardware PMUs due to the limited availability of signal interface devices in the RTDS. The remaining eight PMUs are software PMUs simulated in a

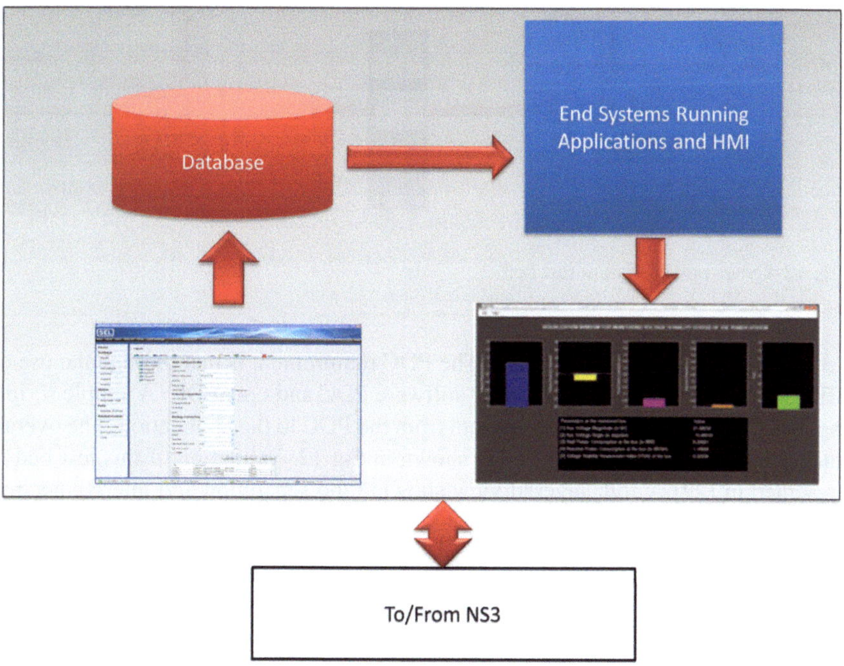

Fig. 11 Energy management system view for transmission system

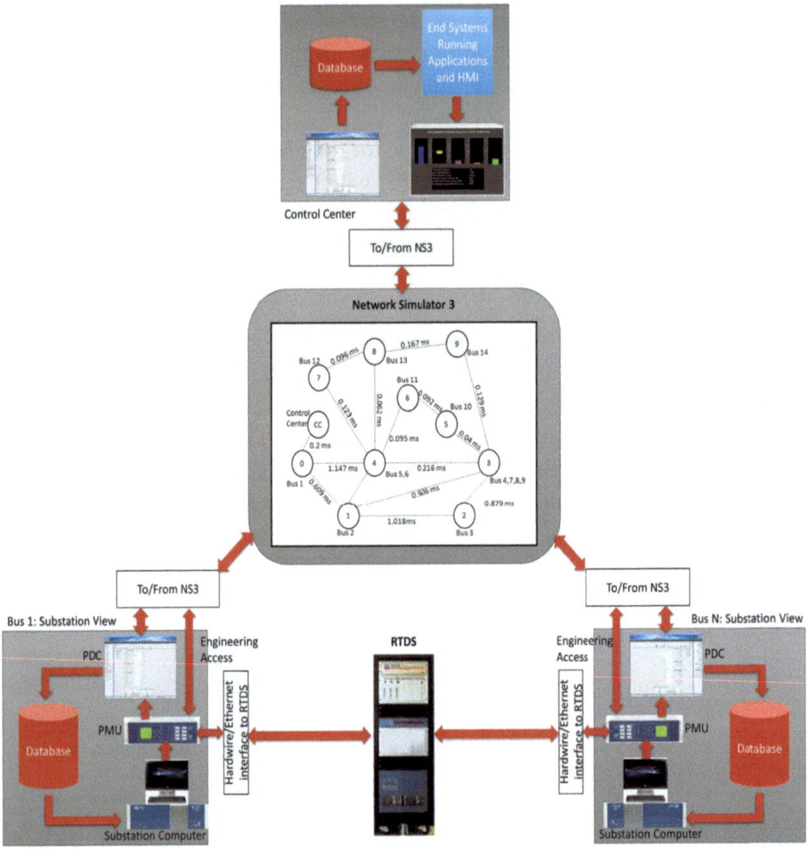

Fig. 12 Cyber-power system test bed

dedicated expansion card in RTDS. The PDC requirement is met through the use of SEL 3373 hardware PDC, SEL 5073 software PDC and openPDC. A simple scripting interface is provided to deliver data from the PDC to the application. The overall integrated cyber-physical test bed is shown in Fig 12. Validation of this test bed is presented in [22] A four layered view showing the separation and interconnection of the different layers is shown in Fig 13.

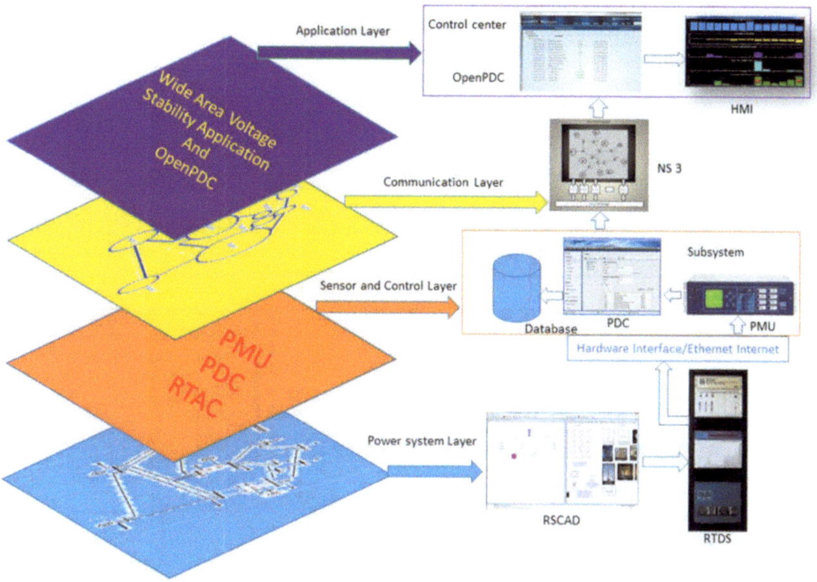

Fig. 13 Four layer view of the cyber-power test bed

4 Applications of Cyber-Power Test Bed

This section provides examples of applications for cyber-physical analysis using different configurations of cyber-power test bed. In addition to the application testing, the test bed can also be used for device testing such as PMUs and PDCs [23].

4.1 Local Voltage Stability Monitoring Algorithm

The test bed setup for the local voltage stability monitoring algorithm (LVSMA) is shown in Fig 14. Only the implementation of LVSMA is discussed here and further details about the implementation can be found in [24].

The requirements of the algorithm are stated as follows:

1) The Local Voltage Stability Monitoring Algorithm (LVSMA) will be running in the substations only. A substation computer is required to carry out the analysis locally at the substation. For this purpose, the SEL 3354 substation computer is used.

2) In addition to the local phasor data, the algorithm requires the use of voltage angle of the slack bus to obtain the angles with reference to the slack bus. So, the slack bus voltage angle is sent to each substation in the network.

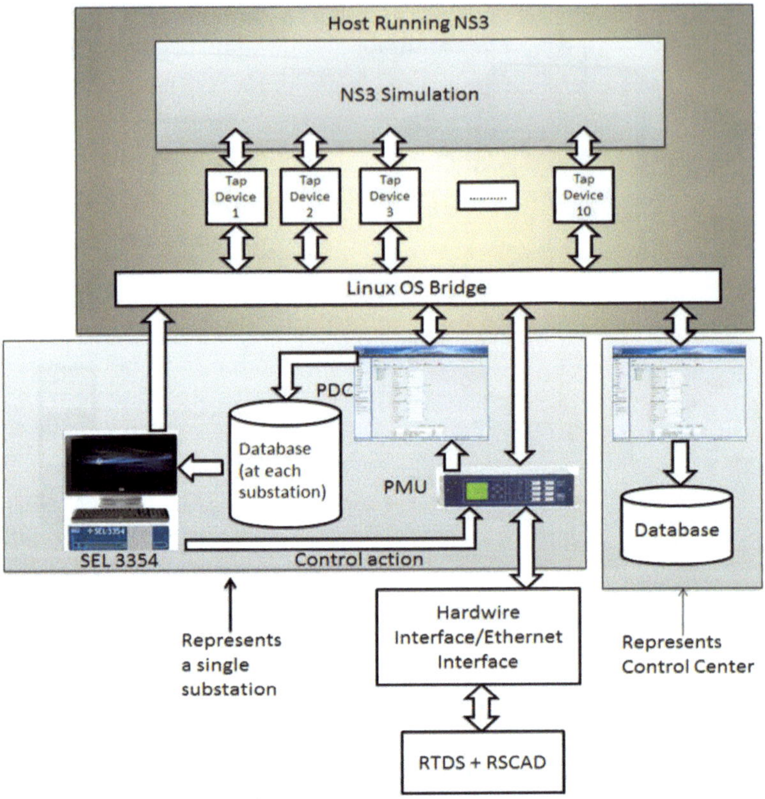

Fig. 14 Test bed setup for testing LVSMA

3) The calculated voltage stability index is then transferred to the control center through the communication network. A control algorithm may be present at the local or the control center. This algorithm is responsible for taking any control actions necessary to avoid voltage collapse.

4) The application running time is in the range of microseconds. Therefore, the application running frequency is restricted only by the availability of data and requirement.

The substation PDC receives the voltage angle from the slack bus through the control center PDC. The scripts for obtaining data from the database at the substation are executed locally in the substation computer. The local VSMA is successfully implemented to compute the index for voltage stability monitoring.

4.2 *Wide Area Voltage Stability Monitoring Algorithm*

The wide area voltage stability monitoring algorithm (WAVSMA) in EMS layer uses a non iterative methodology for computing voltage stability indices and is based on system centric reduced network equivalent method [25]. The algorithm makes use of system wide voltage, voltage angle measurements, and system topological information to compute the distance to the point of voltage collapse (PoC) for the load buses in the power system. An index termed Voltage Stability Assessment Index (VSAI) that ranges between '0' and '1' is used to specify this value. A value near '0' indicates a highly voltage stable load bus and a value near '1' indicates that the load bus is near the point of voltage collapse.

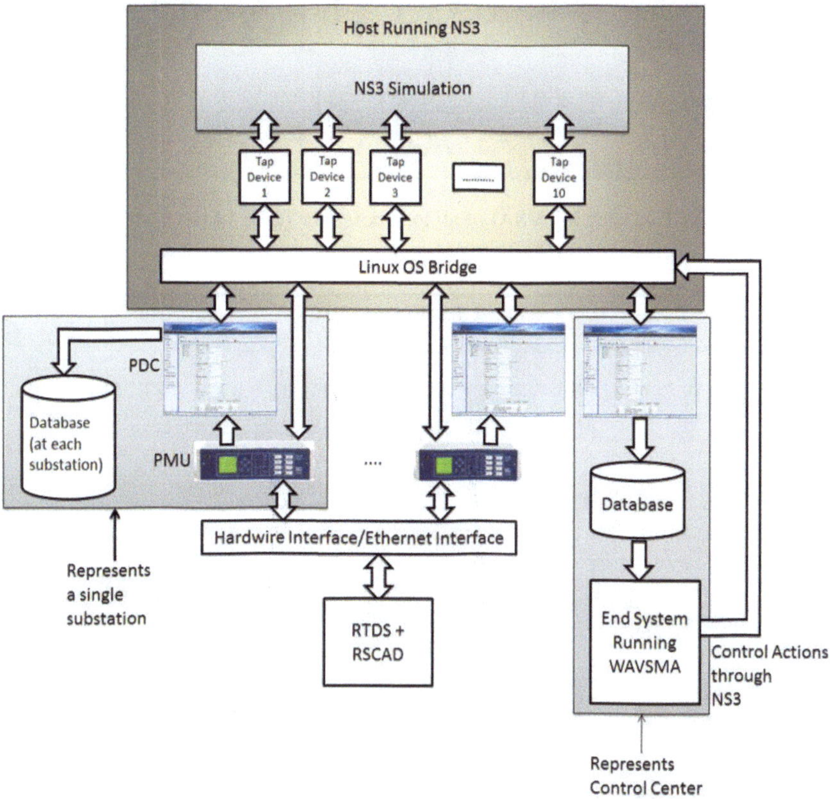

Fig. 15 Test bed setup for testing WAVSMA

As the developed algorithm is non iterative, it is computationally less burdensome than the existing multiple power flow based approaches. Hence it is suitable for the real time monitoring of the system voltage stability. The typical run time of the application is in the range of a few milliseconds, allowing it to process PMU data

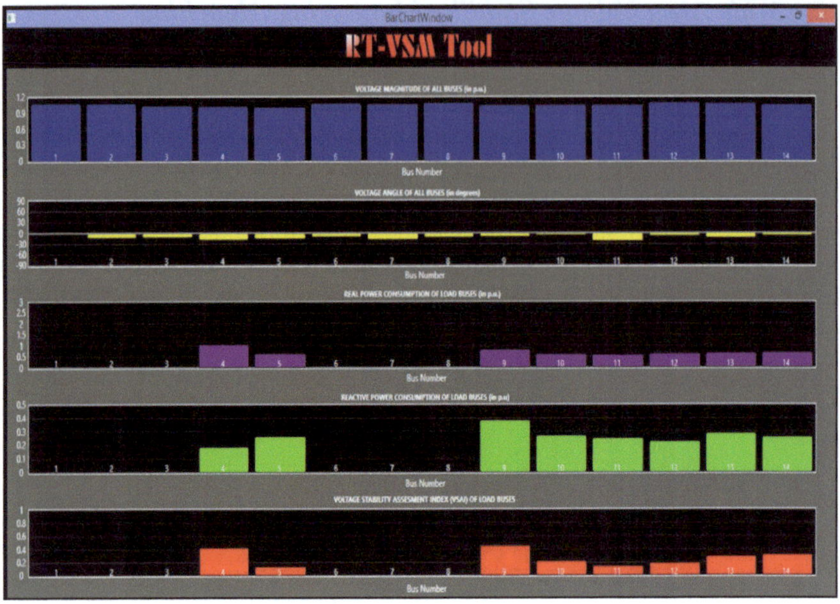

Fig. 16 RT-VSM Tool showing VSAI of all load buses for IEEE 14 bus system

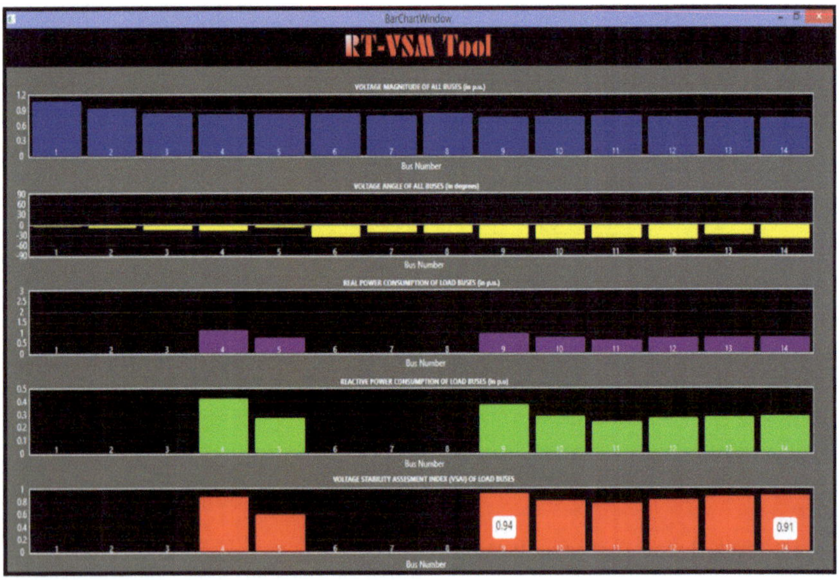

Fig. 17 RT-VSM Tool showing VSAI of all load buses for IEEE 14 bus system following load increase

as available. Additionally, the algorithm can integrate and operate with traditional monitoring systems as well as PMU based systems governing the time step required to run this application.

The requirements of this application are summarized as follows:

1) The WAVSMA tool is assumed to be running at the control center. However, it could be used at different locations as needed. One of the Dell precision workstations available in the test bed is used as the platform for running the tool. 2) The application needs the voltage phasor and voltage angle values only. Breaker statuses are not required if a topology processor is providing the topological information.

3) The WAVSMA calculates the voltage stability assessment index, and based on this any necessary control action may be taken. The control action is relayed to the appropriate substation.

All substation PDCs send the phasor data aggregated at the substation to the control center PDC. At the control center, a script is used to retrieve the data in the format as needed by the application. It is to be noted here that the algorithm running time is small, and the frequency of application execution is restricted generally by the rate of data input. Any control action specified by the application is relayed to the appropriate substation through NS3.

A tool that can enable power system operators to visualize the real time voltage stability condition has been developed using the proposed algorithm and has been named Real Time Voltage Stability Monitoring (RT-VSM) Tool. C# Language and

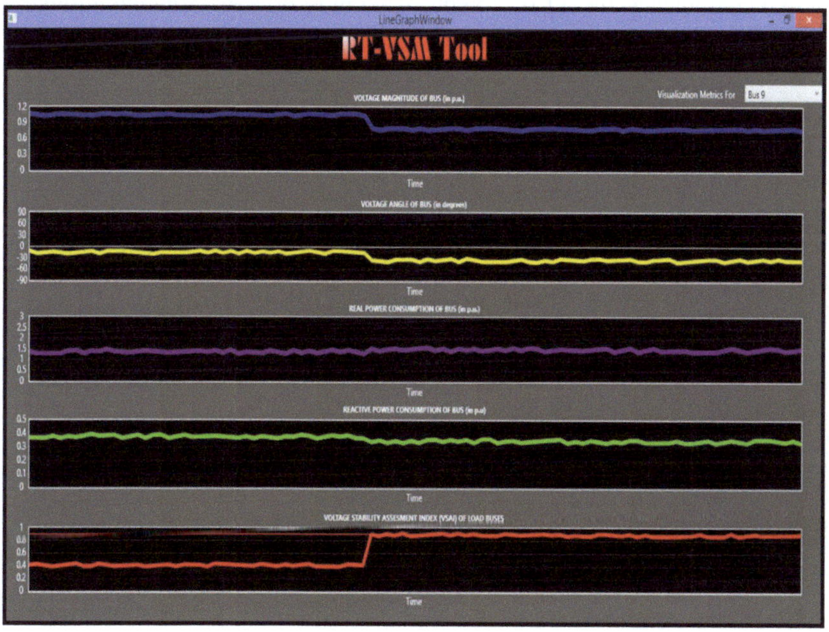

Fig. 18 Real time visualization window of the RT-VSM Tool

XAML have been used to build this tool. Fig 16 shows the visualization window of the RT-VSM Tool when the loading at the load buses in the IEEE 14 bus test case is increased. The increase in load leads to an increase in the voltage stability assessment index indicating a stressed system. The visualization windows of the RT-VSM Tool show that during the stressed case, VSAI of all the load buses increase indicating deterioration of voltage stability. It is found that buses 9 and 14 have VSAI values above the set alarm value of 0.9, as has been indicated in the Fig 17. Fig 18 shows the visualization window of the RT-VSM Tool that tracks the changes in the critical system metrics during the change in system loading. The VSAI of the weakest bus under the given system conditions i.e. Bus-9 changes from 0.44 to 0.94. Further details related to the use of the test bed for this application can be found in reference [26].

4.3 Shipboard Power System Reconfiguration

The Shipboard Power System Reconfiguration Algorithm (SPSRA) implemented in the test bed is a genetic algorithm based application [28]. The algorithm is coded in structured text and implemented in the SEL 3530 RTAC. Power generation, load on the system and breaker status data in the system are the data required in addition to load priority information. DNP data communication is implemented between all devices in the network. Fig 19 shows the test bed setup for SPSRA. The algorithm monitors the state of the power system continuously. In the event of generation loss due to a contingency, the RTAC computes the most effective solution through the

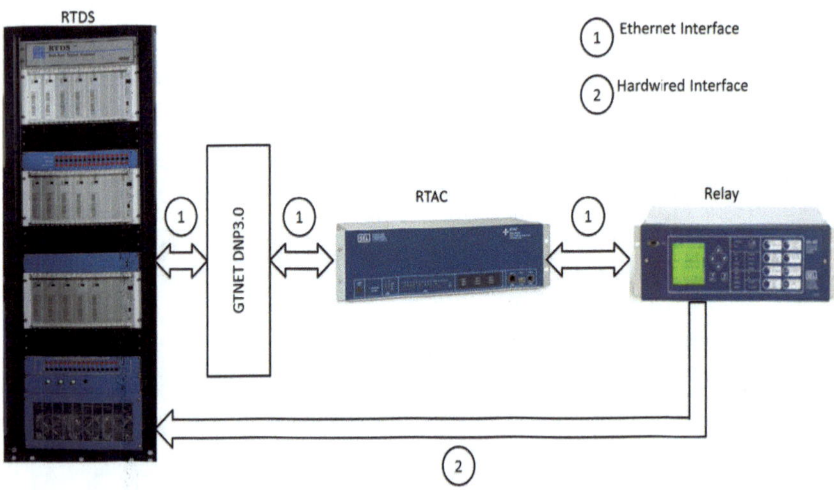

Fig. 19 Test bed setup for testing SPSRA

use of genetic algorithm and performs a reconfiguration of the system based on load priority, restoring power very quickly.

4.4 Aurora Attack Simulation

In 2007, Idaho National Laboratories (INL) demonstrated the Aurora attack involving the opening and closing of breaker associated with the generator in fixed intervals. The assumption is that the sync check element (25) is disabled in the relay

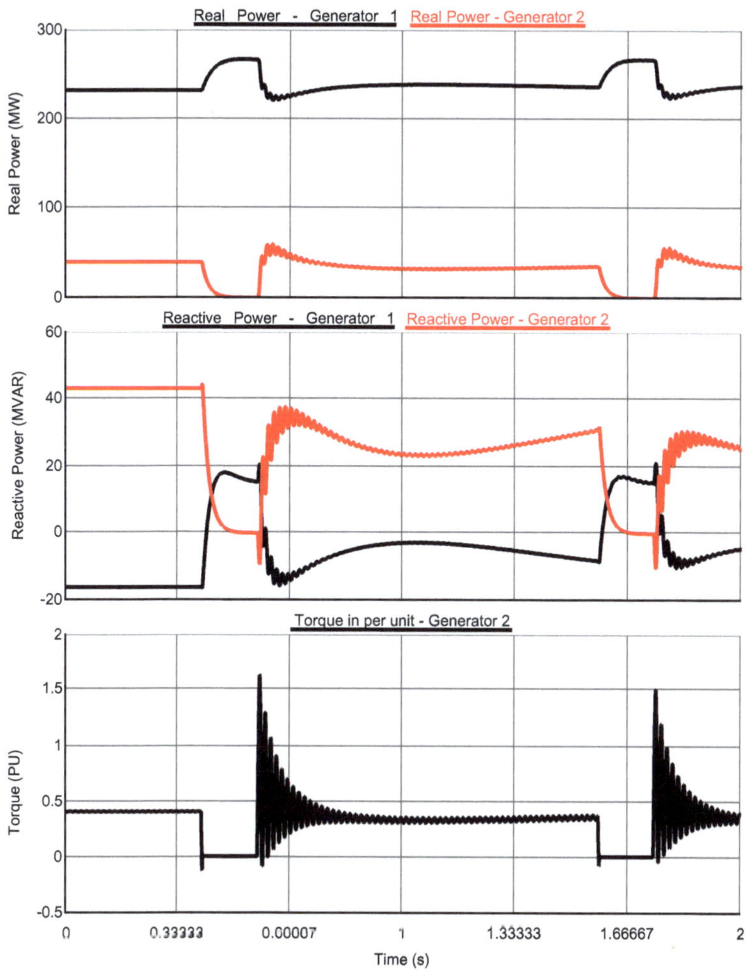

Fig. 20 Plot showing swing in real, reactive power, and electrical torque at Gen 2

leading to the possibility of re-closing the generator without a synchronism check leading to a out-of-sync close.

Aurora attack was re-simulated using the developed cyber-power test bed. The ranking of critical generators for the IEEE 14 bus system is computed based on the generator contingency ranking with incomplete information and cyber vulnerability index for relays [29]. For the purpose of simulating the attack in the test bed, it is assumed that the attacker has access to the network information and IEDs by compromising the private network. The attacker gets access through engineering access to the IED used to control the breaker at Bus 2 of the IEEE 14 bus system. This means that the attacker now has access to the IED settings, which can be modified maliciously. The IED settings is reprogrammed by the attacker to open the breaker for 10 cycles, causing the generator to spin faster upon loss of the load, and reclose the breaker after one second. The one second intervals gives enough time for the system to recover and not cause tripping due to any other relays picking up the induced fault condition in the system. The simulation of this attack can be made realistic by overlaying a security model on the communication network. This requires implementation of a security module that can be integrated with NS3. The impact on the machine due to the attack is depicted in Fig 20. It can be seen that the machine is subjected to high mechanical stress due to the out of sync reclosing and will suffer physical damage if attack is repeated.

5 Summary

In this chapter, the cyber-physical characteristics of the smart grid and need for a cyber-power system analysis using a real time test bed is presented. The different layers of the smart grid and the coupling between these layers are discussed. The modeling and simulation of the power system layer, communication network layer, measurement and data collection layer, and EMS layer using real time digital simulator (RTDS), network simulator-3 (NS3), hardware sensors, and controllers have been presented. Challenges involved in developing such a test bed and guidelines have also been discussed. Additionally, example applications using the developed cyber-power test bed have been successfully implemented and presented. The next step in the test bed development process is to add a security layer, which would reflect a real world scenario. By defining access policies and firewall rules for each private network, the attack-defense mechanism testing can be made possible. Recent developments in NS3 in the implementation of network address translation (NAT) and Netfilter provide basic framework for the implementation of a security layer. These improvements are still in development and need validation and testing before deployment in the test bed.

References

1. Singh, S.N.: Electric Power Generation, Transmission and Distribution. Prentice Hall India Pvt. Limited (2004) ISBN: 9788120321922

2. Padiyar, K.R.: HVDC Power Transmission Systems: Technology and System Interactions, New Age International (1990) ISBN: 9788122401028
3. Bose, A.: Power System Stability: New Opportunities for Control. In: Liu, D., Antsaklis, P.J. (eds.) Stability and Control of Dynamical Systems and Applications. Birkhäuser, Boston (2003)
4. Gungor, V.C., et al., Smart Grid Technologies: Communication Technologies and Standards. IEEE Trans. Ind. Informat. (2011), doi: 10.1109/TII.2011.2166794
5. National Institute of Standards and Technology, NIST framework and roadmap for smart grid interoperability standards 2.0 (2012), `http://www.nist.gov/smartgrid/upload/NIST_Framework_Release_2-0_corr.pdf`
6. Smith, R.: Assault on California Power Station Raises Alarm on Potential for Terrorism (2014)
7. Govindarasu, M., et al.: Cyber-Physical Systems Security for Smart Grid. PSERC (2012), `http://www.pserc.wisc.edu/documents/publications/papers/fgwhitepapers/Govindarasu_Future_Grid_White_Paper_CPS_Feb2012.pdf`
8. Meserve, J.: Stage cyber attack reveals vulnerability in power grid (2007), `http://www.cnn.com/2007/US/09/26/power.at.risk/`
9. Bou-Harb, E., et al.: Communication security for smart grid distribution networks. IEEE Commun. Magazine (2013), doi:10.1109/MCOM.2013.6400437
10. Yilin, M., et al.: Cyber-Physical Security of a Smart Grid Infrastructure. Proceedings of the IEEE (2011), doi: 10.1109/JPROC.2011.2161428
11. Khaitan, S.K., McCalley, J.D., Cyber physical system approach for design of power grids: A survey. In: IEEE Power and Energy Society General Meeting (PES) (2013), doi:10.1109/PESMG.2013.6672537
12. Khaitan, S.K., McCalley, J.D.: Design Techniques and Applications of Cyberphysical Systems: A Survey. IEEE Systems Journal (2014), doi:10.1109/JSYST.2014.2322503
13. Gurusinghe, D.R., et al.: Modeling of a synchrophasor measurement unit in an electromagnetic transient simulation program. In: Proc. Int. Conf. on Power Syst. Transients (2012), doi:10.1109/PESGM.2012.6343954
14. Menike, S., et al.: Implementation of communication network components for transient simulations in PSCAD/EMTDC. In: Int. Conf. on Power Syst. Transients (2013)
15. Anderson, P.M., Fouad, A.A.: Power System Control and Stability, p. 450. Iowa State University Press, Ames (1977)
16. Premerlani, M., Kasztenny, B.: Synchrophasors: Definition, Measurement, and Application (2006), `http://www.gedigitalenergy.com/SmartGrid/Sep06/Synchrophasors_Paper.pdf`
17. Lin, H., et al.: Global Event-Driven Co-Simulation Framework for Interconnected Power System and Communication Network. IEEE Trans. Smart Grid (2012), doi:10.1109/TSG.2012.2191805
18. Mallouhi, M., et al.: A testbed for analyzing security of SCADA control systems. In: IEEE PES Innovative Smart Grid Technologies (2011), doi:10.1109/ISGT.2011.5759169
19. Davis, C.M., et al.: SCADA cyber security testbed development. In: North American Power Symp., Carbondale, Illinois (2006)
20. McDonald, M.J.: Modeling and simulation for cyber-physical system security research. Sandia National Laboratories Development and Applications, SAND2010-0568 (2010)

21. Bergman, D.C., et al.: The virtual power system testbed and inter-testbed integration. In: Proceedings of Cyber Security Exp. Test (August 2009)
22. Vellaithurai, C., et al.: Development and Application of a Real Time Cyber-Power Test Bed. IEEE Trans. on Industrial Informatics (in Review)
23. Biswas, S.S., et al.: Development of a smart grid test bed and applications in PMU and PDC testing. In: North American Power Symp., Champaign, Illinois (September 2012)
24. Biswas, S.S., et al.: Development and real time implementation of a synchrophasor based fast voltage stability monitoring algorithm with consideration of load models. In: IEEE Ind. Applicat. Soc. Annual Meeting (October 2013), doi:10.1109/IAS.2013.6682584
25. Biswas, S.S., et al.: RT-VSMAP: A Real Time Voltage Stability Monitoring and Prediction Algorithm for Electric Power Grids, Patent filed, USPTO: 27158.8052.US01 (2014)
26. Srivastava, A., et al.: Real Time Cyber-Power System Analysis. In: TCIPG Annual Ind. Day Workshop (2013), `http://www.youtube.com/watch?v=_wfbcOWckmM`
27. Dommel, H.W.: Digital Computer Solution of Electromagnetic Transients in Single- and Multiphase Networks. IEEE Trans. on Power Apparatus and Systems (1969), doi:10.1109/TPAS.1969.292459
28. Shariatzadeh, F., et al.: Real-Time Implementation of Intelligent Reconfiguration Algorithm for Microgrid. IEEE Trans. Sustain. Energy (2013), doi:10.1109/TSTE.2013.2289864
29. Srivastava, A., et al.: Modeling Cyber-Physical Vulnerability of the Smart Grid With Incomplete Information. IEEE Trans. Smart Grid (2013), doi:10.1109/TSG.2012.2232318

Cyber Physical Approach to HVDC Grid Control

Lars Nordström and Davood Babazadeh

Abstract. This chapter presents a cyber-physical approach to design of HVDC control system architectures and evolving HVDC grid operation and control modes. In addition, the chapter describes the communication system architectures needed for centralized and distributed operation and control of HVDC grids. Modeling and analysis methods suitable to analyze such systems using graph theoretic concepts, and also the design of distributed control systems utilizing a Multi-Agent approach and its dependence on the information graph theory. The chapter is concluded with a description of an application for distributed control of DC grids utilizing the concepts introduced. The application is presented both with regards to comparison with other design choices and analysis of performance and robustness of the algorithm versus communication metrics.

1 Multi Terminal HVDC and HVDC Grids

This section provides background and overview to the development towards High Voltage DC (HVDC) grids. The section is started with an introduction to some of the driving forces behind the development. Thereafter two key technologies for HVDC grids – the Voltage Source Converter and the DC breaker are presented in some detail. The section is concluded with a discussion on the type challenges that appear for DC system operation in various configurations.

1.1 HVDC Grid Rationale

In the recent power system expansion driven by growing energy demand, more attentions are being put toward integration of large-scale renewable resources.

Lars Nordström · Davood Babazadeh
Department of Industrial Information and Control Systems, School of Electrical Engineering, KTH- Royal Institute of Technology,
SE 100-44 Stockholm, Sweden
e-mail: davood@kth.se, larsn@ics.kth.se

S.K. Khaitan, J.D. McCalley, and C.-C. Liu (eds.), *Cyber Physical Systems Approach to Smart Electric Power Grid*, Power Systems, DOI: 10.1007/978-3-662-45928-7_4

This is tangible by observing the intention of the European Union in producing 20 % of its electric power needs through renewable resources by 2020 [1]. The benefits of integration of renewable resource such as wind or solar have been justified through many researches and to some extent in real-world projects. One of the challenges in increasing the share of renewable production in the power system is the remote location of the resources and consequently the problem in transmitting bulk electric energy to load centers.

Current AC power transmission grids operate close to their limits. Furthermore, the expansion of the grid involves problematic legislative rights-of-way efforts limiting the speed of expansion. Therefore, recently there is a considerable attention drawn to application of HVDC transmission grids on top of, or in complement to, existing AC power systems as a solution to these limitations on integration of renewable resources. Potential benefits of such an HVDC grid includes the possibility to access remote energy sources thereby increasing renewable penetration, improve grid security and decrease congestion in the system [2].

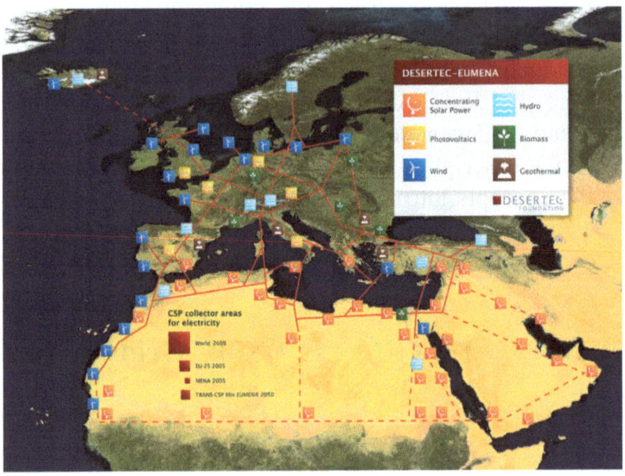

Fig. 1 Schematic of HVDC connections in DESERTEC project

In this regard, there have been several proposals and projects to study the feasibility of such HVDC grids in the European power system from market, technical, as well as political perspectives. For example, the conceptual study of a *Supergrid* proposes a HVDC network connecting offshore and onshore Supernodes that collect the remote renewable energy and transmit it to the existing power grid on land. On prime target application for the Supergrid is to integrate Europe's abundant offshore wind in the North Sea [2]. In parallel, another European project DESERTEC aims to develop a wide renewable energy integration plan to harness sustainable power from remote sites to load centers also see Fig. 1. Although this project covers all kind of renewable resources, but it emphasizes the sun-rich deserts with solar energy capabilities [3].

Besides these proposals, there have been similar projects planned and some of them partially installed. For example, the North-East Agra project in India aims to use series DC links that form a multi-terminal HVDC system to transfer power to loads. The North-East Agra links can transfer around 65000 MW hydro power resources scattered in a large area in the north of India to electrify around 90 million people. Once operational this system will be the world's first multi-terminal Ultra-HVDC link with four terminals [4].

1.2 HVDC Grid Technologies

In order to convert high voltage AC power to DC power, two well-known technologies are available, classical line commutated converter (LCC) and the voltage-source converter (VSC).

Line Commutated Converter (LCC) uses a semiconductor switch called thyristor to rectify AC current. Thyristors similar to diodes only conduct current when the AC voltage over them is positive. The difference however, is that thyristors need to be turned on, or fired, in order to begin conducting. These switches can withstand the AC voltage in either polarity. But current can only flow in one direction and can be limited by adjusting the time the thyristors are turned on. This time, or angle in a sinusoid, at which the thyristors are turned on is called the firing angle, or valve ignition delay angle, and is used to control power flow between the HVDC stations.

Voltage Source Converter technology uses Insular Gate Bipolar Transistors (IGBT) instead of thyristors. The IGBT semiconductor can be controlled both with regards to being turned on or off. In VSC technology, the DC current on the DC link can be flowed in both directions that is a benefit over the LCC technology in which the current can flow in one direction. Considering the bi-directional capability of the DC current flow in VSC, there is no need to change the DC voltage polarity of the converters to change the power flow direction between converters. Compared to LCC technology, it is possible for VSC to be connected to weaker grids which has low short-circuit level. In addition, in contrast to LCC, VSC has two degree of control which makes it capable to control the active power and reactive power separately. One challenge with IGBT based VSC is that they have less overload capability compared to LCC.

Regarding HVDC grids development, the VSC technology is due to its power flow flexibility the suitable solution to build meshed topology grids. On the other hand, in order to transfer higher power capacity in a series link topology (similar to North-East Agra project) LCC technology is more suitable. Since in this chapter, the core of the work refers to meshed HVDC grids, the following sections focus on introducing the VSC technology.

The implementation of meshed HVDC grids brings different challenges which the most important one is the need for DC breakers in the case of fault in the HVDC grid. In HVDC grids, the DC lines require to be equipped with DC breakers to de-energize the faulty line and keep the intact part of the DC grid in operation in order to increase the utilization factor of the whole system.

1.3 VSC Technologies

In contrast to LCC technology, the VSC has an additional degree of freedom to also control reactive power separate from the active power. This freedom comes from controlling the converter using the Pulse Width Modulation (PWM) technology. PWM lets the magnitude and phase of the voltage be controlled spontaneously which allows independent control of active and reactive power.

In literature, two different approaches have been introduced to control the VSC, i.e. direct control and vector control. Direct control is based on controlling the voltage in the VSC. This means by controlling phase angle and amplitude of the voltage transmitted active power and reactive power is controlled. Vector control on the other hand sets the converter to work as a controllable current source. In this approach, the injected current vector is set to follow a reference current vector.

The vector control method has some advantages compared to direct control. This includes better power quality since it is less influenced by grid harmonics and disturbances. Besides, in vector control decoupled control of active and reactive power is possible. Finally it also provides the capability of inherent protection during over-current events [5].

The benefits of vector control method compared to direct method makes it the common control scheme in VSC-HVDC systems. Therefore, a brief description of vector-based control architecture for VSC-HVDC is presented in this section.

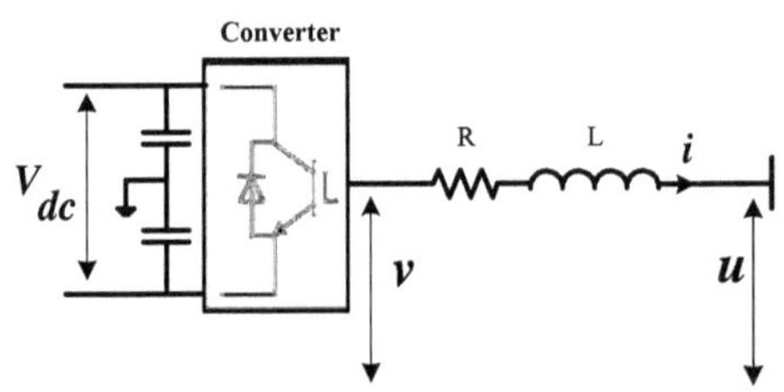

Fig. 2 Basic schematic of VSC

Fig. 2 shows a typical schematic of a VSC station. *R* and *L* are the resistance and inductance of the converter AC side that include the transformer and phase reactor parameters. Based on the schematic, the equation of the AC side in *abc* coordinates can be written as:

$$v_{abc} - u_{abc} = L\frac{di_{abc}}{dt} + Ri_{abc} \tag{1}$$

Where V is the AC voltage of the converter, U is the AC voltage at the point of common coupling (PCC) which connects the station to the AC grid, and i is the current coming/going from/to AC grid. The equation (1) in *abc* coordinates can be transformed to *dq* coordinates presented in equation (2). In these coordinates, both decoupled i_d and i_q currents can be controlled separately by Inner Current Control.

$$\begin{bmatrix} v_d - u_d \\ v_q - u_q \end{bmatrix} = \begin{bmatrix} R & -\omega L \\ \omega L & R \end{bmatrix} \begin{bmatrix} i_d \\ i_q \end{bmatrix} + \begin{bmatrix} L & 0 \\ 0 & L \end{bmatrix} \frac{d}{dt} \begin{bmatrix} i_d \\ i_q \end{bmatrix} \tag{2}$$

The VSC control system includes an inner current controller which helps to control the current in *dq reference frame* and therefore leads to the decoupled control of active and reactive power in the converter. The complete schematic of an inner current controller is shown in Fig. 3. The inner controller follows the reference orders i.e. i_d^* and i_q^* that are set by outer controller.

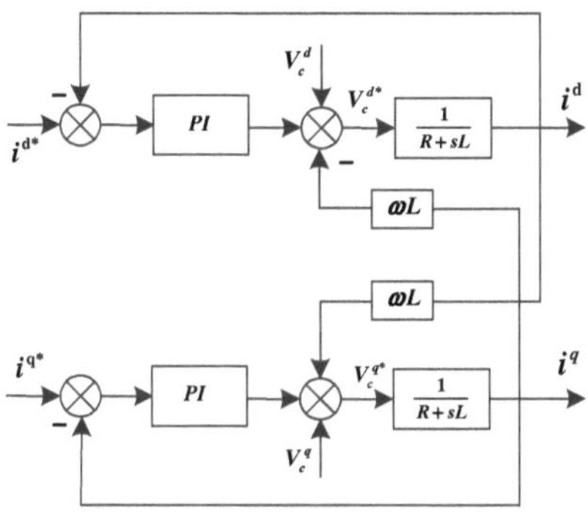

Fig. 3 Block diagram of the complete inner controller

The outer controller provides the inner controller with the reference current values in dq coordinate (i^{d*} and i^{q*}). The outer controller is able to control active power or DC voltage on DC side by controlling i^d (The injected active power in dq coordinates can be written as $P = V_s^q i^d$). The reactive power or AC voltage on AC side can be controlled by controlling i^q (if the voltage is aligned with the q-axis, the reactive power can be written as $Q = V_s^q i^q$). Due to the simplicity of PI controllers, it is normally applied in the outer controller (see Fig. 4).

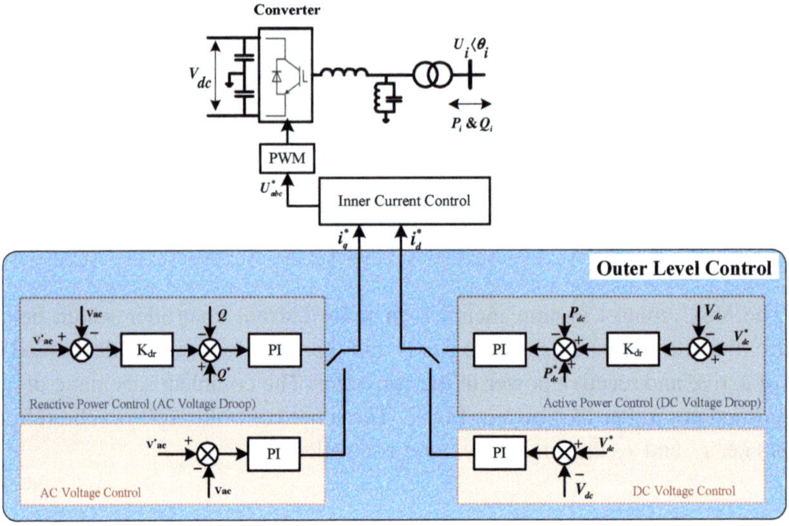

Fig. 4 Block diagram of the complete control system

As mentioned above, DC voltage is one of the parameters in the DC grid operation that can be controlled and therefor helps the power flow control in the DC side. Several control approaches such as Master-Slave, Voltage Margin and Droop have been proposed in literature [7][8][9]. These methods are described in the coming sections.

1.4 HVDC Grid Considerations

As shown above, the overlaid HVDC grid appears to be a promising solution for new demands in expansion of the current power system. However there are still some challenges to be solved to let this vision come to reality. The configuration of the grid, the grid protection and the ancillary services of the HVDC grid are some of those issues. Inspired by these challenges, some researchers have stepped further to examine the extra functionality of the HVDC grid in form of ancillary services [10][11].

1.4.1. HVDC Grid Topologies

One of the key parameters that has vital impact on the architecture of control and protection systems is without doubt the topology of the HVDC network. Different terminologies have been used for the overlaid HVDC connections based on the configuration of the network. To set a base for the discussions in this chapter, we start with the introduction of different topologies for HVDC connections and then clarify the terms "HVDC Grid" and "Multi-Terminal DC (MT-DC)".

The topology of the HVDC network has an indisputable correlation with the reliability of the system and also cost of the total development. This issue becomes even more important as the architecture of the control and protection system and the complete cyber physical system is fully dependent on this topology.

The first choice of topology for a grid is the point-to-point topology (PP) which comprises several point to point HVDC connections. This topology makes it possible to use just AC breakers in the case of failure in the DC line. On the other hand such a topology cannot fulfill the single failure criterion, since losing one link removes the whole connected generation. In a star topology (ST), a central star node connects all the nodes together. The dependence on the central node and reliability of this configuration is one of its drawbacks. Any fault at the central node can damage the whole DC system [12]. The general ring (GR) topology forms a ring with the lines connected to all the nodes. Lines must be designed for maximum capacities since they are needed to transfer the whole power of the system when the ring is opened. In this case, opening two DC circuit breakers placed on the end of a line can isolate a DC fault. Fast communication is however required to coordinate DC breakers to de-energize just the faulted line and keep the rest of the DC system up running.

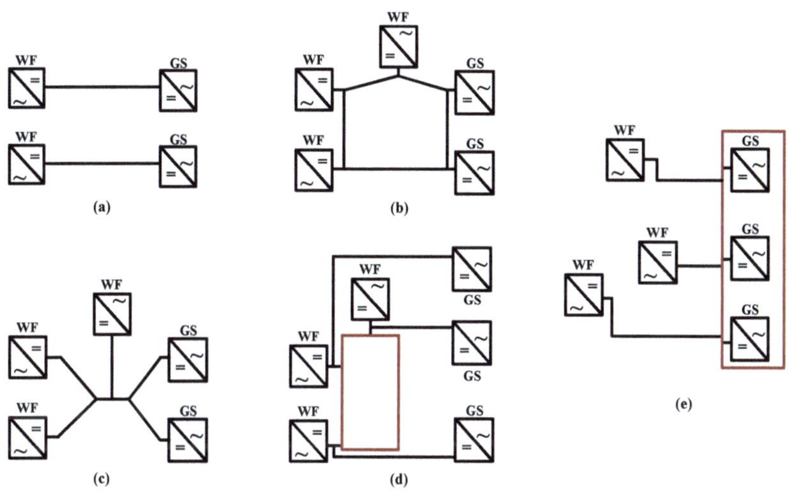

Fig. 5 HVDC grid topologies for off-shore integration (a) point to point topology, (b) general ring topology, (c) star topology, (d) wind ring topology, (e) substation ring topology

Another topology called wind farms ring (WR) provides an opportunity to minimize the costs by decreasing the number of breakers on the DC side. In this topology, there is a ring shaped grid connecting wind farms to each other [12]. Besides, each wind farm is connected separately to the AC grid. In this configuration, breakers are located on the links which connect ring DC grid to farms. This configuration enables the system to control the power flow between production and AC area in a more flexible manner. This topology is one of the strong candidates to be considered for future DC grid connected to offshore plants. Still fast communications is a requirement for this topology in order to coordinate the DC breakers in fault situations. Similar to this configuration, a ring can be considered for the AC side substations. In a substations ring topology (SSR), the wind farm converter must be isolated at fault instead of the AC side converter which is the case in wind ring topology. Therefore communication is still a considerable issue in this topology.

Considering all types of abovementioned topologies, the term MTDC is used for those kinds of the topologies that connect a number of converter stations in radial or parallel manner such as star topology or PP topology. This means there is no mesh inside such topology. The term "HVDC Grid" distinguishes the meshed network architecture from rest of the topologies such as WR or SSR.

1.4.2 Protection of HVDC Grids

One of the time-sensitive constraints in the domain of DC grid is the protection system for fault conditions which can cause many challenges in terms of economic and technical aspects of the DC grids developments. In DC grids, short circuit faults either line to ground or line to line leads to overcurrent faults. DC over voltage faults can occur due to open circuit in the grid or loss of a converter.

Converse to conventional HVDC which do not experience a large overcurrent during the faults due to its large DC smoothing reactance, the discharge of DC link capacitor in VSC-HVDC can lead to high levels of overcurrent. Therefore the protection system should be designed in such a way that it detects and isolates a faulty part of the system within the range of few milliseconds. A reliable and fast protection scheme is one the significant parameters in the reliability of the meshed DC grid.

The protection schemes proposed for DC faults e.g. [13][14] have addressed algorithms that do not use DC breakers. In [13], the method called handshaking method proposed which blocks all the HVDC stations in the fault condition and AC breaker is used to clear it. In [14], a method called a diode clamping method is proposed for small-scale system in which DC circuit breakers are not economically necessary. While in all those schemes, using the AC breakers causes the whole DC system to be de-energized for a moment, DC breakers with a fast clearing time can be used to disconnect just the faulty part and keep the intact parts of the DC grid in operation which increases the availability of the whole system. Innovative protection schemes that consider the topology of the grid and circuit breakers must be developed. These schemes will in turn have an impact on the architecture of the architecture of the communication network.

1.4.3 HVDC Grid Ancillary Services

Besides the transmission of active power from generation plants to load centers in normal operation, the HVDC grids are able to offer a reliable infrastructure for connected AC areas to exchange active power or control local reactive power thereby providing ancillary services to the AC areas. These services includes frequency support, voltage support and rotor angle stability in the form of avoiding loss of synchronism between generators or damping of electro-mechanical oscillations. In this section, a summary of such ancillary services is given.

Frequency Control

The change in the balance between loads and generations in an AC system leads to deviation in the frequency of the system. Such a deviation can be an indication to take control actions. This control action includes increasing or decreasing power production on the generation side or shedding loads on the demand side. Considering hybrid AC/DC grid i.e. including meshed DC grid on top of the AC grid, several studies such as [11] have suggested to take this possibility of the power injection control between DC/AC areas to control the frequency of the AC areas. Therefore, in the case of frequency deviation in one AC area, the remaining connected AC areas contribute in supporting the frequency through the overlaid HVDC grid. This ancillary function can prompt new schemes in the design of control system in both primary and secondary frequency control actions.

Voltage Support

The support of voltage profile at the connecting nodes on the converter can be implemented by injecting reactive power to AC areas. The voltage deviation especially over-voltage situation during faults can be managed independently by reactive power injection. This support function relies on fast actions of local TSO (the TSO controls the AC area connected to that converter) to dispatch reactive power reserves [11][15].

Oscillation Damping

The interactions of rotating machines in the power system can create oscillation modes. One of the vital control actions in the operation of power system is to damp these oscillations. In the new power system structure, the integration of large wind farms can introduce the power oscillation between the rotors of synchronous machine which is the new source of instability. Several studies suggest suitable control methods for point to point HVDC links in order to increase rotor angle stability. Most of the proposed schemes for point to point links can also be used in the case of HVDC grids [11]. This ancillary service can therefore also be considered during the design of upper control level for HVDC grids.

2 Communication and Control for HVDC Grid Control

In this section the operational and control principles that can be used for HVDC grids are provided. The importance of communication infrastructure on some the control schemes are explained.

2.1 HVDC System Operation

The current HVDC grids which are in the planning phase, such as abovementioned ones i.e. the Swedish South-West link or Indian North-East Agra are being developed by a single transmission system operator (TSO). One of the key challenges to be solved for larger HVDC Grids is the interaction between interconnected TSOs and ownerships of the HVDC grid. In case of an overlaid HVDC grid connecting separate AC areas involving several TSOs, different architectures can be proposed for the operation of the HVDC grid. These include the independent and the integrated architecture [11]. In the independent architecture, the DC grid is operated by a separate TSO called the DC grid operator, while in the integrated architecture, one of the connected TSOs is responsible for the control of the HVDC grid. In both architectures, the optimal power flow (OPF) must be calculated in order to run the system in the most optimal situation. The difference in the two architectures in the terms of the power flow calculation is in the required information, boundary of the system, and also objective (cost) function.

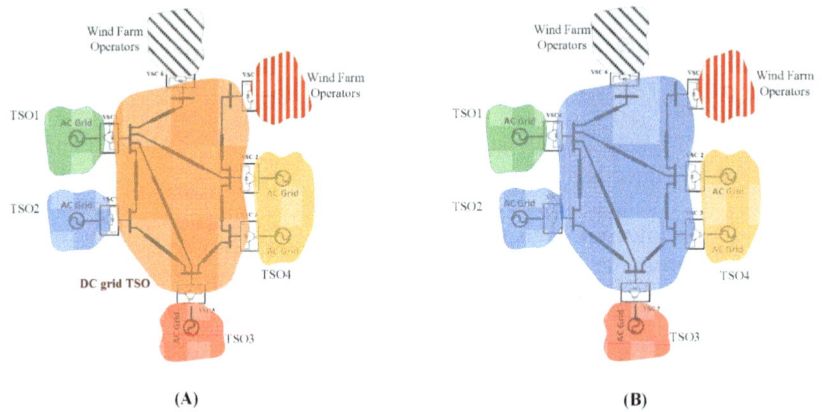

Fig. 6 Operational Architecture of HVDC grids, (A) Independent DC TSO (B) Integrated TSO

Independent Operation

As it is shown in Fig. 6, an independent TSO can operate the DC grid which lies between different AC areas. The DC grid TSO must follow the connection rules and policies set by connected AC TSO while tries to increase its operational benefits [11]. The functions such as optimal power flow, primary/ secondary frequency control and oscillation damping should be defined and implemented by this TSO. Based on the choice of design, these functions can be implemented in centralized or distributed manners. For both designs, the DC TSO needs to have permission to access the corresponding information from other AC TSOs.

Integrated Operation

In the second architecture, one of the TSOs takes the responsibility of operating the HVDC grid. Implementation of some the functions such as optimal power flow or rotor angle stability-related services are simpler in this architecture. In this case, there should be clear policies and agreement on the sharing of benefits from the operation of HVDC grids. The centralized and distributed approaches are again two choices for the implementation of functions in this architecture.

2.2 Communication and Control

As mentioned in previous sections, the HVDC converters are able to control active power or DC voltage on the DC side. The ability of controlling these two parameter give rise to different types of control schemes to operate the HVDC grid in reliable and efficient ways. Based on the required information in each scheme, some of the methods need fast communication while others are dependent on just local information. Master-Slave, Voltage Margin and Droop are three well-known proposals for control of HVDC grid.

Master-Slave scheme
In this scheme, one terminal is responsible for controlling the DC voltage and other converters operate at constant active power mode. This single converter is also responsible for compensate any imbalance in the HVDC grid power flow. Therefore it should be designed for large power deficits as well. The drawback of this design is that the system can collapse if the grid loses this dedicated DC control converter.

Voltage Margin scheme
Similar to the Master-Slave scheme, the Voltage Margin method assigns one converter to be responsible to keep DC voltage of the grid at a desired level and the other converters operate as constant power converter. In the case of losing or reaching the limit in this converter, another converter takes over the responsibility and works as the slack bus in the grid. As shown in Fig. 7, at current operating point, terminal 1 works on constant power mode and terminal 2 controls the DC voltage [16].

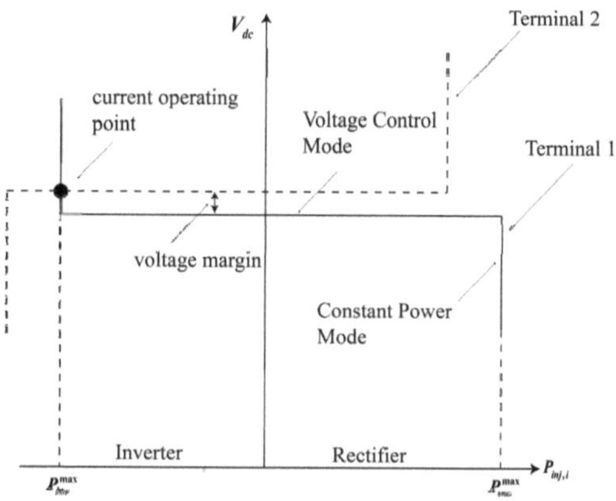

Fig. 7 Voltage Margin Scheme for two terminal system

Droop Scheme
Similar to frequency indication in the case of AC power mismatches, the DC voltage deviates if there is any imbalance in the power injection and extraction on the

DC side. This DC voltage deviation which is a local indication of power mismatch can be used as a control signal to increase or decrease the active power in converters. However there is a difference between frequency in AC and voltage in DC grid; the frequency deviation is almost similar in entire grid but the DC voltage deviation is different in each node. The DC voltage deviation has a direct relation with the topology of the grid.

Based on this characteristic of the DC grid, the Droop method lets some converters change their injected active power proportional to the local DC voltage deviation. The characteristics of converters for a two-terminal system are shown in Fig. 8 [16]. In this scheme, if there is a power loss in the system, the voltage drops in the entire DC grid, and the converters uses this value to inject more power to the grid or extract less power. This process is also valid for the reversed situation i.e. a power increase in the system. The problem with this method is the power sharing in the case of disturbances can not be distribute fairly. Besides, the droop parameter must be reassigned for the new operating point after each disturbance. This recalculation should be done by a centralized master controller. This controller/SCADA can be run by either the independent TSO or the integrated TSO.

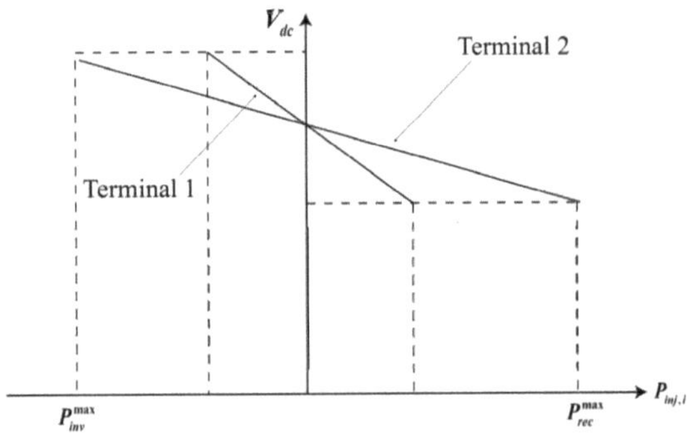

Fig. 8 Droop scheme for two terminal HVDC system

Power Flow Control

The power flow control in the HVDC grid can be carried out by calculating the proper set-point for the converter either on DC voltage control or on constant power control. This calculation needs to be implemented in either a centralized controller or distributed entities. In both cases, the calculator(s) needs the information from either the entire grid or the neighboring HVDC station. Power flow control based on type of implementation i.e. centralized or distributed requires different communication infrastructures. Note that in the case of droop control, the need of power flow controllers is more accentuated since the new stable point after disturbance does not guarantee fair and optimal power flow. Due to the slow dynamic of power flow control compared to voltage control, it can take up to seconds to determine the set-points of the converters.

Need of reliable ICT system
Operation of HVDC grid connecting multi AC areas is a complicated process that required robust control strategies and reliable supporting systems. The ICT system as one of the important supporting systems plays an important role in some of the control and protection architectures [17]. Here, we summarize the functions and architectures that rely on these supporting systems and then in further section the significance of studying and modeling of such a Cyber-Physical System will be emphasized.

As we have seen, the protection system in some grid topologies such as Point to Point, Grid Ring or Station Ring requires fast communication links to isolate the faulty DC line and keep intact areas operational. When it comes to control schemes, Master-Slave is the method needing fast communication to send power set-points. In the case of Voltage Margin or even Droop control, the new parameters for the new operating points should be calculated either centralized or distributed and sent to converters as soon as possible. In some literature this is referred to as power flow control. This function still requires communication between the neighboring nodes in the case of distributed flow control or between nodes and SCADA in the case of centralized approach.

3 HVDC Grids as Cyber-Physical Systems

This section describes the modeling of agent-based control for HVDC grids from a Cyber-Physical System perspective. The term Cyber-physical system (CPS) is recently being used to refer to systems in which the computational entities including control/communication units (cyber) as well as physical processes are strongly coupled [20]. First we describe how the control of power flow in the HVDC grid can be carried out by assigning a set of agents for the HVDC converters to perform the local tasks aligned with the global goals. The characteristic of this cyber physical system is thereafter described. Finally we describe the co-simulation testbed used to evaluate the multi-agent control scheme.

There are several research works that study the CPS concept from different perspectives such as design techniques or application of CPS to design power systems [21]. A cyber physical system model that could be scheduled and controlled to achieve the desired reference values and minimizing the power consumption of the given system has been proposed in [22]. The integration of ICT supporting system with traditional power system is more noticeable nowadays by implementation of new technologies like PMU-based monitoring systems. Such an integration of cyber elements to the operation and control of the power grid can increase the stability of the system. Modeling the power system as physical system interacting with such cyber devices has been further discussed in [23][24][25].

3.1 HVDC Control as a Cyber-Physical System

Voltage Droop and Voltage Margin are two approaches for control of DC voltage that have been described above. In both methods, a disturbance will shift the operating point of converters and as a result the power flow in both AC grid and DC grid will change. Such a new operating point cannot guarantee fair distribution of power in the grid. Therefore there can is a need of centralized or distributed coordinator(s) to allocate the power sharing in the HVDC grid. Since the converters in HVDC grid are dynamic and must coordinate locally or globally for certain control functions such as power sharing, the problem can be formulated in the context of Multi-Agent System (MAS) control. MAS contain a number of controllers called agents interacting with dynamic units. These agents are set to reach global goals. This concept is applicable in many areas such as formation flights, sensor networks, energy networks and distributed computation.

Such agent-based control of HVDC grid involves the interaction of physical power systems and distributed decision makers, which are basically computational/communication units. Therefore, such an interaction can be studied from a Cyber Physical System perspective. Studying such integrated systems of systems requires comprehensive knowledge in different domains such as software engineering, communication and networking, control theory and also electronic design.

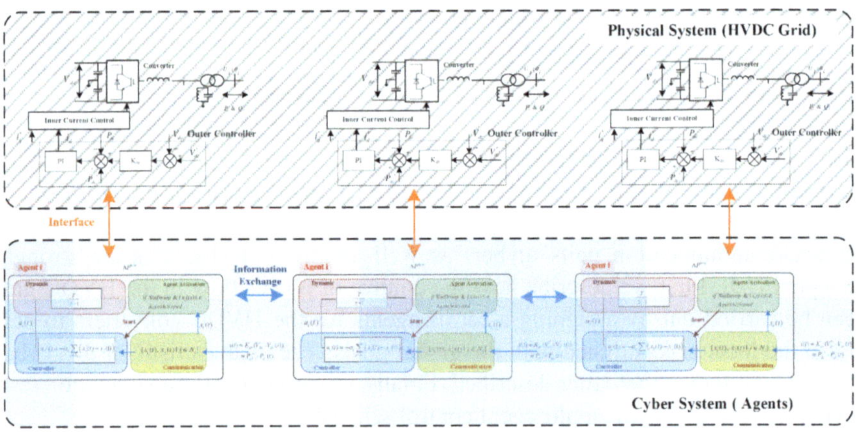

Fig. 9 Agent based control in Cyber-Physical System Framework

As shown in Fig. 9, the control model and process has been separated into two parts. The physical system consists of power system, both AC and DC grids, and the corresponding local control system for each converter station. In this study, the local control system including the inner and outer control level are presented as part of the physical system. However one is possible to consider this local control system as part of cyber system as well, but due to simplicity in the modeling of the system, this assumption has been made in this study.

3.2 Modeling the Communication System

This section explains the multi-agent control scheme developed. The agent's setup, system dynamics and the information link between agents are studied with different tools. Graph theory is one of the tools often used to model the agent's data exchange interactions [26]. Consider a graph *G= {V, e}* consisting of a set of vertices or agents *V= {1,..., N}* and edges *e*. Nodes *i* and *j* are adjacent if a link or edge is between them. The adjacency matrix *A* that shows the adjacency between the nodes in the graph *G* can be defined.

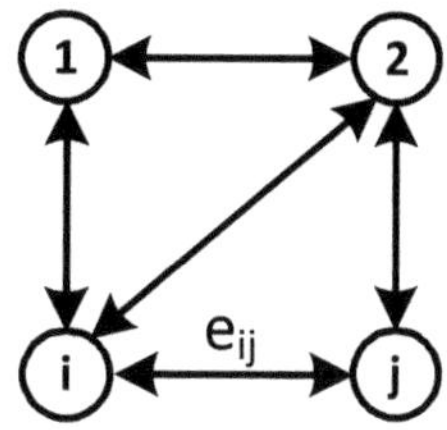

Fig. 10 Information graph G with links e

The distance between two nodes i.e. *d(i,j)* is the shortest path with least number of links that connect nodes *i* and *j*. The degree matrix *D* with the elements of d_i is a diagonal matrix which elements are the cardinality of agent *i* neighbor set $N_i = \{j \in v : (i, j) \in \varepsilon\}$. The Laplacian matrix L is equal to the difference between degree matrix and adjacency matrix *(L=D-A)*. Consider an undirected graph (see Fig. 10), since the Laplacian matrix is symmetric and positive semi-definite i.e. the sums of the elements in each row is zero, the first eigenvalue of the matrix is equal to zero ($\lambda_1(\mathrm{G}) = 0$). For a connected graph, Laplacian matrix *L* has exactly one zero eigenvalue and the eigenvalues increases by the order i.e. $0 = \lambda_1(g) \leq \lambda_2(g) \leq \ldots \leq \lambda_N(g)$. The second smallest eigenvalue $\lambda_2(\mathrm{G})$ of L shows how well the graph *G* is connected. Therefore it is also known as the algebraic connectivity [27].

3.3 Control Algorithm

As mentioned earlier, in the case of any disturbance or power mismatch in a typical DC grid, some converters can be assigned to contribute in power sharing by using DC voltage droop control (distributed DC slack). The assignment of droop parameters for different converters is a real-time problem that must be recalculated for any new system status based on converter capacities, market issues or line limitation. If droop parameters are not reassigned correctly, the next disturbances can drop the voltage level of the entire system to the minimum or maximum limit. In this situation, other converters that work on power constant mode will produce more to recover their DC voltage. Instead of centralized controller to assign the new droop parameter for the converters and decide the fairly power sharing in the new state, this section purposes a distributed agent-based scheme for power sharing and recover the DC voltage level after the disturbance.

As it is shown in Fig. 11, the agent set-up consists of four parts. The communication part is responsible for receiving measurements from the neighbors. The difference in current DC power measurements and pre-defined references is the state of the agent. The controller part is responsible to create the distributed control law. In this study, we use $u_i(t) = -\sum_{j\in N_i}\left(x_i(t) - x_j(\mathrm{t})\right)$ as the distributed control law, where x_i is the local state variable and x_j is the state variable of neighbors. This can be changed based on the designer's interest. The dynamic block is the agent's dynamic behavior that here, is a single integrator agent ($\dot{x} = u$). The last block is the Agent Activation that senses a new operating point, DC voltage drops close to limits, and sends the commands to start the agent or freeze the droop control.

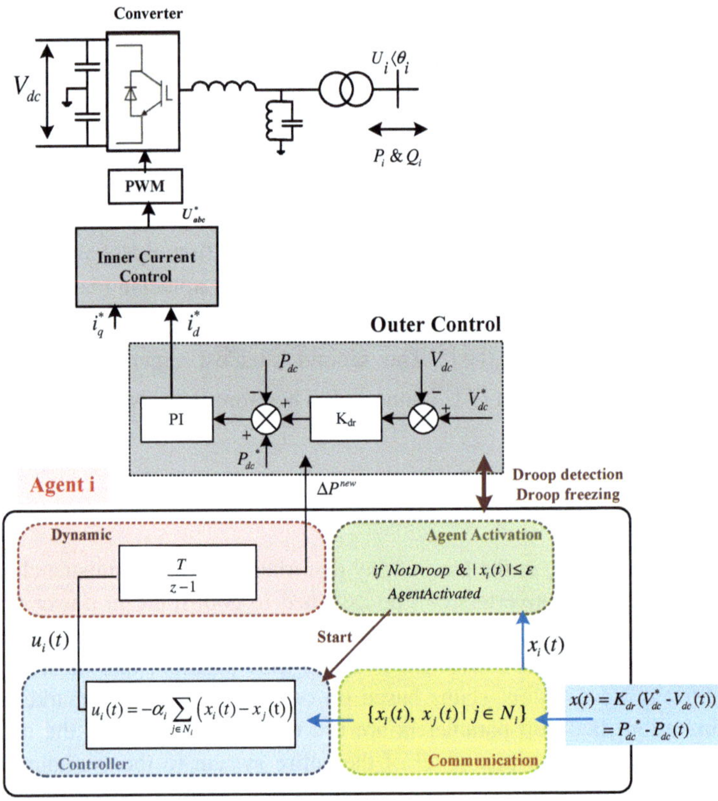

Fig. 11 Agent-based control set-up

The agents start the information sharing either when the DC voltage drop/increase is close to limit or when they are manually set to power consensus. So, when the DC voltage drop/increase is recognized, the agent freezes the DC droop control, increases the voltage level by pre-disturbances value, and finally starts the communication with other agents to converge to agreement. The agreements value is the average of initial value of the agents' state after start i.e. $\Delta v_i(\mathrm{t}) \rightarrow \frac{1}{N}\sum_{i \in v} \Delta v_i(\mathrm{t}^{+}{}_{disturbance})$.

Proof:

If we consider the set of single integrator agents that are connected through an undirected graph g, the closed loop system becomes $\dot{x} = -L(g)x(t)$, *where L is the Laplacian of the graph. The solution of the dynamic for each agent contains the exponential term with the decay rate depending on this matrix i.e.* $x(t) = \mathrm{e}^{-L(g)t}\, x(0)$. *Assume U is the matrix consisting of normalized and orthogonal eigenvectors of matrix L, and V the diagonal matrix comprises of eigenvalues of the matrix. Then it is possible to expand the exponential terms of the system response using the spectral factorization of the Laplacian as follows:*

$$\mathrm{e}^{-L(g)t}\,\mathrm{x}(0) = (\mathrm{e}^{-(UV(\mathrm{g})\mathrm{U}^T)t})\,\mathrm{x}(0) = (U\,\mathrm{e}^{-V(\mathrm{g})t}\,U^T)\,\mathrm{x}(0)$$
$$= \mathrm{e}^{-\lambda_1(g)t}\,u_1^T x(0)u_1 + \mathrm{e}^{-\lambda_1(g)t}\,u_2^T x(0)u_2 + \ldots. + \mathrm{e}^{-\lambda_1(g)t}\,u_n^T x(0)u_n$$

As mentioned above, for the connected graph the first eigenvalue is zero and higher orders have the positive values. So, the exponential terms with positive eigenvalue (negative rate) will decay by time. The smallest positive eigenvalue $\lambda_2(\mathrm{g})$ *dominates the slowest rate of convergence in the solution. Based on system solution, when the time goes to infinite, the respond converges to the average value of the initial states. The detail of the proof can be found in [28].*

$$x(\mathrm{t}) \rightarrow u_1^T x(0)u_1 = \frac{\mathbf{1}^T x(0)\mathbf{1}}{n}$$

4 Use case – 7 Terminal HVDC Grid

This section presents a case study acting as proof of concept of the proposed Multi-agent control scheme. The section begins with a presentation of a DC grid used for the study including all relevant parameters. Thereafter the co-simulation platform used to study the control scheme is presented. Finally the results of the study are presented at the end of the section.

4.1 System under Study

In order to validate the suggested agent-control algorithm in the previous section, a 7-terminal DC grid has been developed in a real-time simulator. Since there is not any standard model or real system available for DC grid studies, the chosen model is designed based on the system data for a 7-terminal DC grid analyzed in [29]. The multi-terminal HVDC system consists of seven converter stations. As shown in Fig. 12, the converter stations are connected through a meshed DC grid. The line parameters are defined in TABLE I.

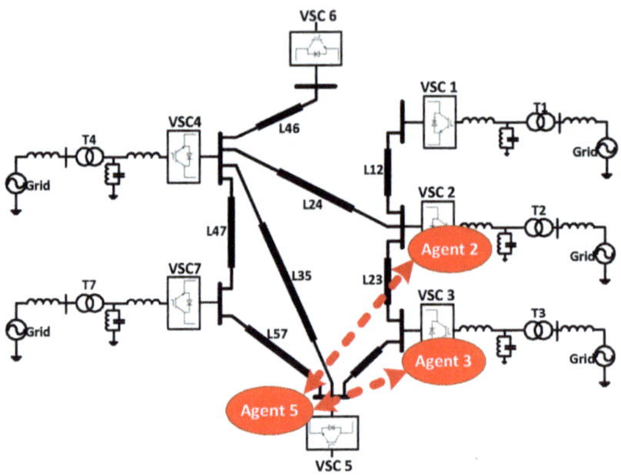

Fig. 12 7-terminal VSC-HVDC transmission grid

The average model has been used to model the VSC stations. So the dynamic of the switching and corresponding effects are not considered in this study since the control scheme is developed for higher system control. Each converter is connected to a separate strong AC grid. The modeled VSC stations can control active power or DC voltage on DC side, and reactive power or AC voltage on AC side. The power ratings and control modes of the stations are provided in TABLE II.

Table 1 DC grid line parameters

Lines	$R(\Omega)$	$L(mH)$	Length (Km)
L12	2.577	22.5	213
L23	3.00	26.2	248
L24	2.50	21.9	207
L35	4	25.0	331
L45	1	8.76	83
L46	2.5	21.9	207
L47	3.5	30.5	289
L57	2	17.4	165

Table 2 Converters Rating and modes

Converter	Capacity (MW)	Control mode	Agent Control
1	200	Active Power	NA
2	300	DC droop Voltage	Implemented
3	150	DC droop Voltage	Implemented
4	200	Active Power	NA
5	300	DC droop Voltage	Implemented
6	100	Active Power	NA
7	50	Active Power	NA

As shown above, three converters (2, 3 and 5) out of seven terminals are interacting with the agents. And in this specific set-up the tree connection is considered between these three agents in the system. Converter 5 is able to exchange the data with both converters 2 and 3. The reflected graph is an undirected graph meaning that both nodes at the end of each edge are able to send/receive the data to/from other node.

4.2 *Simulation Platform*

The Power System Management and Information eXchange (PSMIX) Platform is a real-time co-simulation test-bed which enables the studies regarding the design, testing and implementation of real-time control and operation applications in power system (see Fig. 13). This real-time platform reflects the characteristic of the supporting ICT system and the physical process, as well as the interfacing devices or systems as close as possible to the real life scenarios. PSMIX is a general real-time architecture that can be re-arranged for different studies from distribution grid control scenarios to wide area control of transmission grid. This platform includes real-time power system simulator, real-time communication network simulator, applications, and software-based or real interfacing devices/measurement. The main factor in the development of any such real-time platform is accuracy and performance of the software-based models of the real devices and the implementation of industrial automation protocols such as synchronized phasor measurement units [25]. For this study, the PSMIX Platform is configured to support the modeling of HVDC grid and its supporting control and communication system. The detail information of the components is described as follows.

OPAL-RT eMEGAsim Simulator

The eMEGAsim is a commercial real time simulator which combines OPAL-RT electrical circuit solvers, SimPowerSystem and RT-LAB distributed processing software and hardware for high speed real-time simulations of a Power system for both steady state and transient analysis. This simulator can be customized to meet I/O requirements enabling the Hardware-in-the-Loop (HIL) simulations. The DC grid presented above has been modeled and simulated in this simulator.

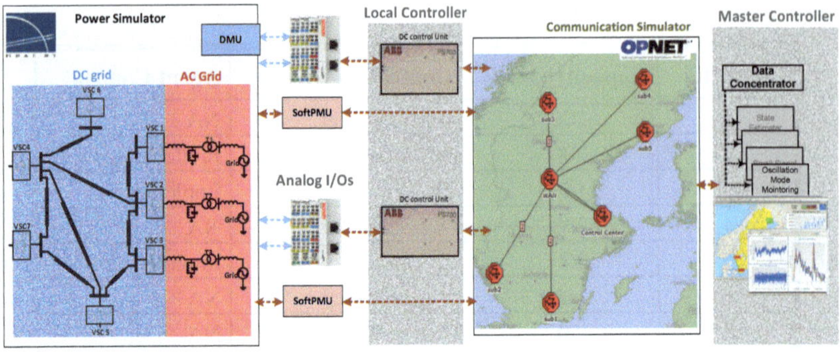

Fig. 13 PSMIX Platform Architecture

Measurement Units

The HIL feature of the OPAL-RT simulator enables the simulated power grid to interact via analog I/Os. This HIL simulation test platform provides the ability for more realistic study of the real world systems. Since the HVDC controller is able to communicate with specific analog I/Os, a special DC measurement unit (DMU) has been developed inside the OPAL-RT simulator to send/receive the DC voltage and DC power measurement with specific accuracy to/from analog I/Os. This device takes in the analog input in the range of 0V to 10V and digitizes it with 16-bits resolution. To provide the input to the simulated power grid 4-channel analog output device is used. This device generates the voltage signal within the range of -10V to 10V. Both I/O devices are mounted on an EtherCAT coupler which provides the means of communication between I/Os via the EtherCAT protocol providing sufficient performance. For the AC side, a Software-based Phasor Measurement Unit (SoftPMU) has been developed (See [25] for detail information). The specific method proposed in this chapter only uses the DMU for control purposes.

HVDC Industrial Controller

ABB's MACH3 system is a high performance control and protection system includes an industrial computer called PS700 which runs Windows embedded integrated with INtime reliable Real time operating systems (RTOS). PS700 communicates with the analog devise via EtherCAT protocol. It can communicate with other HVDC industrial control systems via Ethernet. HiDraw studio is a graphical programming environment based on C++ that is used to program the HDVC station control systems hardware. When the project in HiDraw is compiled, first the C++ code is generated then this C++ code is compiled and released for executing in INtime. The agent logic has been developed in C++ codes in HiDraw.

OPNET Communication Network Simulator

OPNET is a communication system modeler which provides comprehensive development environment for modeling communication networks and distributed systems. The behavior of the simulated communication network can be analyzed by performing discrete event simulations. OPNET contains the System-in-the-loop (STIL) module enabling the connection of the simulation model with live network hardware by providing interfaces and modules [30]. For the platform presented in the chapter each STIL module inside the simulation environment is assigned to a specific network adapter on the machine. There are four such physical network adapters available. Each of the three PS700 has been connected to these adapters to simulate three HVDC control stations located at different geographical locations.

Master Controller: KTH PowerIT Platform
The application component of the PSMIX platform consists of openPDC as the phasor data concentrator and the KTH PowerIT as the application hosting platform that connects to openPDC to receive the synchronized measurements [25]. Besides, it is able to receive the DC grid information i.e. DC voltage and active power from the converters (PS700 controller) using industrial defined Raw Ethernet protocol. Several applications have been implemented in PowerIT platform, such as average frequency visualization and electro mechanical mode estimation for AC grid, and monitoring and control application for HVDC grid. Note that in this particular agent-based control scheme, there is no need of a centralized application to be run on the PowerIT platform.

4.3 Agent Logic and Information Exchange Modeling

The agents' logic has been modeled using HiDraw language, then compiled and implemented in HVDC controller (PS700). The Communication network model considered for this case study consists of just three subnets. This communication network mirrors the information exchange graph of the agents that is a tree graph. Since just three agents are assigned for the power sharing coordination, the communication networks model consists of three subnets. Each subnet represents an HVDC substation and is connected to corresponding HVDC controller through a physical Network Adaptor and SITL gateway. Therefore, the generated traffic by the real controller can be passed to the communications simulator. In addition, the SITL gateways must be configured properly to filter other packets and receive only the relevant packets. The initial network model between the agents is built based on the empirical data from an industrial project [31]. This network is a dedicated network that uses the fiber optic links. The link bandwidth is considered to be 24 Mbps.

4.4 Simulation Results

Scenarios
Several scenarios have been implemented to evaluate the performance of agent-based control schemes in different conditions. In the original set-up, three converters of 1, 4, 6 and 7 are set to control active power, and other three of 2, 3 and 5

are set to DC droop voltage control. The droop coefficients in the first scenario are set to 0.75, 1 and 1.5. In the agent-based set-up, the three converters of 2, 3 and 5 use the proposed schemes to bring back the DC voltage to normal limit and share the power mismatch equally in terms of the power capacity in per unit by using on local and neighboring information. In the first step, these two different set-ups are compared for the case that the droop control is not sufficient for series of events (scenario I). In the second step, the droop control and agent-based power sharing are evaluated for normal operational scenarios in which the voltage limit is not hit (Scenario II). Besides, the performance of agent-based control scheme is evaluated for different bit error rate (BER) on the communication link (scenario III). In scenario I, the active power injection in converter 6 is reduced to 0.6 pu after 5 seconds and to zero after 21 second. In scenario II and III, a disturbance is introduced to the HVDC grid by decreasing the active power injection in converter 6 to 0.4 pu after 5 seconds. And, the second disturbance is introduced in 21 seconds by decreasing the injection of converter 6 to zero. Note that the per unit values for each converter is based on the station power capacity, not the system base. Besides, to evaluate the performance of the real-time co-simulation platform, the first scenario is also fully implemented just in the real-time power simulator. This scenario is called "simulation scenario" from now. The difference of this scenario and "co-simulation scenario" is the environment of modeling. In "simulation scenario", in contrast to co-simulation scenario, the physical power system, control system and communication system are modelled in just one real-time simulator i.e. OPAL-RT simulator. In this scenario, the communication network is modeled with a simple delay block. Note that a tree graph is considered to exchange the information as a minimum connectivity requirement.

Results

Fig. 14 depicts the result of scenario I. The results show that after the first disturbance at 5 seconds, the voltage drop is significant if the Droop Control is used. But still the system remains reliable. Now, with the second disturbance, the voltage level of the system drops to the limits of some converters such as converter 4. Consequently, this converter starts in power sharing even if it is set to just follow its scheduled power (power constant control). On the other hand, the agent-based control first brings the voltage level to pre-disturbance situation and then shares the power mismatch fairly.

The result of power sharing in both cases i.e. droop based and agent-based (scenario II) are presented in Fig. 15. The agents ordering the HVDC converters are sharing the information to come to average power contribution. The result shows that it takes around 5 seconds for the converters to reach the average value. It can be seen that the consensus value is basically the average of power outputs in the droop control scheme.

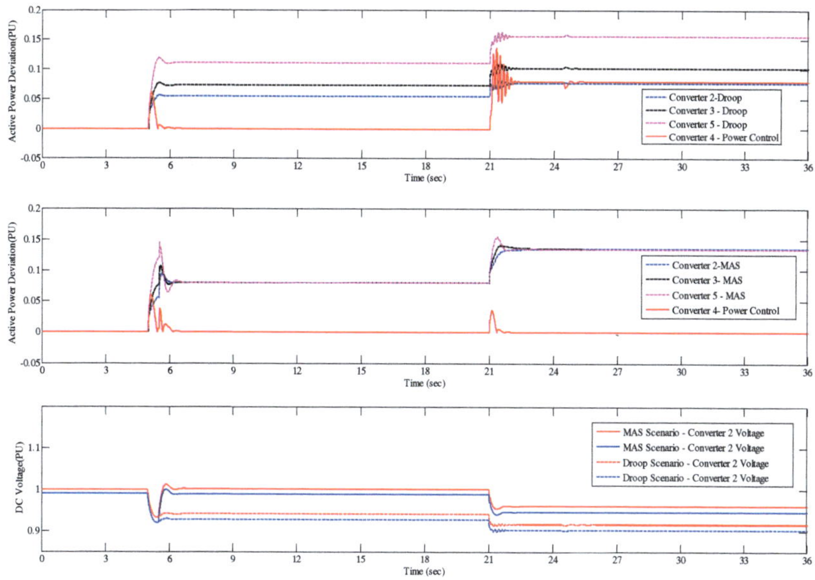

Fig. 14 Comparison of Droop Control and Agent-base Control

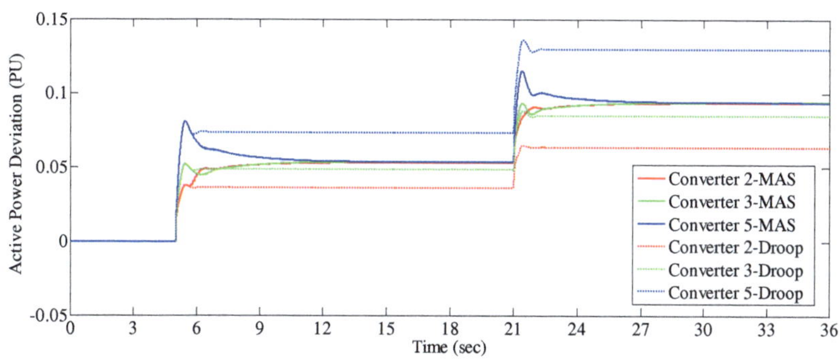

Fig. 15 Consensus-based versus droop-based power sharing

Fig. 16 shows the result of scenario II. As the benefit of having communication simulator in the simulation loop, the BER can be studied on the control scheme. As shown here, during the first disturbance the BER is around 0.01 percent and the controller cannot recognize the information, and when BER reaches approximately 0.004 percent the agents recognize the information and try to reach the consensus value.

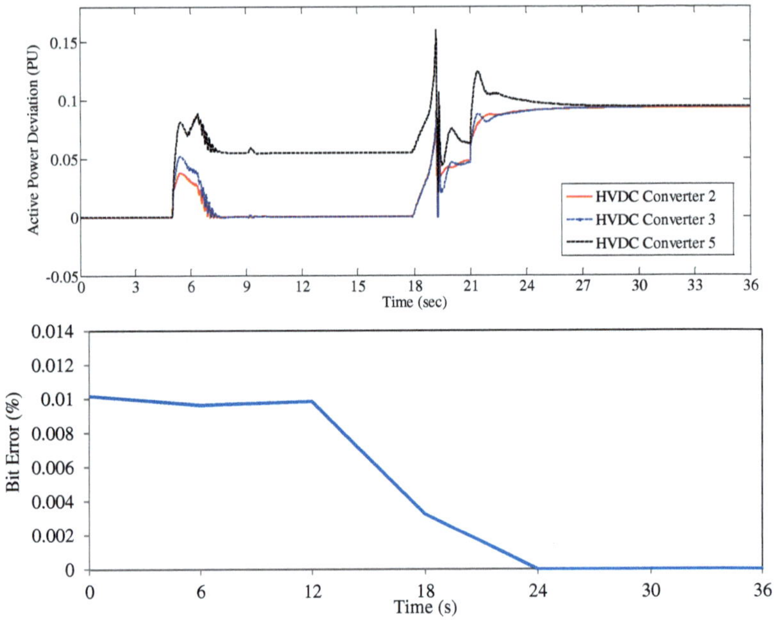

Fig. 16 Impact of BER on the control performance

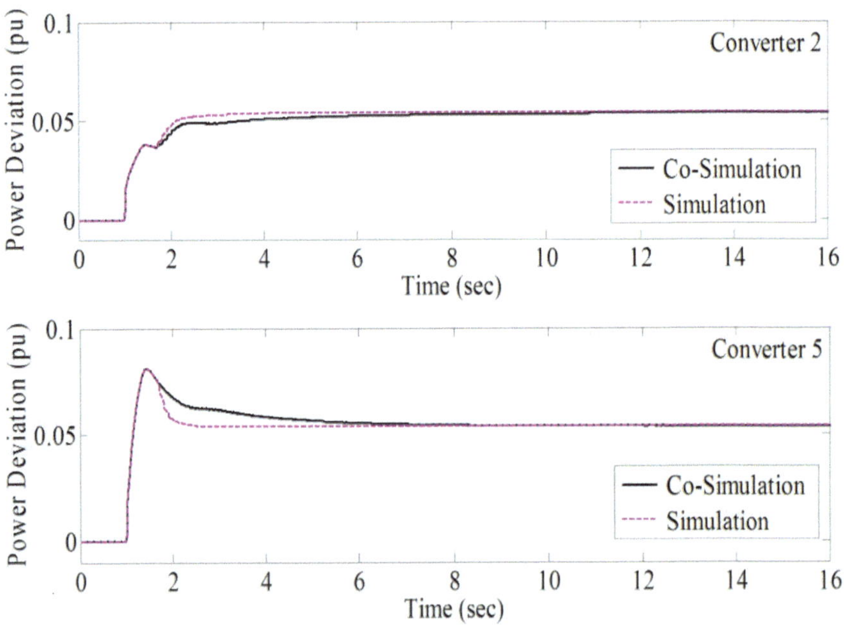

Fig. 17 Result of Simulation and co-Simulation

The comparison of simulation and co-simulation implementation is presented in Fig. 17. The result shows that some dynamic of the real controller and errors in A/D conversion cannot capture in the pure simulation. However the study shows that both results in steady state and in their trend follow each other. It can be concluded that when the studying cyber physical systems, it is uncertain to use a pure simulation to model the whole system.

5 Conclusion

The integration of large-scale renewable energy sources, the inter-connection of grids to neighboring national grids and the limitation in the expansion of new AC transmission line draw the power system planners' attention to move toward HVDC technology with its various benefits such as the visual impact, controllability and lower power loss. Different researches have been carried out to evaluate design and control choices for operating the HVDC grids.

This work presented an agent-based method to coordinate the power sharing in an overlaid HVDC grid. The method has been compared with the existing Voltage Droop Control. The proposed method uses the distributed control law based on local and just neighboring converters to contribute the power between the converters in the case of power deficiency. In contrast to droop control, it does not need to recalculate the coefficient for new operation point. This method regardless the type of DC grid operational architecture (Independent DC TSO or integrated TSO) can be implemented to control the power flow in distributed manner in the HVDC grid. The results showed the advantage of the proposed method. However, the performance of the control scheme can be degraded by communication network parameters.

The study and evaluation of this control scheme has been tackled from the cyber physical system (CPS) perspective. The agent/computational entities and information exchange between them has been modelled as a cyber-system interfacing the physical power system. A real-time co-simulation platform has been developed to evaluate the impact of IT supporting system (cyber system) on the operation of the HVDC grid. In future, this platform using distributed decision makers can be used for more advances functions such as distributed optimal power flow calculation.

References

[1] European Commission, The 2020 climate and energy package, ec.europa.eu, `http://ec.europa.eu/clima/policies/package/indexen.htm` (accessed: May 22, 2014)

[2] Van Hertem, D., Ghandhari, M., Delimar, M.: Technical limitations towards a SuperGrid — A European prospective. In: 2010 IEEE International Energy Conference and Exhibition (EnergyCon), December 18-22, pp. 302–309 (2010)

[3] Erdle, S.: The DESERTEC Initiative; Powering the development perspectives of southern Mediterranean countries. Discussion Paper, German Development Institute (2010)
[4] North-east agra - a 800 kv transmission superhighway multiterminal system with 8,000 mw converter capacity, nea800 brochure (March 2011), http://search.abb.com/library/Download.aspx?DocumentID=POW0071&LanguageCode=en&DocumentPartId=&Action=Launch (accessed: April 22, 2014)
[5] Tamiru Woldeyesus Shire; VSC-HVDC based Network Reinforcement. Master thesis, Delf University of Technology, Netherlands (2009)
[6] Agelidis, V.G., et al.: Recent Advances in High-Voltage Direct-Current Power Transmission Systems. In: IEEE International Conference on Industrial Technology, ICIT 2006, pp. 206–213 (2006)
[7] Haileselassie, T.M., Uhlen, K.: Impact of DC Line Voltage Drops on Power Flow of MTDC Using Droop Control. IEEE Trans. Power Syst. 27(3), 1441–1449 (2012)
[8] Beerten, J., Cole, S., Belmans, R.: Generalized Steady-state VSC MTDC Model for Sequential AC/DC Power Flow Algorithms. IEEE Transactions on Power Systems 27(2), 821–829 (2012)
[9] Jun, L., et al.: Operation and Control of Multiterminal HVDC Transmission for Offshore Wind Farms. IEEE Transactions on Power Delivery 26(4) (2011)
[10] Mehdipour Pirbazari, A.: Ancillary services: definitions, markets and practices in the world. In: IEEE PES Transmission and Distribution Conference and Exposition (2010)
[11] Phulpin, Y., Ernst, D.: Ancillary services and operation of multi-terminal HVDC grids. In: Proc. of International Workshop on Transmission Networks for Offshore Wind Power Plants, Aarhus, Denmark, pp. 1–6 (October 2011)
[12] Gomis-Bellmunt, O., Liang, J., Ekanayake, J., King, R., Jenkins, N.: Topologies of multi-terminal HVDC-VSC transmission for large offshore wind farms. Electric Power Systems Research 81(2), 271–281 (2011)
[13] Tang, L., Ooi, B.: Locating and isolating dc faults in multi-terminal dc systems. IEEE Transactions on Power Delivery 22(3), 1877–1884 (2007)
[14] Yang, J., et al.: Multi-terminal DC Wind Farm Collection Grid Internal Fault Analysis and Protection Design. IEEE Transactions on Power Delivery 25(4), 2308–2318 (2010)
[15] Bell, K., et al.: Economic and technical criteria for designing future off-shore HVDC grids. In: IEEE Innovative Smart Grid Technologies Conference Europe (2010)
[16] Nazari, M., Ghandhari, M.: Application of Multi-Agent Control to Multi-Terminal HVDC Systems. In: EPEC, Canada (2013)
[17] Babazadeh, D., Chenine, M., Nordstrom, L.: Survey on the Factors Required in Design of Communication Architecture for Future DC Grids. In: IFAC (May 2013)
[18] Quaglia, D.: Cyber-Physical Systems: Modeling, Simulation, Design and Validation. In: Mediterranean Conference on Embedded Computing, MECO (2013)
[19] Susuki, Y., et al.: A Hybrid System Approach to the Analysis and Design of Power Grid Dynamic Performance. Proceedings of IEEE (2012)
[20] Khaitan, S., McCalley, J.: Design Techniques and Applications of Cyber Physical Systems: A Survey. To appear in IEEE Systems Journal (2014)
[21] Khaitan, S., McCalley, J.: Cyber Physical System Approach for Design of Power Grids: A Survey. In: IEEE PES GM 2013, Vancouver, BC, July 21-25, pp. 1–5 (2013)
[22] Cheng, S.-T., Chang, T.-Y.: Cyber Physical System Model Using Genetic Algorithm for Actuators Control. In: CECNet (2012)

[23] Stefanov, A., Liu, C.-C.: ICT Modeling for Integrated Simulation of Cyber-Physical Power Systems. In: IEEE PES Innovative Smart Grid Technologies Europe (ISGT Europe), Berlin (2012)
[24] Bottura, R., Babazadeh, D., Zhu, K., Borghetti, A., Nordström, L., Nucci, C.: SITL and HLA Co-simulation Platforms: Tools for Analysis of the Integrated ICT and Electric Power System. In: IEEE Eurocon 2013 (2013)
[25] Babazadeh, D., Chenine, M., Kun, Z., Al-Hammouri, A., Nordström, L.: A Platform for Wide Area Monitoring and Control System ICT Analysis and Development. In: IEEE Grenoble PowerTech 2013 (2013)
[26] Seyboth, G.S., Dimarogonas, D.V., Johansson, K.H.: Event-based broadcasting for multi-agent average consensus. Automatica 49(1), 245–252 (2013)
[27] Beard, R.W.: Consensus seeking in multiagent systems under dynamically changing interaction topologies. IEEE Trans. Automat. Contr. 50(5), 655–661 (2005)
[28] Olfati-Saber, R., Murray, R.M.: Consensus problems in networks of agents with switching topology and time-delays. IEEE Trans. Autom. Control 49(9), 1520–1533 (2004)
[29] Veilleux, É., Ooi, B.-T.: Multi-Terminal HVDC Grid with Power Flow Controllability. In: Cigre 2012 (2012)
[30] OPNET System-In-The-Loop (SITL) Module, `http://www.opnet.com/solutions/network_rd/system_in_the_loop.html`
[31] Gonzalez, R., Karlsson, A.: Impact of Communication Network Quality on Control and Operation of Multiterminal HVDC Systems. Bachelor Thesis, KTH, Royal Institute of Technology (2014)

Smart Buildings in the Smart Grid: Contract-Based Design of an Integrated Energy Management System

Mehdi Maasoumy, Pierluigi Nuzzo, and Alberto Sangiovanni-Vincentelli

Abstract. In a supply-following "smart" grid scenario, buildings can exploit remotely controllable thermostats and "smart" meters to communicate with energy providers, trade energy in real-time and offer frequency regulation services, by leveraging the flexibility in the energy consumption of their heating, ventilation and air conditioning (HVAC) systems. The realization of such a scenario is, however, strongly dependent on our ability to radically re-think the way both the grid and the building control algorithms are designed. In this work, we regard the grid as an integrated, distributed, cyber-physical system, and propose a compositional framework for the deployment of an optimal supply-following strategy. We use the concept of assume-guarantee contracts to formalize the requirements of the grid and the building subsystem as well as their interface. At the building level, such formalization leads to the development of an optimal control mechanism to determine the HVAC energy flexibility while maximizing the monetary incentive for it. At the grid level, it allows formulating a model predictive control scheme to optimally control the ancillary service power flow from buildings, while integrating constraints such as ramping rates of ancillary service providers, maximum available ancillary power, and load forecast information. Simulation results illustrate the effectiveness of the proposed design methodology and the improvements brought by the proposed control strategy with respect to the state of the art.

1 Introduction

Energy consumers do not usually pay enough attention to *when* they use energy. Demand for electricity tends to rise especially at times when it seems natural to use it; so natural, in fact, that we all tend to use energy at the same time and in similar

Mehdi Maasoumy · Pierluigi Nuzzo · Alberto Sangiovanni-Vincentelli
Department of Electrical Engineering and Computer Sciences,
University of California, Berkeley, CA 94720
e-mail: {maasoumy,nuzzo,alberto}@eecs.berkeley.edu

S.K. Khaitan, J.D. McCalley, and C.-C. Liu (eds.), *Cyber Physical Systems Approach to Smart Electric Power Grid*, Power Systems, DOI: 10.1007/978-3-662-45928-7_5

ways. Such a common practice can easily lead to peaks in electricity demand that are traditionally met by operating extra power plants for limited portions of time, a solution which is generally expensive and environmentally unfriendly.

A more environmentally-friendly alternative to such a demand-following strategy is offered by *supply-following* (SF) programs, where utilities provide *incentives* to encourage consumers to reduce their demand during peak periods and use electricity at a less congested time. Another solution would be to significantly increase the penetration of Renewable Energy Sources (RESs), thus avoiding the introduction of expensive power plants. Several states in the United States and countries around the world have set ambitious targets for penetration of RESs by the next few years. The State of California, as an example, has targeted a 33% RES portfolio by 2020 [9]. However, a large-scale power grid requires continuous power balance between supply and demand; the power flows through individual transmission lines and facilities should also be controlled by continuously adjusting generation or load. Such an instantaneous matching becomes challenging due to the volatility, uncertainty, and intermittency of RESs, and makes their integration into the grid extremely difficult. The situation is even worsened by the uncertainties and randomness in the demand, due to short-term random switching of millions of individual loads, or longer-term (e.g. daily or seasonal) fluctuations in load and weather patterns.

While electricity storage is widely believed to be a solution to these problems by partially absorbing the variability associated with RESs and providing the extra power required at the peak energy demand hours, it is considered as an expensive and not environmentally-friendly solution. On the other hand, there is an emerging consensus that the transition from a demand-following strategy to a supply-following one will finally be enabled by the deployment of advanced metering infrastructure (AMI) devices in the *smart grids*. In a smart grid, new functions, denoted as *ancillary services*, are performed by the entities that generate, control, and transmit electricity in support of the basic services of generating capacity, energy supply, and power delivery. In this context, *smart buildings* can play a significant role. Buildings have inherent flexibility in the way their heating, ventilation and air conditioning (HVAC) systems consume electricity while respecting the occupants' comfort. This flexibility could be used to reduce costs if the electricity price is time-varying, or could be traded (i.e., sold to the utility) to be used for ancillary services.

Such flexible loads with thermal storage capabilities, also denoted as Thermostatically Controlled Loads (TCLs), are deemed to play an important role in regulating the grid frequency and, consequently, in enabling deep penetration of RESs. It has been reported that about 20% of the total electricity consumption in the United States is used by residential TCLs such as air conditioners, heat pumps, water heaters, and refrigerators [1, 2]. Recently, [19, 20] have shown that flexible loads such as TCLs are good candidates for providing ancillary services since their aggregate flexibility can be controlled very fast, and sums up to tens of Gigawatts of power, only in the United States. Modeling, estimation, and control of aggregated heterogeneous TCLs for ancillary services have been discussed in [12]. TCLs are particularly well-suited for Direct Load Control and Demand Response programs that require loads to both decrease and increase power consumption because they are capable of storing thermal energy, much like a battery stores chemical energy. Despite several challenges in using loads for system services, key advantages include: (i) reduction in the overall grid emissions [32]; (ii) instantaneous response of loads to operator requests, versus slow response of generators to significant output changes [11]; and (iii) less variability associated to a very large number of small loads with respect to that of a small number of large generators [11]. Therefore, while ancillary services have been conventionally supplied by generation units, the increasing need for more energy storage capacity for frequency regulation, as well as more agile sources of ancillary services, makes it attractive to also use energy reserves on the demand side. It may soon be the case that the only technical impediment to reliable utilization of loads for system services is the development of the necessary models and control strategies and the development of inexpensive and scalable communication and sensing infrastructure [36].

To fully exploit the potential of buildings as service providers, we need to fundamentally re-design the way both the building HVAC system and the grid are controlled. This chapter addresses the problem of developing models and control algorithms for the deployment of a supply-following strategy in a smart grid, from the perspective of a hierarchical, distributed, cyber-physical system. The smart grid is intrinsically distributed, since different control algorithms must be executed in parallel on different components (e.g. buildings, energy providers) to achieve a common goal. On the other hand, the buildings can be abstracted as a load for the grid, which highlights the hierarchical nature of the system, where the designer is allowed to define the global behavior (via the grid controller) *together with* the local behavior of the "plant" (via the building controller).

Historically, very few, high-capacity reserves, such as industrial plants [33], have been used to provide ancillary services on the demand side. However, when a "swarm" of widespread and smaller capacity reserves are available, these service providers are better managed by intermediate entities called *aggregators*. The role of an aggregator is to provide appropriate incentives for a swarm of buildings at the right time, bundle the resulting capacity, and sell it in the wholesale market for frequency regulation. In this chapter, to simplify, we abstract into one *grid* agent all the players beyond the aggregator, such as the wholesale market players and the generation units, and denote as *buildings* the demand-side service providers that deal with the aggregator. We then focus on grid and buildings as the two sides of the supply-demand spectrum, by abstracting all the intermediate entities involved in the chain from power generation to power consumption.

To address the challenges originating from this distributed and hierarchical system, we resort to a Contract-Based Design (CBD) methodology. CBD has recently emerged as a compositional paradigm for the design of complex systems, emphasizing the concept of interface and requirement formalization to facilitate system integration and provide formal support to the whole design flow [25, 31]. We use assume-guarantee contracts to formalize the requirements of the buildings, the grid, and their interface. Based on this formalization, we build on top of the supply-following scenario introduced in [19] and [20]. However, while in [19] and [20] the grid and building control schemes are derived and investigated separately, in this chapter, we provide an *integrated design framework* for Model Predictive Control (MPC) synthesis, which can combine and subsume both the approaches in [19] and [20]. At the building level, we develop an optimal control mechanism to determine the HVAC flexibility while maximizing the monetary incentive for it in a receding horizon fashion. At the grid level, we formulate a model predictive control scheme to optimally control the ancillary service power flow from buildings, while integrating constraints such as ramping rates of ancillary service providers, maximum available ancillary power, and load forecast. We use MPC as a convenient framework that allows optimizing a desired cost-function over a finite time horizon while, at the same time, satisfying a set of constraints.

The advantage of our contract-based methodology with respect to previous works is threefold: (i) it enables compositional design of the building and the grid MPC schemes, so that they can be independently implemented, while still guaranteeing that their integration is correct; (ii) it allows extending the approaches in [19] and [20] to highly distributed architectures, including a large number of control areas and buildings, in a scalable way; (iii) it supports automatic synthesis of embedded control software directly from assume-guarantee specifications.

The remainder of the chapter is organized as follows. Section 2 and Section 3 provide background information on the supply-following scenario of interest, and on our contract-based design methodology, thus setting the stage for our formulations. Section 4, Section 5 and Section 6 detail the main steps of our methodology, i.e. contract-based requirement formalization, generation of the model library, and MPC synthesis. Simulation results, in Section 7, illustrate the effectiveness of our design methodology and the improvements brought by the proposed control strategy with respect to the state of the art. Finally, Section 8 draws some conclusions.

2 A Supply-Following Scenario for Smart Buildings

In this section, we provide an overview of the Supply Following (SF) scenario considered in this chapter, by focusing on commercial buildings. Compared to residential buildings, commercial buildings typically have larger HVAC systems and therefore consume more electricity. In fact, commercial buildings account for more than 35% of electricity consumption in the US. Moreover, more than 30% of them have adopted a Building Energy Management System (BEMS) technology which facilitates the communication with the grid operators to provide flexibility. The majority of these buildings are also equipped with variable frequency drives, which in coordination with the BEMS, can modulate the HVAC system power consumption at intervals of the order of seconds. About 15% of electricity consumption in commercial buildings is related to the fans of the HVAC systems. Fans are the main drivers, moving the conditioned air from the air handling units (AHU) to the rooms for climate control. For instance, the main supply fans that feed one of the buildings on the U.C. Berkeley campus, Sutardja Dai Hall, can spin at variable speeds, with the maximum rated power of 134 kW, proportional to the cube of the fan speed, which is about 14% of the maximum power consumed in the whole building. Moreover, the power consumed by the fans can be directly controlled upward or downward, thus making it an ideal candidate for ancillary services.

We refer to [15, 18, 16, 22, 28] for more information about the physics and control of HVAC systems. Moreover, we refer to [19, 20] for at-scale experiments on a real building, and a discussion on the feasibility of the proposed SF approach. We only observe here that modulating the fan speed of HVAC systems for extended periods of time with the existing control algorithms can lead to discomfort and does not allow optimizing the amount of flexibility provided by a building [17]. Hence, we propose to re-design the control algorithm, and consider commercial buildings whose HVAC systems are controlled by a Model Predictive Control (MPC) scheme running an optimal control problem at each time step k. Typically, the MPC aims at minimizing the total energy cost (in dollars). In an SF scenario, such cost must account for the reward received from the utility because of the building flexibility in energy consumption [19]. We refer to Table 1 and Table 2, later in the chapter, for a summary of the variables and parameters used in the description below.

To quantify the building *flexibility*, we adopt as a natural metric the difference between the upper and lower power envelopes that can be consumed without violating any constraints, i.e., at each time step k,

$$\text{Flexibility}(k) \triangleq P_f(e_k^u) - P_f(e_k^l) \tag{1}$$

where e_k^u and e_k^l are, respectively, the upper and lower bounds on the air mass flow of a building, and $P_f(.)$ returns the power consumption as a function of the air mass flow. Moreover, following the approach of [19], we assume that a *commercial contract* is stipulated between the utility and the building manager, whose duration, in terms of time steps, is $t_{ce} - t_{cs} = H^c$, where t_{cs} and t_{ce} are the commercial contract start and end times, respectively. The commercial contract has a limited duration because of the limited accuracy of the predicted flexibility by the building far ahead in time. Based on such commercial contract, the BEMS declares a lower envelope $\mathbf{e^l} = [e_{t_{cs}}^l, \ldots, e_{t_{ce}}^l]$, an upper envelope $\mathbf{e^u} = [e_{t_{cs}}^u, \ldots, e_{t_{ce}}^u]$ and a baseline $\mathbf{u}^* = [u_{t_{cs}}^*, \ldots, u_{t_{ce}}^*]$ air mass flow profiles for the duration of the contract. The utility is then allowed to select any power trajectory $P_f(\mathbf{u}) = [P_f(u_{t_{cs}}), \ldots, P_f(u_{t_{ce}})]$ such that, for all $k \in \{t_{cs}, \ldots, t_{ce}\}$, $P_f(e_k^l) \leq P_f(u_k) \leq P_f(e_k^u)$. However, the SF contract is deterministic, since the utility and the building operator both know how much money they have to pay or they receive from the beginning of the commercial contract. The utility charges the building operator for the baseline power consumption $P_f(\mathbf{u}^*)$, irrespective of the deviations due to flexibility signals, at a rate $\boldsymbol{\pi}^e$. On the other hand, the utility rewards the building operator for its declared flexibility, by providing both a downward flexibility rate $\underline{\boldsymbol{\beta}}$ and an upward flexibility rate $\overline{\boldsymbol{\beta}}$ with respect to the baseline power.

Clearly, by obeying the utility power consumption signals, the building may consume more or be in a worse state at the end of the H^c time slots with respect to a conventional demand-following protocol. The flexibility declared by the building operator would then be significant only if: (i) it is enough to be effectively exploited for frequency regulation services [20], and (ii) the reward from the utility is

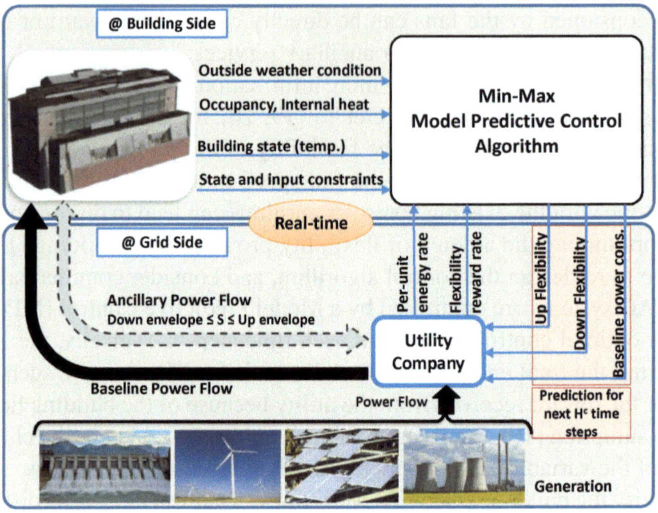

Fig. 1 Schematic of the proposed grid architecture and contractual framework

appropriate for the building. The schematic of the entire system architecture is shown in Fig. 1. The solid-line arrows correspond to the baseline power flow. The ancillary power flow is represented by a dashed-line arrow. For the architecture in Fig. 1 and the commercial contract summarized above, we can state our energy management integrated control problem as follows:

Integrated Energy Management Problem Statement. *Given* the real-time state of the buildings (e.g. indoor temperature, occupancy, internal heat, outside weather condition), a set of building temperature and control requirements, the real-time state of the grid (e.g. frequency deviation, generation and load forecast), a set of frequency regulation requirements, the per-unit energy price, the upward and downward flexibility rewards, and the duration of the commercial contract, *design* an optimal control strategy to determine the baseline power consumption, downward and upward building power envelopes, and grid flexibility signals for the buildings, while satisfying both the building and grid requirements.

In our scenario, the grid essentially controls the building consumption for the next H^c time slots, by sending flexibility signals (similar to frequency regulation signals) to be tracked by the HVAC fans as frequently as every few seconds.

In the next sections, after an overview of Contract-Based Design (CBD), we present a design framework that leverages a formalization of the control goals above and a library of models to generate the MPC schemes at both the building and the grid levels.

3 Contract-Based Design of Cyber-Physical Systems

The notion of formal contracts originates in the context of assume-guarantee reasoning. Informally, a contract is a pair $\mathcal{C} = (A, G)$ of properties, assumptions and guarantees, respectively representing the assumptions on the environment and the promises of the system under these assumptions. The essence of contracts is a compositional approach, where design and verification complexity is reduced by decomposing system-level tasks into more manageable subproblems at the component level, under a set of assumptions. System properties can then be inferred or proved based on component properties.

Compositional reasoning has been known for a long time, but it has mostly been used as a verification mean for the design of software. Rigorous contract theories have then been developed over the years, including assume-guarantee (A/G) contracts [5] and interface theories [4]. However, their concrete adoption in CPS design is still in its infancy [25]. Examples of application of A/G contracts have only been recently demonstrated in the automotive [6] and consumer electronics [26] domains. The use of A/G contracts for control design in combination with platform-based design (PBD) [30] was advocated in [25, 31], while in [27, 24, 10], a methodology was introduced that used contracts to integrate heterogeneous modeling and analysis frameworks for synthesis and optimization of CPS architectures and control protocols. The design flow was demonstrated on a real-life example of industrial interest, namely the design of system topology and supervisory control for aircraft electric power systems (EPS).

3.1 Contracts

We summarize the main concepts behind our methodology by presenting a simple contract model centered on the notion of platform *component*. A platform component $\mathcal{M}$ can be seen as an abstraction representing an element of a design, characterized by a set of *attributes*, including: *variables* (input, output and internal), configuration *parameters*, and *ports* (input, output and bidirectional); a *behavioral model*, uniquely determining the values of the output and internal variables given the values of the input variables and configuration parameters, and a set of *non-functional models*, i.e. maps that allow computing non-functional attributes of a component, corresponding to particular valuations of its input variables and configuration parameters. Components can be connected together by sharing certain ports under constraints on the values of certain variables. In what follows, we use variables to denote both component variables and ports. A component may be associated with both implementations and contracts. An *implementation* M is an instantiation of a component $\mathcal{M}$ for a given set of configuration parameters. In the following, we also use M to denote the set of behaviors of an implementation, which assign a history of "values" to ports. Behaviors are generic and abstract. For instance, they could be continuous functions that result from solving differential equations, or sequences of values or events recognized by an automata model.

A *contract* $\mathcal{C}$ for a component $\mathcal{M}$ is a pair of assertions (A,G), called the *assumptions* and the *guarantees*, each representing a specific set of behaviors over the component variables [5]. An implementation M satisfies an assertion B whenever M and B are defined over the same set of variables and all the behaviors of M satisfy the assertion, i.e. when $M \subseteq B$. An implementation of a component satisfies a contract whenever it satisfies its guarantee, subject to the assumption. Formally, $M \cap A \subseteq G$, where M and $\mathcal{C}$ have the same variables. We denote such a *satisfaction* relation by writing $M \models \mathcal{C}$. An implementation E is a legal *environment* for $\mathcal{C}$, i.e. $E \models_E \mathcal{C}$, whenever $E \subseteq A$. Two contracts $\mathcal{C}$ and $\mathcal{C}'$ with identical variables, identical assumptions, and such that $G' \cup \neg A = G \cup \neg A$, where $\neg A$ is the complement of A, possess identical sets of environments and implementations. Such two contracts are then *equivalent*. In particular, any contract $\mathcal{C} = (A,G)$ is equivalent to a contract in *saturated form* (A,G'), obtained by taking $G' = G \cup \neg A$. Therefore, in what follows, we assume that all contracts are in saturated form. A contract is *consistent* when the set of implementations satisfying it is not empty, i.e. it is feasible to develop implementations for it. For contracts in saturated form, this amounts to verify that $G \neq \emptyset$. Let M be any implementation, i.e. $M \models \mathcal{C}$, then $\mathcal{C}$ is *compatible*, if there exists a legal environment E for M, i.e. if and only if $A \neq \emptyset$. The intent is that a component satisfying contract $\mathcal{C}$ can only be used in the context of a compatible environment.

Contracts associated to different components can be combined according to different rules. Similar to parallel composition of components, *parallel composition* ($\otimes$) of contracts can be used to construct composite contracts out of simpler ones. Let M_1 and M_2 two components that are composable to obtain $M_1 \times M_2$ and satisfy, respectively, contracts $\mathcal{C}_1$ and $\mathcal{C}_2$. Then, $M_1 \times M_2$ is a valid composition if M_1 and M_2 are *compatible*. This can be checked by first computing the contract composition $\mathcal{C}_{12} = \mathcal{C}_1 \otimes \mathcal{C}_2$ and then checking whether $\mathcal{C}_{12}$ is compatible. To compose multiple views of the same component that need to be satisfied simultaneously, the *conjunction* ($\wedge$) of contracts can also be defined so that if $M \models \mathcal{C}_1 \wedge \mathcal{C}_2$, then $M \models \mathcal{C}_1$ and $M \models \mathcal{C}_2$. Contract conjunction can be computed by defining a preorder on contracts, which formalizes a notion of *refinement*. We say that $\mathcal{C}$ refines $\mathcal{C}'$, written $\mathcal{C} \preceq \mathcal{C}'$ if and only if $A \supseteq A'$ and $G \subseteq G'$. Refinement amounts to relaxing assumptions and reinforcing guarantees, therefore strengthening the contract. Clearly, if $M \models \mathcal{C}$ and $\mathcal{C} \preceq \mathcal{C}'$, then $M \models \mathcal{C}'$. On the other hand, if $E \models_E \mathcal{C}'$, then $E \models_E \mathcal{C}$. Mathematical expressions for computing contract composition and conjunction can be found in [5].

3.2 Design Flow

In [27, 24], we introduced a design methodology that addresses the complexity and heterogeneity of cyber-physical systems by using assume-guarantee contracts to

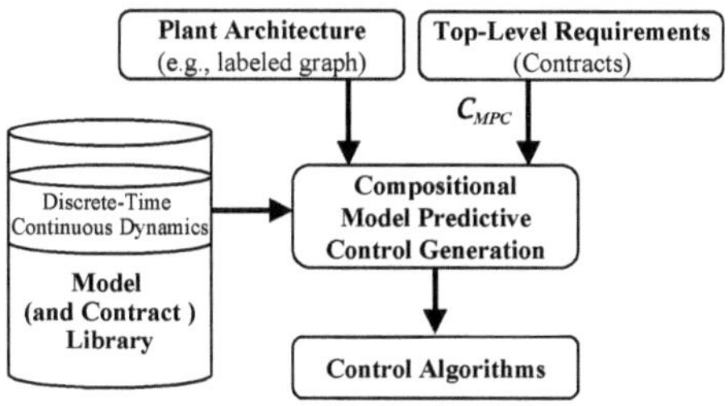

Fig. 2 Contract-based model predictive control synthesis flow

formalize the design process and enable realization of control protocols in a hierarchical and compositional manner. Given the architecture of the physical plant to be controlled, the design is carried out as a sequence of refinement steps from an initial specification to a final implementation, including synthesis from requirements and mapping of higher-level functional and non-functional models into a set of candidate solutions built out of a library of components at the lower level. Initial top-level requirements are captured as contracts and expressed using linear temporal logic (LTL) [29] and signal temporal logic (STL) [23] formulas to enable requirement analysis and early detection of inconsistencies. Requirements are then refined into a controller architecture by combining reactive synthesis steps from LTL specifications with simulation-based design space exploration steps. We have demonstrated our approach on the design of embedded controllers for aircraft electric power distribution. In this work, we adapt and extend our methodology to the design of MPC algorithms. In our design flow, pictorially represented in Fig. 2, platform component design and characterization is completely orthogonalized from system specification and algorithm design.

Requirement Formalization. In the top-down phase of the design process, top-level system requirements are formalized as contracts. Responsibilities of achieving requirements are split into those to be established by a system (guarantees) and those characterizing admissible environments (assumptions). In a distributed control setting as in our application, the requirements of a controller C can be expressed as a contract $\mathcal{C}_C = (A_C, G_C)$, where A_C encodes the allowable behaviors (i.e. trajectories, or sequences of valuations over a set of variables) of the environment (e.g. physical plant, or other controllers) and G_C encodes the required behaviors of C. To define $\mathcal{C}_C$, as shown Fig. 2, we can leverage a discrete time abstraction of the continuous behaviors of the components. We then express A_C and G_C as either first order difference equations involving the component variables and parameters (time varying properties), or arithmetic constraints on real variables that must hold at each time step (time invariant properties). The algebra of contracts can then be

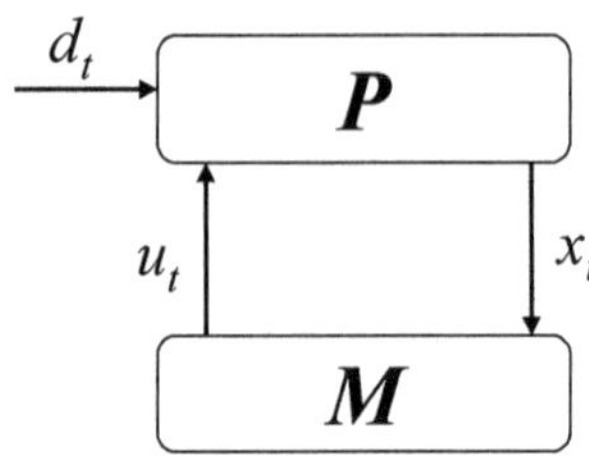

Fig. 3 Generic feedback control scheme

implemented by simply combining constraints via conjunction or disjunction to express, respectively, intersections or unions of behaviors. Examples of this approach will be provided in Section 4.

Platform and Contract Library Generation. In the bottom-up phase of the design process, a library of components (and contracts) is generated to model (or specify) both the plant architecture (e.g. the power system or the building) and the controllers. Components can be hierarchically organized to represent the system at different levels of abstraction, e.g. *steady-state* (static), *discrete-event* (DE), and *hybrid* levels. At each level of abstraction, components are also capable of exposing multiple, complementary *views*, associated with different design concerns (e.g. safety, performance, reliability) and with models that can be expressed via different formalisms (e.g. graphs, linear temporal logic, differential equations), and analyzed by different tools. Such models include non-functional (performance) metrics, such as timing, energy and cost. As detailed in Section 5, in this work, we model our platform components by adopting the same discrete-time abstractions and formalisms we use for requirements. System behavioral models are expressed using difference equations, while performance and cost models are polynomial functions of the component variables and parameters.

Mapping Functions to Implementations. System design (synthesis) is cast as a set of problems mapping functionality (specifications or requirements) over implementations. A mapping problem can be solved by casting an optimization problem that uses information from both the system and the component levels to evaluate global tradeoffs among components or minimize a cost function.

For an MPC scenario, let $\mathcal{C}_{MPC}$ be the contract formalizing the requirements of the closed loop architecture in Fig. 3, where M is the controller and P the plant, and let H be the MPC horizon. We assume that the system dynamics are described by the following difference equation at each time step t:

$$x_{t+1} = p(x_t, u_t, d_t) \quad \forall\, t \in \mathbb{N}, \tag{2}$$

where x_t is the system state, u_t the control input, and d_t an external uncontrolled input (disturbance). We denote as $s = (x, u, d)$ the set of system variables. A system

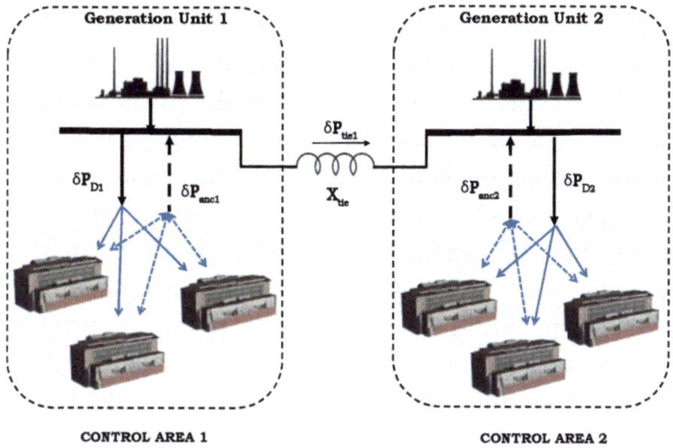

Fig. 4 A two-area model of the system architecture considered in this chapter

behavior or trajectory $\sigma = s_0, s_1, s_2, \ldots$ is a sequence of valuations over s, for all $t \in \mathbb{N}$. We also assume that the assumptions and guarantees in $\mathcal{C}_{MPC}$ can be represented as follows:

$$A_{MPC} = \{\sigma | \alpha_t(s_t) \leq 0 \;\; \forall\, t \in \mathbb{N}\} \qquad G_{MPC} = \{\sigma | \gamma_t(s_t) \leq 0 \;\; \forall\, t \in \mathbb{N}\}, \tag{3}$$

where $\alpha_t(.)$, $\gamma_t(.)$ are generic real functions (constraints) parameterized by t. Finally, let $C(u,x)$ be a real function providing the system cost in terms of system state and control. Then the optimal control problem aiming at minimizing the cost over time horizon H while satisfying the system dynamics and contract can be formulated as follows:

$$\min_{\mathbf{u}_t} \quad \sum_{k=0}^{H-1} C(u_{t+k}, x_{t+k}) \tag{4a}$$

$$\text{subject to:} \quad x_{t+k+1} = p(x_{t+k}, u_{t+k}, d_{t+k}), \quad \forall\, k \in \{0, \ldots, H-1\} \tag{4b}$$

$$\gamma_{t+k}(x_{t+k}, u_{t+k}, d_{t+k}) \leq 0, \quad \forall\, k \in \{0, \ldots, H-1\} \tag{4c}$$

$$\alpha_{t+k}(x_{t+k}, u_{t+k}, d_{t+k}) \leq 0, \quad \forall\, k \in \{0, \ldots, H-1\} \tag{4d}$$

where $\mathbf{u}_t = (u_t, u_{t+1}, \ldots, u_{t+H-1})$. Both contract assumptions and guarantees are captured as optimization constraints. The resulting optimal control algorithm executes the optimization problem (4) in a receding horizon fashion, and is returned as the final design.

4 System Requirement Formalization

We consider a control area network as the one shown in Fig. 4. In particular, for the sake of illustration, we use a simplified two-area model, and abstract the complexity of the full power transmission and distribution subsystem. Commercial buildings have HVAC systems controlled by an MPC scheme, while the grid utilizes a hierarchical control scheme composed of a high-level MPC framework on top of the low-level classical automatic generation control (AGC), detailed in Section 5.2. The MPC schemes are based on discrete-time models of the system dynamics. Let τ_G and τ_B be the sampling times for the grid and building dynamics, respectively. Generally, τ_G is much smaller than τ_B since the grid must be able to send control signals as fast as every second, while τ_B may range from 15 min to 1 h.

We also assume that H^m and H are the time horizons of the MPC scheme adopted, respectively by the building subsystem (B-MPC) and the grid (G-MPC), with $H^m \gg H^c \gg H$, where H^c is the length of the commercial contract. The choice of H^m depends on how far in the future the predicted values of the building model parameters have an acceptable accuracy and the inputs to B-MPC (e.g. cost of energy) are available. Typical values for H^m range from a few hours to a few days. For instance, by assuming $\tau_B = 1$ h for the building dynamics, H^m may range from 3 to 72 time slots. Similarly, the choice of H depends on the accuracy of the grid model and grid load forecast. Finally, typical values for H^c can range from one to a few building time slots τ_B. In particular, we pick the contract start time t_{cs} and end time t_{ce} such that $t_{ce} - t_{cs} = H^c$, $t_{cs} \geq 1$ and $t_{cs} + 1 \leq t_{ce}$, in units of τ_B. In our simulations, we use $H^m = 24$ h, $H^c = \tau_B = 1$ h, $\tau_G = 1$ s and $H = 60$ s.

A summary of all the variables and parameters used to formalize the requirements and generate the MPC schemes is provided in Table 1 and Table 2. The overall set of requirements for the integrated energy management system can be formalized in terms of contracts as follows.

4.1 Building Contract

The building contract $\mathcal{C}_B$ can be expressed as follows.

Assumptions. Each building receives from the grid the vector of *electric energy prices* $\boldsymbol{\pi} = [\pi^e_t, \ldots, \pi^e_{t+H^m-1}]$ as well as the *prices of non-electric cooling and heating energy*, $\pi^{ne,c}$ and $\pi^{ne,h}$, respectively, every H^c time slots. Similarly, it receives a pair of *rewards* vectors for providing upward flexibility ($\overline{\boldsymbol{\beta}}_t$) and downward flexibility ($\underline{\boldsymbol{\beta}}_t$). Finally, the building receives air flow control signals w_t from the grid, with the only assumption that they are bounded by the air flow flexibility as defined below, i.e. at each time t, we have $\underline{\varphi}_t \leq w_t \leq \overline{\varphi}_t$.

Guarantees. The building must satisfy a set of *temperature requirements*, expressed as predicates on the building states of the form "x_{t+k} should be in $\mathcal{X}_{t+k}$ for all times $t+k$ where $k \in \{1, \ldots, H^m\}$". Similarly, *air mass flow requirements* can be formalized as a set of constraints $\mathcal{U}_{t+k}$ on the building inputs for all $k \in \{0, \ldots, H^m - 1\}$, where

$$\mathcal{X}_t := \{x \mid \underline{T}_t \leq x \leq \overline{T}_t\} \tag{5}$$

$$\mathcal{U}_t := \{u \mid \underline{U}_t \leq u \leq \overline{U}_t\}, \tag{6}$$

Table 1 Building and Building/Grid interface variables

Variable	Definition
	Building Requirements
H^m	Prediction horizon for the building MPC
$\mathscr{X}_t$	Set of permissible states at time t
$\mathscr{U}_t$	Set of permissible inputs at time t
	Building Model
τ_B	Sampling time for discretizing the building continuous dynamics
d_t	Disturbance at time t (e.g. outside temperature, occupancy, solar radiation)
T_t^{out}	Outside air temperature at time t
T^s	Supply air temperature exiting air handling unit (AHU)
COP_h	Coefficient of performance of heating system
COP_c	Coefficient of performance of cooling system
x_t	State of the system at time t
u_t	Input to the system at time t
w_t	Grid control uncertainty variable for the building control problem
P_f, P_h, P_c	Power consumption of fan, heating and cooling systems
	Building/Grid Interface Variables
H^c	Horizon (length) of the commercial contract
π_t^e	Per-unit price of electric energy at time t [$/kWh]
$\pi^{ne,c}$	Per-unit price of non-electric cooling energy [$/kWh]
$\pi^{ne,h}$	Per-unit price of non-electric heating energy [$/kWh]
t_{cs}	Commercial contract start time
t_{ce}	Commercial contract end time
$\{\overline{\beta}_t, \underline{\beta}_t\}$	Reward paid from the grid to the building for upward flexibility ($\overline{\beta}_t$) and downward flexibility ($\underline{\beta}_t$) at time t [$/kWh]
e_t^u	Upper envelope for safe air mass flow
e_t^l	Lower envelope for safe air mass flow
$\{\overline{\varphi}_t, \underline{\varphi}_t\}$	Upward ($\overline{\varphi}$) and downward ($\underline{\varphi}$) flexibility of the building at time t (in air flow)
$\{\overline{\psi}_t, \underline{\psi}_t\}$	Upward ($\overline{\psi}$), and downward ($\underline{\psi}$) flexibility of building at time t (in power)
$C_{\text{hvac}}(u_t, \pi_t^e)$	Total HVAC energy consumption cost at time t
$R(\Phi, \mathscr{B})$	Total reward from the grid to the building for flexibility

$\underline{T}_t$ and $\overline{T}_t$ are the upper and lower temperature limits, and $\overline{U}_t$ and $\underline{U}_t$ are the upper and lower feasible air mass flow rates at time t.

We say that the building offers a *flexibility* $\Psi := \{\underline{\boldsymbol{\psi}}, \overline{\boldsymbol{\psi}}\}$ in fan power or equivalently a flexibility $\boldsymbol{\Phi} := \{\underline{\boldsymbol{\varphi}}, \overline{\boldsymbol{\varphi}}\}$ in air mass flow, including downward flexibility $\underline{\boldsymbol{\varphi}}$ and upward flexibility $\overline{\boldsymbol{\varphi}}$, from the contract start time t_{cs} to the contract end time t_{ce} if there exist two trajectories $\mathbf{e^l} = \mathbf{u} + \underline{\boldsymbol{\varphi}}$ and $\mathbf{e^u} = \mathbf{u} + \overline{\boldsymbol{\varphi}}$, that satisfy:

$$\underline{\varphi}_k \leq 0, \quad \overline{\varphi_k} \geq 0 \qquad \forall k \in \{t_{cs}, \ldots, t_{ce}\} \tag{7}$$

$$f(x_k, u_k + \underline{\varphi}_k, d_k) \in \mathscr{X}_{k+1} \qquad \forall k \in \{t_{cs} - 1, \ldots, t_{ce} - 1\} \tag{8}$$

Table 2 Power system variables

Variable	Definition
H	Prediction horizon for the grid MPC
λ	Bound on the rate of change of the power supplied by the buildings
τ_G	Sampling time for discretizing the grid continuous dynamics
P_M	Mechanical power input
P_M^o	Desired real power generation
P_G	Generated real electric power
δP_G	Increase in demand (at rated generator MVA)
V_t	Terminal voltage
P_D	Load (power demand)
δP_D	Input disturbance due to load changes
δP_C	Speed changer position feedback control signal
ω	Angular speed and frequency
ω_o	Rated (desired) frequency
D	Damping coefficient. Range: 0.01 - 0.1
M	Machine inertia constant. Range: 100 - 1000 [MW s]
R	Speed regulation constant. Range: 0.05 [p.u.]
T_i	Time constant for power system components. Range: {0,0.01-10} [s]
K_i	Fraction of total mechanical power outputs associated with different operating points of the turbine. Range: {0,0.1-1}

$$f(x_k, u_k + \overline{\varphi}_k, d_k) \in \mathscr{X}_{k+1} \qquad \forall k \in \{t_{cs}-1, \ldots, t_{ce}-1\} \tag{9}$$

$$u_k + \underline{\varphi}_k \in \mathscr{U}_k \qquad \forall k \in \{t_{cs}, \ldots, t_{ce}\} \tag{10}$$

$$u_k + \overline{\varphi}_k \in \mathscr{U}_k \qquad \forall k \in \{t_{cs}, \ldots, t_{ce}\} \tag{11}$$

where f captures the building dynamics, and d_k is an estimate of unmodelled disturbances, as detailed in Section 5.1. If the BEMS declares $\underline{\boldsymbol{\varphi}}$ and $\overline{\boldsymbol{\varphi}}$, then the utility can choose any fan power (and consequently air flow $\hat{u}_k$) for all time steps $t_{cs} \leq k \leq t_{ce}$ as long as $u_k^* + \underline{\varphi}_k \leq \hat{u}_k \leq u_k^* + \overline{\varphi}_k$, where u_k^* is the *baseline* air mass flow. Hence, we "center" the flexibility around u^*.

4.2 Grid Contract

To express the grid contract $\mathscr{C}_G$, we refer to the grid system architecture in Fig. 5, where commercial buildings are abstracted as a load for the power system.

Assumptions. Buildings can provide both positive and negative power flow to the grid for frequency regulation purposes. When there is a power deficit, buildings will temporarily use less power, and when there is as surplus of power, they will temporarily use the extra power.

To formalize such a behavior at the grid level, we abstract the building subsystem in terms of: (i) aggregate baseline power demand P_D, and (ii) aggregate flexibility for ancillary services δP_{anc}. In practice, the grid operator predicts both the long-

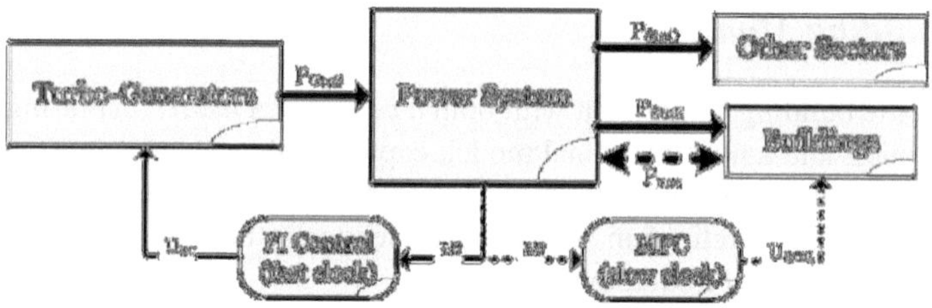

Fig. 5 Schematic of the power system showing its interconnections with the turbo-generators, the building subsystem, and other sectors, along with the control architecture. Thick arrows represent the power flow while thin arrows represent frequency and control signals. Dashed arrows indicate the additional signals and power flows proposed in this work, on top of the state-of-the-art AGC.

term power demand and its short-term deviations from historical data (e.g. weather patterns) by using machine learning algorithms. Therefore, in this formulation, we assume that the power demand P_D, and any variation δP_D, are known parameters, internal to the power system model. Similarly, the upper bound on the ramping rate of the ancillary service power $\lambda > 0$ is treated as a constant parameter of the power system model. Then, the only assumptions of the grid operator on the characteristics of the ancillary service signal from the buildings can be expressed in terms of *maximum capacity* $\max(\delta P_{\mathrm{anc}_k}) > 0$, and a *minimum capacity* $\min(\delta P_{\mathrm{anc}_k}) < 0$ at each time k. Such bounds can be directly derived by the flexibility declared by the buildings at the beginning of the commercial contract. Finally, we assume that u_{sc} in Fig. 5 is also constant, since it is regulated by a local PI controller external to the G-MPC problem.

Guarantees. The utility guarantees that its power consumption signals to the buildings will be within the assumed maximum and minimum capacity boundaries above. Moreover, the recent Federal Energy Regulatory Commission (FERC) Order 755 requires scheduling coordinators to procure and compensate more for regulation resources with faster ramping rates. This ramping rate constraint can be formalized by requiring that $|\Delta P_{\mathrm{anc}}| = |\delta P_{\mathrm{anc}_{k+1}} - \delta P_{\mathrm{anc}_k}| \leq \lambda$ holds at each time step k, i.e. the *rate of change* of the power supplied by the buildings must be guaranteed to be limited by λ.

5 System Model Library

We describe the models of the different components of the grid, starting with the building subsystem. These models will be used together with the contracts in Section 4 to formulate the MPC optimization problems.

5.1 Building Model

We equip the building component with both a *behavioral model*, capturing the system dynamics, and a non-functional model, capturing the electric power consumption as a function of the air mass flow u_t and the outer temperature T_t^{out} at time t. By assuming a discretization step τ_B as in Section 4, the building dynamics are regulated by a difference equation of the form

$$x_{t+1} = f(x_t, u_t, d_t) \tag{12}$$

where x_t represent the system state, i.e. the temperatures of different rooms or zones in the building, u_t is the air mass flow to the thermal zones, and d_t is an estimate of the unmodelled disturbances, e.g. outside temperature or building occupancy [21]. The function $f(.)$ is generally non-linear, which makes it more difficult to handle in an optimization framework. Therefore, we adopt a linearized expression of the state update equation using the forward Euler integration formula with time-step τ_B as:

$$x_{t+1} = Ax_t + Bu_t + Ed_t. \tag{13}$$

The building HVAC *power consumption* is the summation of fan power, cooling power and heating power. With the assumption of no recirculation of air and constant air mass flow, the three contributions can be calculated as follows:

$$P_f(u_t) = c_1 u_t^3 + c_2 u_t^2 + c_3 u_t + c_4 \tag{14}$$

$$P_h(u_t, T_t^{out}) = c_p u_t (T^s - T_t^{out})/COP_{\mathrm{h}} \tag{15}$$

$$P_c(u_t, T_t^{out}) = c_p u_t (T_t^{out} - T^s)/COP_{\mathrm{c}}, \tag{16}$$

where T^{out} is the outside air temperature, constants c_{1-4} are fan parameters, c_p is the specific heat of air, COP_{h} and COP_{c} are the performance coefficients for the heating system and the cooling system, respectively, and the supply air temperature T^s is considered constant. To move the coolant fluid around, heating and cooling systems use pumps which consume electric power. However, we assume that electric power consumption of pumps is negligible compared to the non-electric heating and cooling powers of these systems [16, 18].

5.2 Grid Model

The power system model, based on [8, 34, 13], consists of a governor, turbine, and generator, interconnected as shown in the block diagram of Fig. 6. Electric power is generated by the turbo-generators, and is fed to the power system. The power system transmits and distributes the power to the end users. δP_C is a control input which acts against increase or decrease in the power demand to regulate the system frequency. δP_D denotes the fluctuations in the power demand, which are here considered as an exogenous input (disturbance), while the aggregate flexibility from all the buildings participating in the SF program is lumped into δP_{anc}. As discussed in Section 4.2,

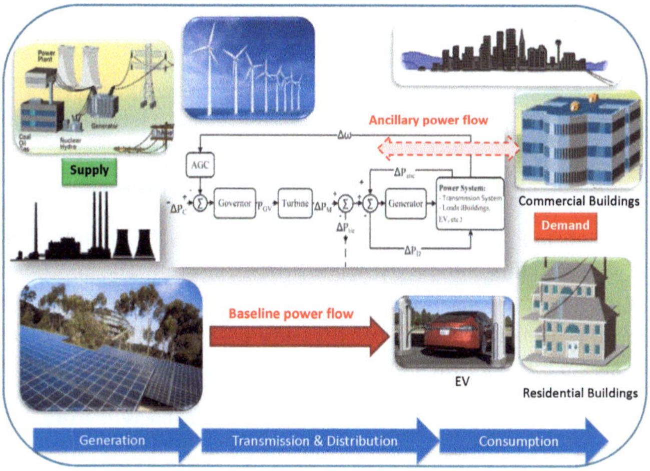

Fig. 6 Block diagram of power system and its relation to governor, turbine, generator, and the AGC signal for each control area

we assume that both the power demand P_D and its variations δP_D are known from historical data. The variables of the power system are listed in Table 2 together with a short description for each of them. Our model relies on the following simplifying assumptions:

- The resistance of the transmission lines is ignored;
- The transmission line between areas i and j is characterized by a reactance $X_{\text{tie}_{ij}}$;
- Reactive power flows are ignored;
- The voltage V_i of bus i is considered constant.

At steady state, we have: $\omega = \omega_o$, $V_t = V_t^o$, and $P_M = P_G = P_M^o$, where ω_o, V_t^o, and P_M^o are the nominal values for rated frequency, terminal voltage and mechanical power input. We are interested in modeling the incremental changes with respect to the steady-state condition.

Governor. The overall input-output transfer function of the governor is given by

$$F_{Gov}(s) = \frac{(1+sT_2)}{(1+sT_1)(1+sT_3)}. \tag{17}$$

Typical values for the time constants depend on whether governors are mechanical-hydraulic or electro-hydraulic, with or without steam feedback [8].

Turbine. The input-output transfer function of a turbine is given by

$$\frac{\delta P_M}{P_{GV}} = K_1F_1 + K_3F_1F_2 + K_5F_1F_2F_3 + K_7F_1F_2F_3F_4, \tag{18}$$

where F_1, F_2, F_3, and F_4 are transfer functions corresponding to steam chest, piping system, re-heaters, and cross-over mechanisms, respectively, and are given by

$$F_1(s) = \frac{1}{1+sT_4}, \qquad F_2(s) = \frac{1}{1+sT_5}, \tag{19}$$

$$F_3(s) = \frac{1}{1+sT_6}, \qquad F_4(s) = \frac{1}{1+sT_7}. \tag{20}$$

The basic time constant associated with steam turbines is T_4 (*steam chest*). For non-reheat steam turbines, this is the only needed time constant. The coefficients K_1, K_3, K_5, and K_7 represent fractions of the total mechanical power outputs associated with *very high*, *high*, *intermediate*, and *low* pressure components, respectively. Typical values of the steam turbine time constants and fractions are reported in [8].

Generator. The dynamics of the generator is given by the following transfer function

$$F_{Gen} = \frac{1}{D+sM}, \tag{21}$$

where constants D and M represent the damping coefficient and the inertia of the governor, respectively.

Two Area System Model. The components above can be interconnected to generate a model for the system in Fig. 4, including two areas connected by a tie line with reactance X_{tie}. P_{tie} is the power flow on the tie line from area 1 to area 2. A positive δP_{tie} represents an increase in power transfer from area 1 to area 2. This in effect is equivalent to increasing the load of area 1 and decreasing the load of area 2. The model can be directly extended to n areas, under the assumption that non-negligible power transfers can only occur between area i and its nearest neighbors $i-1$ and $i+1$. In the two-area model, the superscripts refer to the control area ($i, j \in \{1,2\}$) and the subscripts index the state in each area:

$$\frac{dx_1^i}{dt} = \frac{(-D^i x_1^i + \delta P_M^i - \delta P_D^i - \delta P_{\text{tie}}^{ij} + \delta P_{\text{anc}}^i)}{M_x^i} \tag{22a}$$

$$\frac{dx_2^i}{dt} = \frac{(x_3^i - x_2^i)}{T_7^i} \tag{22b}$$

$$\frac{dx_3^i}{dt} = \frac{(x_4^i - x_3^i)}{T_6^i} \tag{22c}$$

$$\frac{dx_4^i}{dt} = \frac{(x_5^i - x_4^i)}{T_5^i} \tag{22d}$$

$$\frac{dx_5^i}{dt} = \frac{(P_{GV}^i - x_5^i)}{T_4^i} \tag{22e}$$

$$\frac{dx_6^i}{dt} = \frac{(x_7^i - x_6^i)}{T_3^i} \tag{22f}$$

$$\frac{dx_7^i}{dt} = \frac{(-x_7^i + \delta P_C^i - x_1^i/R^i)}{T_1^i} \tag{22g}$$

where

$$\delta P_M^i = K_1^i x_5^i + K_3^i x_4^i + K_5^i x_3^i + K_7^i x_2^i \tag{23}$$

$$P_{GV}^i = (1 - T_2/T_3)x_6^i + (T_2/T_3)x_7^i \tag{24}$$

In (22), the first state variable represents the frequency increment, $x_1^i = \delta\omega_i$. The differential equations for the seven state variables are derived using the mathematical models in (17)–(21). When the time constant representing the system pole is zero, the corresponding differential equation becomes an algebraic equation. For instance, when $T_5 = 0$, the equation $\frac{dx_4^i}{dt} = 1/T_5(x_5^i - x_4^i)$ turns into $x_5^i = x_4^i = 0$.

The real power transferred from bus i to bus j can be approximated as $P_{\text{tie}}^{ij} \approx V_i V_j b_{ij} cos(\theta_i - \theta_j)$. Since here we are concerned with incremental changes in all variables, the incremental change in P_{tie}^{ij} is given by $\delta P_{\text{tie}}^{ij} = v_{ij}(\theta_i - \theta_j)$ where at the nominal operating points, θ_i^o, $i = 1, 2$, the *transmission line stiffness* coefficient v_{ij} is given by

$$v_{ij} = -V_i V_j b_{ij} cos(\theta_i^o - \theta_j^o). \tag{25}$$

In terms of the incremental state variables used, we have:

$$\delta P_{\text{tie}}^i = \sum_{j=1}^{n} v_{ij}(x_8^i - x_8^j), \tag{26}$$

where the state variable x_8^i is the integral of the frequency increment of area i, i.e.,

$$\frac{dx_8^i}{dt} = x_1^i. \tag{27}$$

The state space model (22)-(27) can be written in compact form as follows:

$$\frac{dx(t)}{dt} = A'x(t) + B_1'u^{\text{sc}}(t) + B_2'u^{\text{anc}}(t) + E'd(t). \tag{28}$$

States are stored in x, input signals to the speed changers are $u^{\text{sc}} = [\delta P_{C_1}\ \delta P_{C_2}]^T$, the ancillary inputs from the buildings are $u^{\text{anc}} = [\delta P_{\text{anc}_1}\ \delta P_{\text{anc}_2}]^T$, and the exogenous inputs (disturbance), modeling the variations in demand are denoted by $d = [\delta P_{D_1}\ \delta P_{D_2}]^T$.

We discretize the state space dynamics using the forward Euler scheme. We show the result on the equation for dx_1^i/dt. The discretized dynamics for the other states can be obtained in a similar way. At time t_n we approximate the derivative of x_1^i by

$$\frac{dx_1^i(t_n)}{dt} \approx \frac{x_1^i(t_n + \delta t) - x_1^i(t_n)}{\delta t} \tag{29}$$

Hence the discretized version of (22a) is

$$x_1^i(t_{n+1}) \quad = \quad \left(1 - \frac{D_x^i \delta t}{M_x^i}\right) x_1^i(t_n) \quad + \quad \frac{\delta t}{M_x^i}\left[\delta P_M^i - \delta P_D^i - \delta P_{\text{tie}}^{ij} + \delta P_{\text{anc}}^i\right],$$

where $t_{n+1} = t_n + \delta t$ and $\delta t = \tau_G$ is the discretization time step. The discrete-time state-space model is obtained as

$$x_{n+1} = Ax_n + B_1 u_n^{\text{sc}} + B_2 u_n^{\text{anc}} + Ed_n. \tag{30}$$

We use this state update equation for the G-MPC problem in Section 6.3.

Automatic Generation Control. The AGC is the main control function of a utility's energy control section. The purpose of an AGC is to track the load variations while maintaining the system frequency, net tie-line interchanges, and optimal generation level close to scheduled values [8]. This function is referred to as Load-Frequency Control. A secondary objective is to distribute the required change in generation among units to minimize operating costs [13]. In the case where several areas are interconnected, each will perform its own AGC independent of the others.

In the classical AGC, a simple PI control is utilized to regulate the frequency of the grid. The Area Control Error (ACE) is defined as

$$ACE^i = \delta P^i_{tie} + \beta^i x^i_1, \tag{31}$$

where $\delta P^i_{\text{tie}} = P^i_{\text{tie}} - P^i_{\text{tie,scheduled}}$, and β^i is the bias coefficient of area i. The standard industry practice is to set the bias β^i at the so-called Area Frequency Response Characteristic which is defined as $\beta^i = D^i + 1/R^i$. The integral of the ACE is then used to construct the speed changer position feedback control signal δP^i_C. A new state x^i_9 is then defined as

$$\frac{dx^i_9}{dt} = ACE^i. \tag{32}$$

Consequently the control input δP^i_C is given by

$$\delta P^i_C = -K^i x^i_9, \tag{33}$$

where K^i is the feedback gain. The MPC scheme proposed in Sec. 6.3 controls the available ancillary service from commercial buildings to improve on the classical AGC practice. This optimization-based control framework is utilized as a *higher-level* control in a "hierarchical" fashion on top of the *low-level* classical AGC control, as visualized in Fig. 5.

6 Energy Management System Optimal Control

We first describe the communication protocol followed by the grid and the building subsystem over time. Then, based on this protocol, we show how the building and the grid MPC schemes can be generated, respectively, from the contracts in Section 4 and the models in Section 5.

6.1 Grid-Building Communication Protocol

The grid operator and the building subsystem communicate as follows:

1. Buildings and grid agree upon the length H^c of the commercial contract.
2. The utility declares $\boldsymbol{\pi} = [\pi_0^e, \dots, \pi_{H^m-1}^e]$, the vector of prices of electric energy per unit step, the vector $\underline{\boldsymbol{\beta}} = [\underline{\beta}_0, \dots, \underline{\beta}_{H^m-1}]$ of rewards for downward flexibility, and the vector $\overline{\boldsymbol{\beta}} = [\overline{\beta}_0, \dots, \overline{\beta}_{H^m-1}]$ of reward for upward flexibility. If the utility is not willing to commit to the flexibility rates for the time span beyond the next, immediate contract period, e.g. $[\underline{\beta}_{H^c+1}, \dots, \underline{\beta}_{H^m-1}]$, and $[\overline{\beta}_{H^c+1}, \dots, \overline{\beta}_{H^m-1}]$, each building operator can obtain an estimate of these values from historical data. The same can be stated for the prices of electric energy beyond the next contract period $[\pi_{H^c+1}^e, \dots, \pi_{H^m-1}^e]$.
3. Each building operator computes the baseline air mass flow u_k^* and the two envelopes e_k^l and e_k^u, for the time frame $k \in \{0, 1, \dots, H^m - 1\}$, by solving the B-MPC, and declares the envelope $P_f(\mathbf{e^l})$, $P_f(\mathbf{e^u})$ and the baseline $P_f(\mathbf{u}^*)$ power consumption profiles. The B-MPC can be safely solved *independently* of the power control signals received from the grid, since the buildings do not make any assumption on such signals, but the fact that they are confined within the declared power envelopes. It is then possible for the buildings to separately minimize their cost *for all admissible values* of the grid control signals. Such a robust control problem is key to compositional design since it breaks the circularity between the B-MPC and the G-MPC problems.
4. The grid operator aggregates the envelope profiles received from the buildings and repeatedly solves the G-MPC, to obtain the control signals u_j^{anc} for $j \in \{0, 1, \dots, H\}$. These control signals are the aggregation of all the ancillary service powers provided by the buildings. At each time step, the grid operator will disaggregate δP_{anc_i} for each control area i into N pieces proportional to the declared flexibility of each building, such that $\delta P_{\text{anc}_i} = \sum_{k=1}^{N} s_i^k$, where N is the number of buildings participating in the ancillary service program.
5. During the next H^c time slots, the grid operator will send signals s_j^k in each control area, such that $P_f(e_j^{l,k}) \leq s_j^k \leq P_f(e_j^{u,k})$ and the building operator k has to track the signals, i.e., has to consume power equal to s_j^k in time slot j. The flexibility signal s_j^k may arrive as frequently as every few seconds, as mentioned earlier.

6.2 Design of the Building Optimal Control Scheme

Let H^m be the prediction horizon of the B-MPC in terms of building time slots, selected as discussed in Section 4. We design the B-MPC scheme to minimize the building economic cost in terms of baseline power consumption and flexibility over H^m, while satisfying the building constraints and the occupants' comfort requirements as encoded by the building contract $\mathcal{C}_B$ in Section 4.1. At each time t, the predictive controller solves an optimization problem to compute the baseline air mass flow $\mathbf{u}_t = [u_t, \dots, u_{t+H^m-1}]$ to the thermal zones of the building, and the downward and upward mass flow flexibility vectors $\underline{\boldsymbol{\varphi}}_{t+1} = [\underline{\varphi}_{t+1}, \dots, \underline{\varphi}_{t+H^m}]$ and $\overline{\boldsymbol{\varphi}}_{t+1} = [\overline{\varphi}_{t+1}, \dots, \overline{\varphi}_{t+H^m}]$.

The inputs to the optimization problem are the *initial state* x_t (zone temperatures), the set of *electric* energy prices per time slot $\{\pi_t^e, \dots, \pi_{t+H^m-1}^e\}$, the *non-electric* energy prices, such as gas price for heating $\pi^{ne,h}$ and cooling $\pi^{ne,c}$ (which are considered time-invariant), the prediction of the outside temperature and inside heat generation d_t. The cost function consists of the *cost* for the baseline HVAC power consumption, C_{hvac}, minus the *reward* for the flexibility R, computed as follows:

$$R(\Phi, \mathscr{B}) = \overline{\boldsymbol{\beta}}^T \overline{\boldsymbol{\psi}}(\mathbf{u}, \overline{\boldsymbol{\varphi}}) + \underline{\boldsymbol{\beta}}^T \underline{\boldsymbol{\psi}}(\mathbf{u}, \underline{\boldsymbol{\varphi}}) \tag{34}$$

where $\mathscr{B} := \{\underline{\boldsymbol{\beta}}, \overline{\boldsymbol{\beta}}\}$, with $\underline{\boldsymbol{\beta}}$ and $\overline{\boldsymbol{\beta}}$ as in Section 6.1, $\Phi := \{\underline{\boldsymbol{\varphi}}, \overline{\boldsymbol{\varphi}}\}$, and $\underline{\boldsymbol{\psi}}(.)$ and $\overline{\boldsymbol{\psi}}(.)$ are given, at each time step k, by

$$\overline{\psi}(u_k, e_k^u) \triangleq P_f(e_k^u) - P_f(u_k) \tag{35a}$$

$$\underline{\psi}(u_k, e_k^l) \triangleq P_f(u_k) - P_f(e_k^l) \tag{35b}$$

in which $\mathbf{e^l} = \mathbf{u} + \underline{\boldsymbol{\varphi}}$ and $\mathbf{e^u} = \mathbf{u} + \overline{\boldsymbol{\varphi}}$. The total HVAC power consumption cost $C_{\text{hvac}}(u_t, \pi_t)$ is the summation of fan power, cooling power and heating power costs, given by:

$$C_{\text{hvac}}(u_t, \pi_t) = \pi_t^e P_f(u_t) + \pi^{ne,c} P_c(u_t, T_t^{out}) + \pi^{ne,h} P_h(u_t, T_t^{out}) \tag{36}$$

where T^{out} is the outside air temperature, and the three power contributions are calculated based on the building model in Section 5.1.

As discussed in Section 6.1, the building operator aims to minimize the economic cost in the worst-case scenario for the grid signals (environment) following a game-theoretic approach. The result is a robust *min-max* optimization problem. The inner maximization problem derives the worst-case scenario cost and constraints, while the outer minimization problem solves for its arguments $(\mathbf{u}_t, \Phi_{t+1})$, while guaranteeing that the constraints are satisfied for all the values of the uncertain mass flow signals $\mathbf{w}$ from the grid, as long as they are within the range allowed by the building flexibility. Therefore, at time t the building operator solves:

$$\min_{\mathbf{u}_t, \Phi_{t+1}} \max_{\mathbf{w}_t} \sum_{k=0}^{H^m-1} C_{\text{hvac}}(u_{t+k}, \pi_{t+k}) - R(\Phi_{t+k+1}, \mathscr{B}_{t+k+1}) \tag{37a}$$

$$\text{subject to:} \quad x_{t+k+1} = f(x_{t+k}, u_{t+k} + w_{t+k}, d_{t+k}), \quad \forall\, k \in \{0, \dots, H^m - 1\} \tag{37b}$$

$$\forall\, w_t \text{ s.t. } \underline{\varphi}_t \le w_t \le \overline{\varphi}_t \tag{37c}$$

$$\forall\, w_{t+k} \text{ s.t. } \underline{\varphi}_{t+k} \le w_{t+k} \le \overline{\varphi}_{t+k}, \quad \forall\, k \in \{1, \dots, H^m - 1\} \tag{37d}$$

$$\overline{\varphi}_{t+k} \ge 0, \ \forall\, k \in \{1, \dots, H^m - 1\} \tag{37e}$$

$$\underline{\varphi}_{t+k} \leq 0, \ \forall\, k \in \{1, ..., H^m - 1\} \tag{37f}$$

$$x_{t+k} \in \mathscr{X}_{t+k}, \ \forall\, k \in \{1, ..., H^m\} \tag{37g}$$

$$u_{t+k} + w_{t+k} \in \mathscr{U}_{t+k}, \ \forall\, k \in \{0, ..., H^m - 1\} \tag{37h}$$

We observe that $\underline{\varphi}_t$ and $\overline{\varphi}_t$ are computed in the previous time step and are constant values in (37), while $\underline{\varphi}_{t+k}$ and $\overline{\varphi}_{t+k}$ for $k \in \{1, \ldots, H^m - 1\}$ are optimization variables and will be computed in the current time step by solving the optimal control problem. Therefore, B-MPC computes the future flexibility profile (starting from the next time step), based on the current flexibility. For the very first time step, we assume $\underline{\varphi}_0 = \overline{\varphi}_0 = 0$.

The inner optimization problem can be solved analytically. In fact, according to a fundamental theorem on convex functions [7], if a convex function attains a maximum over a closed convex set, then the maximum is achieved at some extreme point of the set. When the building state update equation are linearized as described in Section 5.1, the feasible set for states (the temperature of the rooms in the building) and controls (air mass flow into the thermal zones) is closed and convex, being an intersection of closed half-spaces. The objective function is also convex in $\mathbf{w}_t$, since $\mathbf{w}_t$ does not appear in the cost function. Hence, the min-max problem (37) is equivalent to:

$$\min_{\mathbf{u}_t, \boldsymbol{\Phi}_{t+1}} \quad \sum_{k=0}^{H^m-1} C_{\text{hvac}}(u_{t+k}, \pi_{t+k}) - R(\boldsymbol{\Phi}_{t+k+1}, \mathscr{B}_{t+k+1}) \tag{38a}$$

$$\text{s. t.:} \quad \underline{x}_{t+k+1} = f(x_{t+k}, u_{t+k} + \underline{\varphi}_{t+k}, d_{t+k}) \quad \forall k \in \{0, ..., H^m - 1\} \tag{38b}$$

$$\overline{x}_{t+k+1} = f(x_{t+k}, u_{t+k} + \overline{\varphi}_{t+k}, d_{t+k}) \quad \forall k \in \{0, ..., H^m - 1\} \tag{38c}$$

$$\overline{\varphi}_{t+k} \geq 0, \quad \forall k \in \{1, ..., H^m - 1\} \tag{38d}$$

$$\underline{\varphi}_{t+k} \leq 0, \quad \forall k \in \{1, ..., H^m - 1\} \tag{38e}$$

$$\underline{x}_{t+k} \in \mathscr{X}_{t+k} \quad \forall k \in \{1, ..., H^m - 1\} \tag{38f}$$

$$\overline{x}_{t+k} \in \mathscr{X}_{t+k} \quad \forall k \in \{1, ..., H^m - 1\} \tag{38g}$$

$$u_{t+k} + \underline{\varphi}_{t+k} \in \mathscr{U}_{t+k} \quad \forall k \in \{0, ..., H^m - 1\} \tag{38h}$$

$$u_{t+k} + \overline{\varphi}_{t+k} \in \mathscr{U}_{t+k} \quad \forall k \in \{0, ..., H^m - 1\} \tag{38i}$$

The result of (38) is the nominal power consumption u^*_{t+k} and the maximum available flexibility Φ^*_{t+k+1}, for $k \in \{0, ..., H^m - 1\}$. The building declares u^*_{t+k} and Φ^*_{t+k+1} for $k \in \{0, ..., H^c - 1\}$ to the utility. After H^c time slots, the BEMS collects the updated parameters such as new measurements and disturbance predictions, sets up the new MPC algorithm for $k \in \{H^c, H^c + 1, \ldots, H^c + H^m - 1\}$, solves the new MPC for this time frame and uses only the first H^c values of baseline power consumption and flexibility, i.e. for $k = H^c, \ldots, 2H^c - 1$, and this process repeats.

6.3 Design of the Grid Optimal Control Scheme

We design the G-MPC problem by combining assumptions and guarantees under the responsibility of the grid operator, as expressed by the contract $\mathcal{C}_G$ in Section 4.2. Let $U_k^{\text{anc}} = (u_k^{\text{anc}}, u_{k+1}^{\text{anc}}, \ldots, u_{k+H-1}^{\text{anc}})$ the trajectory of the power control signal of the grid to the buildings for time steps from k to $k+H-1$, where H is the prediction horizon of G-MPC. We aim to minimize the ℓ_2 norm of the ACE signal in areas $i = 1,2,...,n$, by exploiting the ancillary service available from buildings, taking into account the system dynamics and constraints. More formally, at each time step k, we solve:

$$\begin{aligned}
\min_{U_k^{\text{anc}}} \quad & \sum_{i=1}^{n}\sum_{j=0}^{H-1} (\text{ACE}_{k+j}^{i})^2 \qquad (39)\\
\text{s.t.} \quad & x_{k+j+1} = Ax_{k+j} + B_2 u_{k+j}^{\text{anc}} + Ed_{k+j}\\
& \underline{\mu}_{k+j} \leq u_{k+j}^{\text{anc}} \leq \overline{\mu}_{k+j}\\
& |u_{k+j+1}^{\text{anc}} - u_{k+j}^{\text{anc}}| \leq \lambda_{k+j}
\end{aligned}$$

where, for each area i, $\overline{\mu}_{k+j}(i) = \sum_{m=1}^{N} P_f(e_{k+j}^{u,m,i})$ and $\underline{\mu}_{k+j}(i) = \sum_{m=1}^{N} P_f(e_{k+j}^{l,m,i})$. All the constraints of problem (39) should hold for $j = 0,1,\ldots,H-1$. The constraints of the optimization problem are $\overline{\mu}_{k+j} > 0$ for the maximum positive power and $\underline{\mu}_{k+j} < 0$ for maximum negative power provided by the set of buildings in each area. Here, "positive" and "negative" refer to the flow of power from generation to consumption. These values are computed by the buildings and sent to the grid operator periodically, as detailed in Section 6.1. λ_{k+j} is the maximum limit on the rate of change of ancillary service provided by the buildings. Based on the assumptions in Section 4.1, deviations in the power demand from the buildings are known and lumped into the signal d_{k+j}. A robust version of the G-MPC problem to deal with uncertainties in the power demand, following a similar approach as in the B-MPC problem, will be object of future work. Finally, we do not incorporate $B_1 u_k^{sc}$ in the state-space model, since we assume that u^{sc} is constant and regulated by the local PI controller.

7 Simulation Results

To validate the proposed methodology, we simulated the control algorithms in Section 6 by using the building model in [18, 21], developed and validated against historical data. For rapid prototyping, we used YALMIP [14] as an interface to back-end optimization solvers. The non-linear optimization problem in B-MPC was solved using IPOPT [35], while CPLEX [3] was used to solve the quadratic program generated by G-MPC.

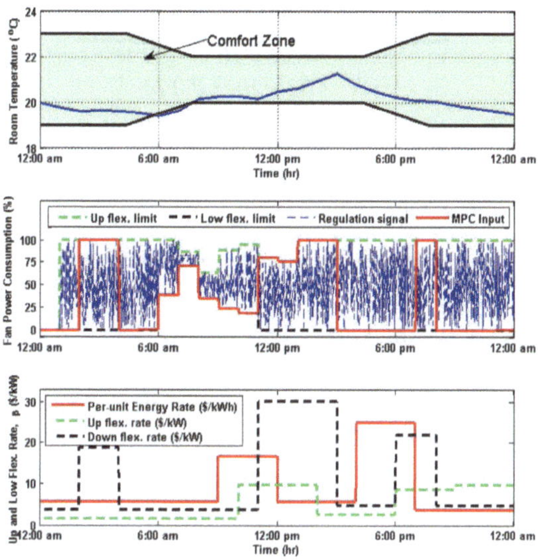

Fig. 7 Per-unit energy rate, upward and downward flexibility reward (bottom), building flexibility (middle), room temperature profile (top). Flexibility signals are sent every minute from the grid.

7.1 Validation of the B-MPC Algorithm

Different reward rates have been considered for upward and downward flexibility at each time step, as shown at the bottom of Fig. 7, under the constraint that downward flexibility is rewarded more than upward flexibility, i.e. $\underline{\beta} > \overline{\beta}$ for most of the day. We performed simulations with a sampling time of 1 hour and a prediction horizon of $H^m = 24$ h. On a 4-core 2.67-GHz Intel processor with 3.86 GB of memory, the mean and standard deviation of solver times were 8.9 s and 5.3 s, respectively. Fig. 7 shows the results when ancillary signals are received from the grid every minute: no building constraint (e.g. temperature comfort zone) is violated for arbitrary values of the fan speed enforced by the grid, as long as the fan power consumption is within the safe envelope calculated by B-MPC. The maximum flexibility (100%) is obtained when the room temperature is far from the boundaries of the comfort zone. The flexibility decreases as the temperature of the room approaches the comfort zone boundary, and reaches its minimum (about 0-15%) when the room temperature is close to the boundaries of the comfort zone, and the reward is small. Fig. 7 shows that the control strategy in (38) can indeed offer HVAC energy consumption flexibility via proper incentives. Table 3 lists the parameters used in our simulations.

Table 3 Simulation parameters used for B-MPC validation

Parameter	Value
c_1	-6.06×10^{-13} $[CFM^{-3}]$
c_2	6.73×10^{-8} $[CFM^{-2}]$
c_3	-1.2×10^{-3} $[CFM^{-1}]$
c_4	59.2
COP_h	3
COP_c	2
c_p	1.0 [kJ/(kg·K)]

7.2 *Validation of the G-MPC Algorithm*

We consider two interconnected control areas with model parameters as in Table 4, and with inter-area stiffness coefficient $v = 1.0$ p.u. The main generation unit for area 1 is a non-reheat turbo generator (TG) system while the main generation unit for area 2 is a hydro TG system. Some metrics such as root mean square (rms) values of frequency and ACE signal are considered to compare the performance of the proposed controller with respect to a traditional scheme. To simplify, we use time-invariant bounds for the maximum and minimum ancillary power $\overline{\mu}_k = \underline{\mu}_k = \mu$ and maximum rate of change of ancillary power λ in the following simulations.

Table 4 Simulation parameters used for G-MPC validation

Control Area	Parameters
Area 1	$T_1 = 0.1$, $T_3 = 0.1$, $T_4 = 1.0$
	$K_1 = 1.0$
	$M = 132.6$ [MW s]
	$D = 0.0265$ [p.u.]
Area 2	$T_1 = 0.2$, $T_3 = 0.3$, $T_4 = 0.1$, $T_5 = 0.5$
	$K_1 = 0.2$, $K_3 = 3$
	$M = 663.13$ [MW s]
	$D = 0.1325$ [p.u.]

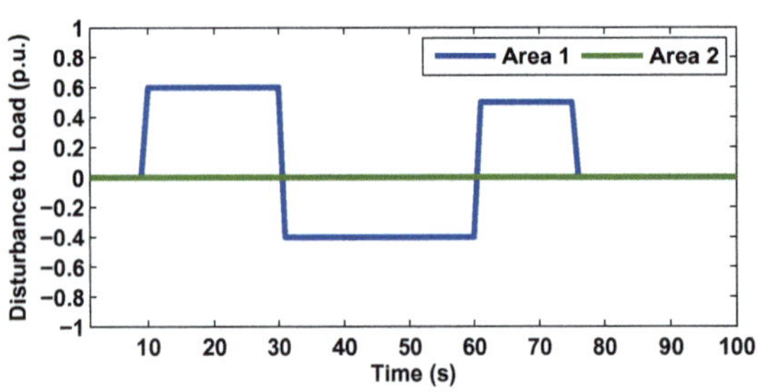

Fig. 8 Load disturbance signal

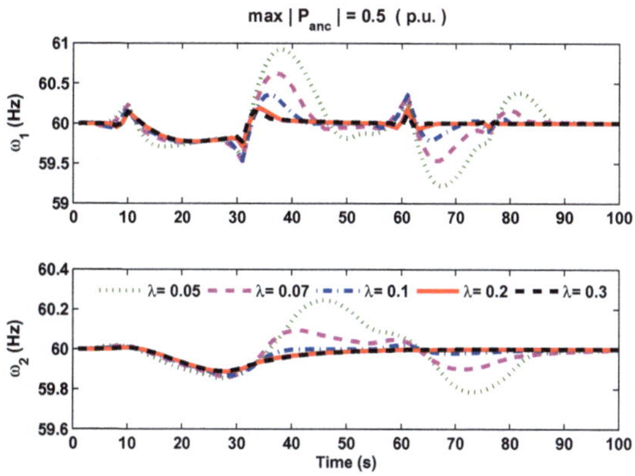

Fig. 9 Frequency of areas 1 and 2 in response to the load disturbance. The prediction horizon is $H = 10$ and the maximum ancillary power is $\max(P_{\text{anc}}) = 0.5$ p.u. Results relate to different values of rate-of-change of ancillary power ($\max|\Delta P_{\text{anc}}| = \lambda$).

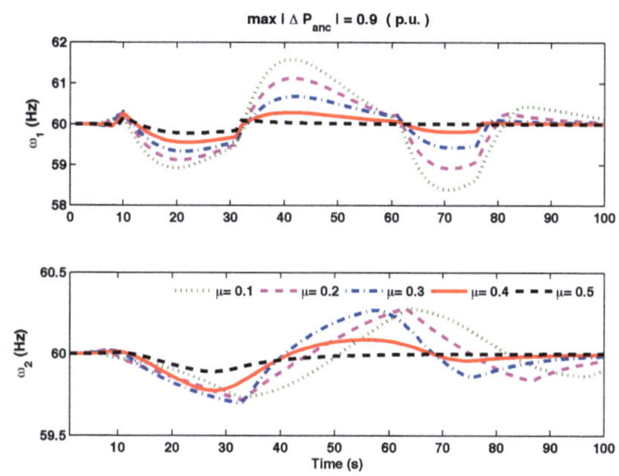

Fig. 10 Frequency trajectories in control areas 1 and 2 ($H = 10$, $\max(\Delta P_{\text{anc}}) = 0.9$ p.u., $\max|P_{\text{anc}}| = \mu$)

We performed simulations for a time horizon of 100 s and a sampling time $\tau_G = 1$ s. The mean and standard deviation solver times were 0.02 s and 0.005 s, respectively, for a prediction horizon of $H = 10$ s. We consider a disturbance signal in the load of area 1, and no disturbance in the load of area 2, as shown in Fig. 8. We assess the performance of the proposed controller considering the following scenarios.

Scenario 1. The maximum ancillary service available in each area is 0.5 per unit (p.u.) of power. We consider a prediction horizon of $H = 10$ time steps. As shown in Fig. 9, by increasing the maximum rate of change of ancillary power (ramping rate for generation units) the resulting frequency deviation decreases. A ramping rate of $\lambda = 0.05$ (p.u./s) is associated with large power generator size, a ramping rate of $\lambda = 0.1$ (p.u./s) is associated with smaller size generators. High ramping rates, such as $\lambda = 0.3$ (p.u./s), are associated with fast ancillary services such as the one provided by building HVAC system fans.

Scenario 2. To investigate the impact of the maximum available ancillary power constraint on the frequency deviations in each area, we relax the constraint on the ramping rate of the ancillary services, by selecting a higher bound $\lambda = 0.9$ (p.u./s). As shown in Fig. 10, by increasing the maximum available ancillary power μ the frequency deviation decreases, thus showing the effectiveness of our control strategy. The disturbance in the load of area 1 affects both the interconnected areas; however, the frequency change in area 1 is larger than the one in area 2.

Simulation results for the scenarios above show that G-MPC can effectively utilize ancillary services from the buildings for frequency regulation purposes.

8 Conclusions

We addressed the problem of optimal design of an integrated energy managing system based on a supply-following strategy. In our framework, assume-guarantee contracts formalize the requirements of both the power grid and the commercial building subsystem, and specify their interface so as to allow for independent implementation of two model predictive control (MPC) schemes in a compositional fashion.

At the building level, we cast a robust optimal control problem to determine the baseline power consumption of the HVAC system and the amount of allowed power consumption flexibility to maximize the building monetary incentive while satisfying its temperature and air mass flow requirements. At the grid level, we optimally control the ancillary service power flow within the buildings' flexibility, while integrating constraints such as ramping rates of ancillary service providers, maximum available ancillary power, and load forecast information. Simulation results show that commercial buildings can profitably provide ancillary services that can be effectively regulated at both the building and the grid levels by the proposed MPC scheme.

As a future work, we plan to further refine the communication interface between grid and buildings to incorporate a more realistic scenario, in which an electricity broker seeks rate offers from suppliers for "bundled" groups of customers and acts on their behalf. We also plan to develop a robust formulation of the grid MPC problem to address the uncertainties associated with imperfect load predictions.

Acknowledgements. This work was supported in part by the TerraSwarm Research Center, one of six centers supported by the STARnet phase of the Focus Center Research Program (FCRP) a Semiconductor Research Corporation program sponsored by MARCO and DARPA.

References

1. Buildings energy data book, http://buildingsdatabook.eren.doe.gov/default.aspx
2. U.S. Energy Information Administration, annual energy review (2010), http://www.eia.gov/totalenergy/data/annual/consumption
3. IBM ILOG CPLEX Optimizer (2012), http://www.ibm.com/software/integration/optimization/cplex-optimizer/
4. de Alfaro, L., Henzinger, T.A.: Interface automata, pp. 109–120. ACM Press (2001)
5. Benveniste, A., Caillaud, B., Ferrari, A., Mangeruca, L., Passerone, R., Sofronis, C.: Multiple Viewpoint Contract-Based Specification and Design. In: de Boer, F.S., Bonsangue, M.M., Graf, S., de Roever, W.-P. (eds.) FMCO 2007. LNCS, vol. 5382, pp. 200–225. Springer, Heidelberg (2008)
6. Benveniste, A., Caillaud, B., Nickovic, D., Passerone, R., Raclet, J.B., Reinkemeier, P., Sangiovanni-Vincentelli, A., et al.: Contracts for System Design. Rapport de recherche RR-8147, INRIA (2012)
7. Bertsekas, D.P.: Nonlinear programming (1999)
8. Debs, A.: Modern power systems control and operation. Kluwer Academic Publishers, Norwell (1988)
9. Helman, U.: Resource and transmission planning to achieve a 33% RPS in California–ISO modeling tools and planning framework. In: FERC Technical Conference on Planning Models and Software (2010)
10. Iannopollo, A., Nuzzo, P., Tripakis, S., Sangiovanni-Vincentelli, A.L.: Library-based scalable refinement checking for contract-based design. In: Proc. Design, Automation and Test in Europe (2014)
11. Kirby, B.: Spinning reserve from responsive loads. United States. Department of Energy (2003)
12. Koch, S., Mathieu, J.L., Callaway, D.S.: Modeling and control of aggregated heterogeneous thermostatically controlled loads for ancillary services. In: Proc. PSCC, pp. 1–7 (2011)
13. Kundur, P., Balu, N.J., Lauby, M.G.: Power system stability and control, vol. 4. McGraw-Hill, New York (1994)
14. Lofberg, J.: Yalmip: A toolbox for modeling and optimization in MATLAB. In: Proceedings of the CACSD Conference, Taipei, Taiwan (2004), http://users.isy.liu.se/johanl/yalmip
15. Ma, Y., Borrelli, F., Hencey, B., Coffey, B., Bengea, S., Haves, P.: Model predictive control for the operation of building cooling systems. In: 2010 American Control Conference (ACC), pp. 5106–5111. IEEE (2010)
16. Maasoumy, M.: Modeling and optimal control algorithm design for HVAC systems in energy efficient buildings. Master's thesis, EECS Department, University of California, Berkeley (2011), http://www.eecs.berkeley.edu/Pubs/TechRpts/2011/EECS-2011-12.html
17. Maasoumy, M., Ortiz, J., Culler, D., Sangiovanni-Vincentelli, A.: Flexibility of commercial building HVAC fan as ancillary service for smart grid. In: IEEE Green Energy and Systems Conference, Long Beach, USA (2013)
18. Maasoumy, M., Pinto, A., Sangiovanni-Vincentelli, A.: Model-based hierarchical optimal control design for HVAC systems. In: ASME Dynamic System Control Conference (DSCC) (2011)

19. Maasoumy, M., Rosenberg, C., Sangiovanni-Vincentelli, A., Callaway, D.: Model predictive control approach to online computation of demand-side flexibility of commercial buildings HVAC systems for supply following. In: Proc. IEEE American Control Conf. (2014)
20. Maasoumy, M., Sanandaji, B.M., Sangiovanni-Vincentelli, A., Poolla, K.: Model predictive control of regulation services from commercial buildings to the smart grid. In: Proc. IEEE American Control Conf. (2014)
21. Maasoumy, M., Sangiovanni-Vincentelli, A.: Total and peak energy consumption minimization of building HVAC systems using model predictive control. IEEE Design and Test of Computers (2012)
22. Maasoumy Haghighi, M.: Controlling energy-efficient buildings in the context of smart grid: A cyber physical system approach. UCB, EECS-2013-244 - PhD Thesis (2013), `http://www.eecs.berkeley.edu/Pubs/TechRpts/2013/EECS-2013-244.html`
23. Maler, O., Nickovic, D.: Monitoring temporal properties of continuous signals. In: Lakhnech, Y., Yovine, S. (eds.) FORMATS/FTRTFT 2004. LNCS, vol. 3253, pp. 152–166. Springer, Heidelberg (2004)
24. Nuzzo, P., Finn, J.B., Iannopollo, A., Sangiovanni-Vincentelli, A.L.: Contract-based design of control protocols for safety-critical cyber-physical systems. In: Proc. Design, Automation and Test in Europe (2014)
25. Nuzzo, P., Sangiovanni-Vincentelli, A.: Let's get physical: Computer science meets systems. In: Bensalem, S., Lakhneck, Y., Legay, A. (eds.) From Programs to Systems. LNCS, vol. 8415, pp. 193–208. Springer, Heidelberg (2014)
26. Nuzzo, P., Sangiovanni-Vincentelli, A., Sun, X., Puggelli, A.: Methodology for the design of analog integrated interfaces using contracts 12(12), 3329–3345 (2012)
27. Nuzzo, P., Xu, H., Ozay, N., Finn, J., Sangiovanni-Vincentelli, A., Murray, R., Donze, A., Seshia, S.: A contract-based methodology for aircraft electric power system design. IEEE Access 2, 1–25 (2014), doi:10.1109/ACCESS.2013.2295764
28. Oldewurtel, F., Parisio, A., Jones, C., Morari, M., Gyalistras, D., Gwerder, M., Stauch, V., Lehmann, B., Wirth, K.: Energy efficient building climate control using stochastic model predictive control and weather predictions. In: 2010 American Control Conference (ACC), pp. 5100–5105. IEEE (2010)
29. Pnueli, A.: The temporal logic of programs. In: 18th Annual Symposium on Foundations of Computer Science, pp. 46–57. IEEE Computer Society Press (1977)
30. Sangiovanni-Vincentelli, A.: Quo vadis, SLD? Reasoning about the trends and challenges of system level design. Proceedings of the IEEE 95(3), 467–506 (2007)
31. Sangiovanni-Vincentelli, A., Damm, W., Passerone, R.: Taming Dr. Frankenstein: Contract-Based Design for Cyber-Physical Systems. European Journal of Control 18(3), 217–238 (2012)
32. Strbac, G.: Demand side management: Benefits and challenges. Energy Policy 36(12), 4419–4426 (2008)
33. Todd, D., Caufield, M., Helms, B., Starke, M., Kirby, B., Kueck, J.: Providing reliability services through demand response: A preliminary evaluation of the demand response capabilities of Alcoa Inc. ORNL/TM 233 (2008)
34. Vittal, V., Bergen, A.: Power systems analysis. Prentice Hall (1999)
35. Wächter, A., Biegler, L.T.: On the implementation of an interior-point filter line-search algorithm for large-scale nonlinear programming. Mathematical Programming 106(1), 25–57 (2006)
36. Woo, C.K., Kollman, E., Orans, R., Price, S., Horii, B.: Now that California has AMI, what can the state do with it? Energy Policy 36(4), 1366–1374 (2008)

Decision-Support Tools for Renewables-Rich Power Systems: A Stochastic Futures Approach*

Jiayi Jiang, Sandip Roy, Juhua Liu, and Vaibhav Donde**

Abstract. The growing penetration of intermittent renewables (primarily wind and solar generation) in deregulated electric power systems is introducing significant challenges in forecasting generation and scheduling units. At the same time, the pervasive integration of cyber- tools in the control room provides unique opportunities for leveraging data sources like weather forecasts, computational resources, and visualization tools for real-time decision-making. Here, we introduce a framework and algorithm set for day-ahead generation scheduling, or unit commitment, that takes advantage of the close tie between cyber- and physical- resources in the electric power grid. First, we use a class of stochastic automata models known as influence models to forecast relevant spatio-temporal environmental parameters (wind speeds/direction, cloud cover), and in turn simulate probabilistic wind and solar generation futures across a wide area. These models can be parameterized in real time to statistically match publicly-available ensemble forecast products, yet can be tailored to provide generation futures at appropriate spatial and temporal resolutions for scheduling. The models also permit rapid selection of representative renewable-generation futures, and are able to capture local variability and spatial/temporal correlation in the generation profiles. Second, a new method for unit scheduling for the day-ahead market, which uses the probabilistic wind/solar generation futures, is proposed and

Jiayi Jiang · Sandip Roy
Washington State University

Juhua Liu
ABB US Corporate Research

Vaibhav Donde
Pacific Gas & Electric

* This research described here was primarily supported by the ABB Research Grant Program.

** Work performed while at ABB US Corporate Research.

S.K. Khaitan, J.D. McCalley, and C.-C. Liu (eds.), *Cyber Physical Systems Approach to Smart Electric Power Grid,* Power Systems, DOI: 10.1007/978-3-662-45928-7_6 developed in a preliminary way. A novelty in this approach is a pre-selection step that can provide operators with situational awareness of critical (sensitive) units. The generation-scheduling and unit-commitment tools are demonstrated on a small-scale example, which is concerned with wind generation in the Columbia River Gorge of Washington State on a historical weather day.

1 Introduction

Electric-power-system operation requires coordinated scheduling and dispatch of generation units across a wide area, to match generation with demand. In many modern deregulated systems [1-13], scheduling and dispatch are achieved at three different time horizons. First, the on/off schedules and tentative hourly dispatch levels of generators are set by the transmission system operator (TSO) or independent system operator (ISO), usually via a binding market, on the day ahead. Second, refined dispatch levels are obtained via an hourly market mechanism which uses an economic dispatch. Finally, local small-scale mismatches are corrected for at a fast time scale, usually on the order 5-10 minutes.

The research described here is primarily concerned with generation scheduling for the day-ahead market. Historically, unit scheduling (as well as longer-term generation-resource planning) was done by human operators, who largely drew on experience to develop on/off schedules for a limited number of generator units. As electric power networks have become increasingly complex and computing technologies have improved, automation for *unit commitment* have been developed and integrated into transmission system operation. These unit commitment technologies, used in tandem with experience-driven decision-making, have proved valuable for wide-area management in both deregulated and regulated systems.

The last ten years has seen a rapid integration of intermittent renewable generation (primarily, wind and solar generation) into electric power systems worldwide, and the penetration of these intermittent renewables is expected to continue growing rapidly. These new generation technologies hold promise to permit sustainable low-cost power for years to come. However, they also bring forth new challenges in control and management of the power grid across multiple temporal and spatial scales, including specifically for day-ahead unit scheduling. Crucially, intermittent renewable generation trajectories are dependent on environmental parameters (e.g., wind speed and direction, cloud cover, humidity, etc.) which may have significant uncertainty at a 24-36 hour look-ahead horizon. This uncertainty must be accounted for in commitment and dispatch of conventional generation, and hence the unit-commitment problem becomes a stochastic one. In addition, the intermittence and consequent temporal variability in wind and solar generation means that unit schedules may change significantly from day to day. This variability makes experience-driven decision-making more difficult, and also requires flexible scheduling paradigms and improved tools for evaluating system-level performance (including economic performance, security and fault management, etc). As the penetration of intermittent renewables increases, these challenges in day-ahead scheduling will become increasingly prominent.

New tools for scheduling generation for the day-ahead market are needed to meet these challenges. These include tools for 1) forecasting intermittent-renewable generation futures, 2) stochastic unit commitment, and 3) evaluation of power-network performance across renewable-generation futures. Additionally, advances in these directions must be translated into practical decision-support software for the control room. In fact, numerous research efforts are underway in these directions. However, these efforts are still largely academic in nature, and have not yet been translated to implemented software solutions. Our viewpoint is that several barriers remain in obtaining implementable technologies:

1) Forecasts of uncertain environmental futures and consequent generation trajectories are needed, that have sufficient resolution for decision-making yet capture uncertain propagation across a wide area as needed for unit commitment.
2) Techniques for stochastic unit commitment are needed that yield practical, robust, and economically viable schedules across generation futures, yet are computationally attractive for wide-area scheduling.
3) End-to-end solutions are needed, that use realistic environmental forecasts for unit commitment and system performance evaluation.

While the growing penetration of intermittent renewables is complicating generation scheduling, new technologies also provide entirely new capabilities for resource scheduling that have not yet been fully exploited. During the last 20 years or so, a wide array of new computing and communication tools have been introduced in the control room: these include increasingly-powerful computers and sophisticated software for analysis, dedicated communication channels as well as high-speed Internet access, and mobile handheld technologies (cell phones, iPads, etc.), among others. These pervasive cyber- tools can facilitate control and management of the wide-area network across time scales [41,42]. In particular, relevant to the unit commitment problem for the renewables-rich grid, these technologies can allow fast transfer of high-dimensional weather-forecast data to the control room, provide operators with convenient interfaces and displays to evaluate consequences of decisions, simplify wide-area monitoring, permit rapid integration of stakeholders' inputs, and allow intensive computing for weather-impact forecasting and schedule optimization. Indeed, the new cyber-technologies have brought about rapid advances in control room operations, but they have not yet yielded significant improvements in unit commitment for a renewables-rich grid. At its essence, exploiting these technologies for stochastic unit commitment requires an understanding of the tight interface between engineered (electromechanical), natural-world (weather), human (market and operational), and cyber components in the electric power grid.

The research presented here approaches stochastic unit commitment from this "cyber-physical systems" viewpoint, focusing particularly on cyber- solutions to the forecasting and scheduling aspects of the problem. Research efforts in three directions are discussed:

1) Development of an end-to-end operational concept for day-ahead unit scheduling.
2) Motivation for and development of a new generation-forecasting tool, which uses a stochastic automaton model known as the influence model.
3) Exploration of tools for stochastic unit commitment that use the new generation-forecasting model.

The chapter is organized as follows. The end-to-end operational concept is first introduced (Section 2). Next, the new generation-forecasting tool is developed in detail (Section 3), and illustrated using a case study of wind generation in the Columbia River Gorge area of Washington State on a historical weather day. Finally, some initial explorations on using the generation forecasts for stochastic unit commitment are presented (Section 4), and conclusions are given (Section 5).

2 Operational Concept

A *stochastic-futures approach* to generation scheduling for the day-ahead market is considered, see Figure 1. The operational concept has two main parts: a module for determining representative spatiotemporal futures (or time-trajectories of wind and solar generation over a 48-hour horizon (the *generation-forecasting module,* as contained in the green blocks), and a second module for scheduling dispatchable generation units using these wind/solar generation futures (the *stochastic unit commitment module*, pink blocks). These modules require development of new algorithms for generation forecasting and unit commitment, respectively, as well as prototype Matlab software development.

The generation-forecasting module in our approach exploits a new *influence modeling* technology (14-17). This technology leverages ensemble forecast outputs, but also allows interpolation of forecasts to the proper resolution for generation forecasting, represents spatial and temporal correlation in weather/generation, and permits rapid simulation of many generation futures. The module uses the influence-modeling technology as follows (see flowchart): 1) relevant forecast data (wind speeds and directions, cloud cover, humidity) is extracted from an ensemble forecast (e.g., the Short Range Ensemble Forecast or SREF, which is available in the public domain, see 18); 2) influence models for wind speeds and cloud dynamics are built (parameterized) to statistically match the ensemble-forecast data; 3) many possible spatio-temporal futures of wind- and solar- generation are obtained through simulation of the stochastic influence model and mapping of the results into generation profiles; and 4) a few representative futures are chosen using the probabilistic-collocation method

(19,20). The futures produced by the module predict wind/solar generation at each network bus over the full day ahead, at a 15-min. resolution.

Meanwhile, the scheduling module aims to develop day-ahead commitment plans for dispatchable (non-renewable) generation, to minimize an expected performance cost across the representative scenario set while respecting numerous constraints (including ramp-up and ramp-down constraints, and transmission-network constraints upon dispatch). The performance cost for scheduling in our formulation captures dispatch cost, ramp-up and ramp-down costs, reserve-generation usage costs, and line losses.

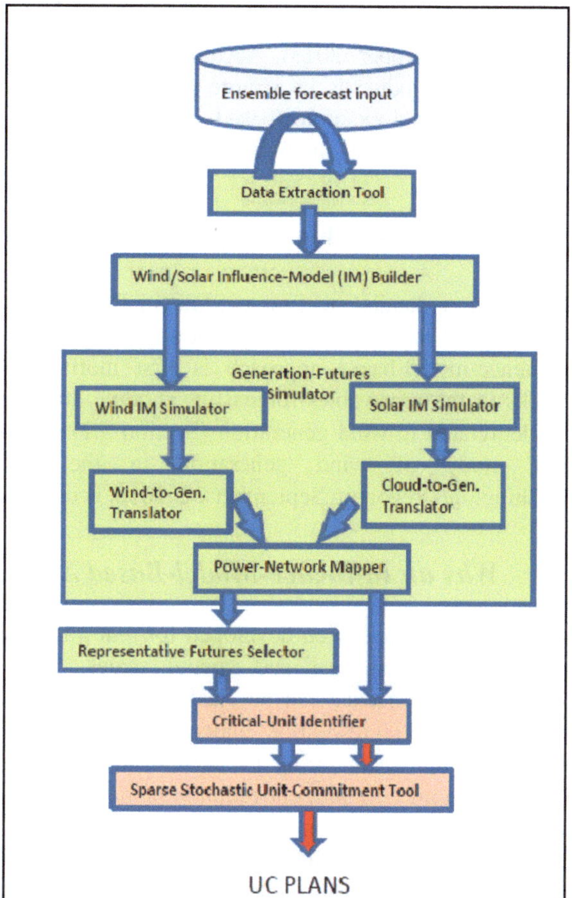

Fig. 1 Operational concept for the stochastic-futures-based unit commitment tool

Since the forward market for the day ahead is a binding one, the module must either provide a single plan or a very small number of alternatives (with human operators choosing one). To combat the computational challenges inherent to stochastic unit commitment (UC) and to provide operators with robust plans, a new two-stage approach to generator scheduling is proposed, which contrasts with existing stochastic UC paradigms (11,21,22). In particular, rather than trying to optimize all generators' schedules (including when they are on-line/off-line and hourly dispatch levels) at once, we instead first **identify critical units** that are difficult to plan. These are units whose optimal schedules of on-line times and/or dispatch are highly sensitive to the future renewable-generation profile. The critical units are identified by pursuing hourly economic dispatch (EDs) for each representative weather-impact future assuming all units are potentially online, and determining the units whose generation levels are sensitive to the weather future. In the second stage, the full scheduling problem is solved using only the critical units' on/off times and dispatch as design variables, while using mean dispatch levels from the first stage for the remaining non-critical units.

Several new algorithms are needed for the stochastic-futures-based UC solution, including for building and simulating the influence model, translating weather futures into renewable-generation futures, choosing representative futures, identifying critical units, and solving the pruned scheduling problem. The algorithms are under development as part of the WSU-ABB collaborative project. Significant progress has been made in developing the algorithms related to generation-forecasting, and initial software implementation has been developed for a case study (on wind-generation in the Pacific Northwest during a cold-front passage in October 2013). Algorithm development for the stochastic-unit-commitment module is in a more preliminary stage. Specifically, for a small-scale constructed example (based on the IEEE 14-bus model), we are pursuing implementation and evaluation of the two-stage approach to unit commitment.

3 The Renewable Generation-Forecasting Module

This section details the first module in the operational concept, an influence-model-based tool for wide-area renewables forecasting for the day ahead. The influence-model-based approach is first motivated (Section 3.1). Then, the blocks in the renewables-forecasting module are described, with a focus on the blocks related to wind generation (Section 3.2). Throughout the development, a case study of wind generation in the Columbia River Gorge of Washington/Oregon on September 22, 2013, is used to illustrate the model.

3.1 Why an Influence-Model-Based Approach?

Many methods have been developed for renewable-generation forecasting, which span multiple temporal and spatial scales. Broadly, renewable-generation forecasting approaches can be classified based on their temporal resolution and look-ahead, spatial resolution, underlying modeling mechanism (physics-based vs. empirical), and their ability to capture uncertainties, among other factors (10). Day-ahead unit commitment requires models with look-ahead horizons of 24-48 hours, preferably with temporal resolutions of 5-15 minutes (which is fine enough to capture hourly generation profiles in some detail). Additionally, the models must be able to provide relatively accurate predictions of generation from wind farms, as well as for solar farms and/or distributed solar generators in a locality. At the same time, the models must be able to provide predictions of such generation across the geographic domain covered by a transmission system operator. Finally, the proposed approach to generation scheduling requires stochastic generation futures. While many deterministic models for generation-forecasting for wind-farms at the time-resolution of interest have been developed, fewer models capture uncertainties and are extensible to wide-area prediction. Among the models that do capture uncertainties, many simply identify error bounds around a nominal forecast, and hence do not naturally provide futures or trajectories of generation.

Among the models in the literature, ensemble-forecast-based approaches are the most relevant to generation-forecasting for day-ahead unit commitment. These approaches translate commercially-available ensemble forecasts for environmental parameters (winds, humidity, etc.) into multiple generation futures. The approaches are relevant and appealing for wide-area forecasting, in that they 1) have the proper look-ahead horizon for forecasting (typically, up to three days), 2) are able to provide predictions across a wide area, and 3) directly yield stochastic futures. The models are also appealing in that they use physics-based representations of environmental processes, and in that they are available in the public domain. However, we believe that the ensemble-forecasting approaches cannot be used directly for day ahead resource scheduling, for several reasons:

1) Ensemble forecasts typically have a temporal resolution of 3 hours at a one-day look-ahead, and a spatial resolution of 15-40 km. The temporal resolution is insufficient for day-ahead scheduling. Likely, environmental conditions may vary sufficiently across a forecast grid square (particularly in complex-terrain regions) to reduce forecast accuracy. Thus, higher-resolution forecasts are needed, particularly in geographic regions with a high density of wind and/or solar generators. From another viewpoint, interpolation of the ensemble forecasts in both space and time is needed.
2) Ensemble forecasts only capture uncertainties in initial conditions. However, wind and solar generation often may be significantly impacted by uncertainties at shorter temporal/spatial scales, which are not forecasted. For instance, beginning and end times for wind events are often highly uncertain even at short time horizons. Likewise, on partly-cloudy days, insolation on solar panels may exhibit significant short-time-scale fluctuations and uncertainties. We note that these smaller-scale fluctuations may exhibit significant temporal and spatial correlation. While a generation-forecasting tool need not capture these smaller-scale patterns exactly, it should be able to account for the resulting uncertainty to some extent.
3) Ensemble forecasts typically only provide a small number of potential weather futures, and hence generation futures. Scheduling potentially may require a larger number of possible generation futures, or at least a representative set that better spans the space of possibilities.

The influence model [14-17] is promising for addressing these needs while still leveraging ensemble forecast products, and hence can provide effective generation forecasts for unit scheduling. Specifically, the model naturally permits simulation at a desired temporal and spatial resolution (including at multiple scales across a region), while matching ensemble forecast probabilities at snapshot times and locations. Additionally, the influence model – which is a stochastic-automaton model – does capture complex uncertainties and patterns in weather evolution, in a way that permits tuning of spatial and temporal correlations. Finally, the model is simple enough to permit rapid simulation and some statistical analysis of many generation trajectories. It is worth noting the influence model has been used to model environmental uncertainties and their impacts in the transportation domain (see [14-16]); this work pursues development of analogous capabilities for generation unit scheduling.

The influence modeling approach to generation forecasting is suited for the modern control room, which has pervasive cyber- technologies. As discussed below, the prediction tool leverages current ensemble forecasting products: the modern control room is designed to access high-volume data (such as ensemble forecasts) through the Internet, and would have the capability to use up-to-date weather data as required in the proposed approach. Additionally, the approach exploits the simulation, analysis, and visualization capabilities available in the modern control room. Thus, it holds promise to provide operators with new, information-rich decision-support and automation for planning under uncertainty.

3.2 *Module Blocks: Overview and Details*

The renewable-generation-forecasting module involves two parallel tracks, one of which simulates possible wind-generation futures and the second of which predicts solar generation futures, for day-ahead resource planning. Here, only the blocks associated with the wind-generation track are discussed. Details on the solar-generation-forecasting track can be found in the companion paper [17]. To begin, let us note that the generation-forecasting-module, as a whole, outputs representative futures of wind and solar generation for each bus in the studied power-system model for the day ahead. To develop these futures, the module uses current ensemble forecast products, as well as archived data (on wind-farm locations and compositions, historical generation profiles, regional solar-generation usage, etc). The blocks comprising the module are envisioned as being implemented in software in the TSO's control room, for use in day-ahead planning. Here, each block's functionality is discussed, and prototype software implementations are illustrated.

3.2.1 Data Extraction Block

The data-extraction block is tasked with downloading weather-forecast data from online ensemble forecast products on a daily basis, for use in renewable-generation forecasting for the day-ahead market. A range of ensemble forecast products are posted on-line in real time, many of them by the United States National Oceanic and Atmospheric Administration (NOAA) and by European counterparts ECMWF [36]. The various ensemble forecasts each use high-resolution deterministic physics-based models for atmospheric dynamics. Multiple ensemble members or futures are produced through randomization of uncertain model parameters and/or initial conditions, with most forecast products including 15-30 ensemble members. The full models are extremely high dimensional and time-consuming to run, usually requiring several hours on a large cluster. Only a subset of the model's states variables are posted to the online server, at a moderate spatial and temporal resolution. Even this lower-resolution filtered output is quite high dimensional, typically requiring tens of gigabytes for storage. The model data on the NOAA servers can be further filtered by the user prior to downloading, permitting extraction of only relevant weather parameters in the geographic region of interest. Specifically, the data can be accessed and parsed via unix script commands. The data is encoded in the *grib2* format, which is commonly used for environmental data sets. Once downloaded, the data can be automatically translated into other common data formats (e.g., csv, plain text, etc.), again using unix scripts. Alternately, several grib2 data readers are available on NOAA's webpage [33], which can be used for display and manual processing of the data.

a)

b)

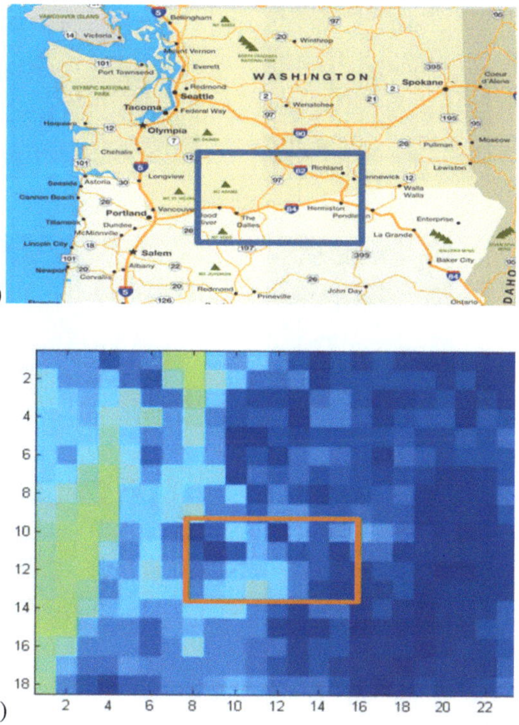

Fig. 2a) A case study of generation forecasting in the Columbia River gorge area of Washington and Oregon is considered; the boxed region has a high density of wind farms, and is the focus area of this case study. b) Snapshot wind map of one ensemble member in the Short Range Ensemble Forecast, with the region of interest indicated. Darker shades of blue indicate low wind speeds, while stronger shades of lime indicate high wind speed.

In the proposed solution, the data-extraction block is responsible for extracting a small subset of the environmental parameters needed for generation-forecasting over the time-horizon of interest, downloading this data to the TSO's local server, and translating it into a convenient form for further processing.

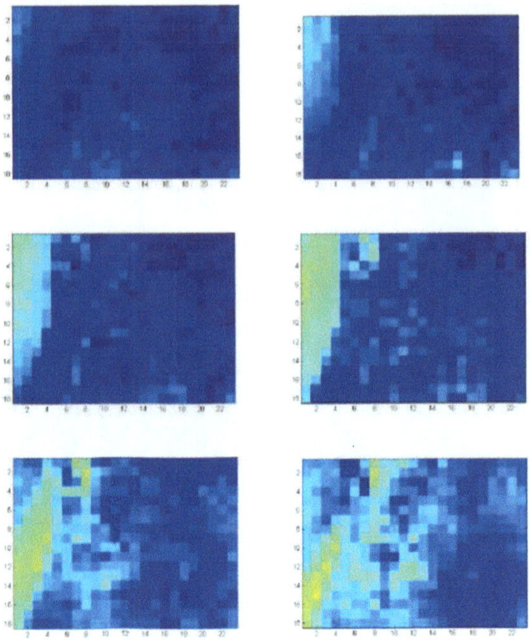

Fig. 3 The time-progression of one ensemble member in the SREF is shown. Specifically, wind speeds predicted by the ensemble member at 3hr intervals across the Pacific Northwest are shown. A cold front is encroaching on the area, leading to a period of increased winds.

Specifically, for wind-generation forecasting, wind speed and direction vari-ables just above ground (10-30 m) are needed, across the geographic region managed by the TSO and over the full day ahead (24 hours, from one midnight until the follow-ing). Solar-generation forecasting is typi-cally more complicated, using temperature, humidity, cloud cover, and possibly other fore-cast data, see [17] for details. The data-extraction block is tasked with downloading this data to the local server, and formatting for use by downstream software blocks.

In this study, the 40km Bias-Corrected Short Range Ensemble Forecast (SREF) for the Continental United States (CONUS) is being used as the forecast data source for the generation-forecasting module. The SREF is appealing for generation scheduling in that the appropriate weather parameters (wind speed, wind direction, etc.) are posted at sufficient frequency (every 6 hours), over an appropriate look-ahead horizon (up to 87 hours into the future) and resolution (3 hrs temporal resolution, 40kmx40km grid squares for the spatial resolution), and with an acceptable delay (forecast becomes available about 2 hours after the initial forecast time). We have chosen to use the 40km bias-corrected model, specifically, because there has been an extensive effort to validate this model version. New higher-resolution versions have recently become available, and may be practical for use in the near future. To permit exploratory study, the DeGrib tool is being used to process SREF data, although basic scripts for automated downloading and processing have also been written. We note that the SREF forecasts are produced at 3Z (3AM Zulu Time), 9Z, 15Z, and 21Z daily: the TSO would use the most recent available forecast, which depends on the TSO's schedule for resolving the day-ahead market. For most markets in the CONUS, this is the 9Z, 15Z, or possibly 21Z forecast.

Case Study: Wind Generation in the Columbia River Gorge

For the case study, wind speed and direction variables were extracted from the 21Z forecast on 9/21/2013, for the period between 6Z and 21Z on 9/22/2013. Based on our focus on generation in the Columbia River Gorge region, the forecast data was extracted for the Pacific Northwest region of the United States (the states of Washington and Oregon, and adjacent areas in Idaho and in the Pacific Ocean). In Figure 2, this region is shown, and the wind speed forecast for one ensemble member at a snapshot time is also displayed; the Columbia River Gorge region is encircled in orange for convenience. Figure 3 compares the wind-speed forecasts at different times for a particular ensemble member, while Figure 4 compares two ensemble members at a snapshot time. During the period of interest, a cold-front is encroaching on the

Pacific Northwest, leading to an increase in wind speeds over the period. While different ensemble members are generally similar, they show noticeable differences in cold-front timing and strength, leading to significant variability in wind speed profiles in the Columbia-River-Gorge region.

3.2.2 The Influence Model Builder

The next blocks in the renewable-generation-forecasting module are tasked with building influence models for wind and cloud cover *(Influence Model Builder)*, which are then used to simulate wind- and cloud- cover profiles (*Influence Model Simulators*), and hence to forecast generation (*Wind-to-Generation Translator* and *Cloud-Cover-to-Generation Translator*). Here, the wind influence model is described, and the algorithm for building specific instances using the ensemble forecasts obtained from the data-extraction block is overviewed. The analogous influence model for cloud cover dynamics is described in [17].

Broadly, the **influence model** is a networked-Markov-chain or stochastic-automaton-network model, which tracks the evolution of discrete statuses across a network of interacting sites. Each site's status evolves in a Markov fashion, via simple interactions with neighboring sites. The model is appealing in that update rule is simple enough to permit rapid simulation and statistical analysis, yet can capture complex spatiotemporal evolution patterns and correlations. The model was originally envisioned as a representation for failure propagation in complex networks [15], but subsequently has been used to model e.g. inter-personal communication patterns, decision-making in sensor networks, and convective-weather evolution, e.g. [20,40].

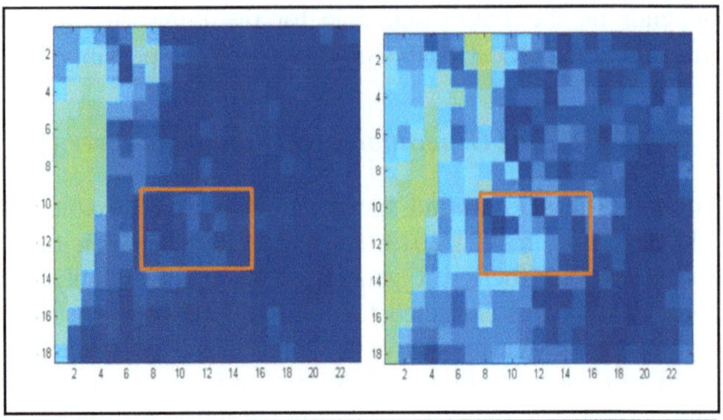

Fig. 4 Two different ensemble members are compared at a snapshot time. Although the predicted wind pattern is similar overall, there is significant variation in the region of interest.

Here, an influence model is considered that forecasts discrete wind levels in N contiguous subregions across an area of interest, at an appropriate temporal resolution for decision-making (e.g., 15 minutes) over the day ahead. Specifically, at each time step, each subregion $i \in 1,..,N$ is modeled as being in one of m statuses (labeled $1,..,m$), which identify different wind speed and direction bins (intervals). For instance, the model may use $m=6$ bins to identify wind speeds between 0 and 30kph, in bins of 5kph each. Alternately, if wind speeds and directions are both tracked, $m=6x4=24$ bins could be used to capture the same wind-speed levels as well as the wind-heading quadrant. The status of subregion i at time k is denoted as $s_i[k]$. These statuses evolve with the time step k based on a simple, Markovian update rule. This update rule captures that forecast wind characteristics in a subregion follow a statistical distribution (as extracted from the ensemble forecasts), but also show persistence over time as well as correlation across space. Specifically, the next-status $s_i[k+1]$ of subregion i is determined via the following two-stage update:

1. Geographical neighbors of subregion i (including the subregion i itself) are viewed as influencing the next status. To define this influence, each neighboring subregion j is modeled as providing an m-element probability vector $\boldsymbol{a}_{ij}(s_j[k], k)$, which depends on its current status $s_j[k]$ and also may vary with the time step k. The vectors $\boldsymbol{a}_{ij}$, as probability vectors, are element-wise non-negative and sum to *1*.
2. A weighted average of the neighbors' probability vectors, $\boldsymbol{a}_i = \sum_{j \in N(i)} d_{ij}[k] \boldsymbol{a}_{ij}(s_j[k], k)$, is computed, where the weights $d_{ij}[k]$ are assumed to be nonnegative and sum to *1*. The probability vector $\boldsymbol{a}_i$ is used to realize the next status of $s_i[k+1]$ of subregion i (independently of all other realizations). That is, the next status is selected stochastically to be one of the discrete possibilities $1,..,m$, with the probability that the status is q given by qth entry in $\boldsymbol{a}_i$.

The time-$(k+1)$ statuses of all sub-regions are determined simultaneously in this fashion, and the process is repeated for each time step. The influence model update is illustrated in Figure 5.

The wind influence model, as defined above, is a stochastic automaton model that produces probabilistic futures of wind trajectories in geographical subregions within an area of interest. Of course, the futures produced by the model crucially depend on the model's parameters, namely the *local influence vectors* $\boldsymbol{a}_{ij}(s_j[k], k)$ and the *network weights* $d_{ij}[k]$. Prior to using the model, these parameters must be selected so that the model produces wind futures that reflect real environmental conditions. Here, the model is parameterized based on the wind forecasts extracted from ensemble forecast products – this is what is meant by "building the influence model".

Several of the previous studies on the influence model have considered parameterization from data or forecasts: broadly, the sparseness of the model often permits parameterization from a fairly limited data set. Most relevant to the research presented here, a method was previously developed for para-meterizing influence models for convective-weather-propagation from ensemble weather forecasts, in the case where the subregions are grid squares. This approach has been adapted to parameterize the wind influence model.

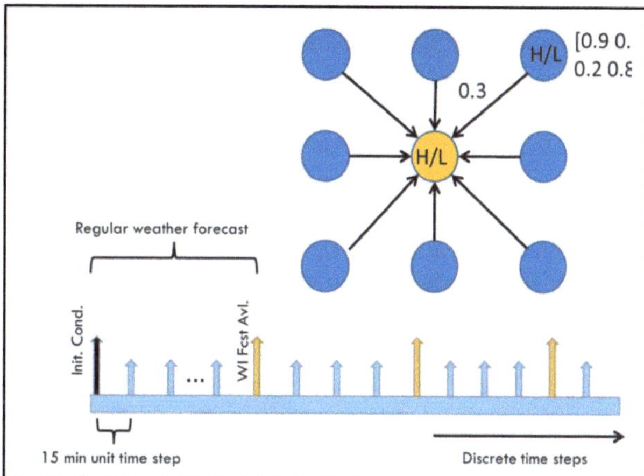

Fig. 5 The influence model update rule is illustrated (top diagram). The influence model is parameterized to statistically match the ensemble forecast at snapshot times, as diagrammed below.

Specifically, let us consider building a gridded influence model using ensemble forecast products at a certain spatial and temporal resolution (e.g., 40km grid squares and 3 hour temporal resolution for the SREF). While the ensemble forecast data has been extracted over a wide area, wind generation is often concentrated in regional clusters – e.g., in the Columbia River Gorge area in the Pacific Northwest. Here, one of these clusters is used to define an area of interest for influence modeling. Within the area, a model with higher resolution than the ensemble forecast may be needed, to capture dynamics and variations at the spatial scale of wind farms, and at an appropriate temporal resolution for unit scheduling. Specifically, a y-fold increase in the spatial resolution (along each dimension) and a z-fold increase in the temporal resolution is assumed. To build the influence model, the ensemble forecast data is first used to determine desired status probabilities for influence-model grid squares at snapshot times (for instance, every three hours if SREF data is used). Specifically, each influence model grid square is located within an ensemble-forecast grid square; the fractions of ensemble members in each status (wind speed/direction bin) in this square can be viewed as desired status probabilities for the corresponding influence model square. In this way, the ensemble forecasts provide local status probabilities across the region of interest at every z time steps. Once these snapshot probabilities have been computed, the technique for influence-model parameterization given in [14-17] can be applied directly. Specifically, the technique allows selection of the influence-model parameters so that the local status probabilities at the snapshot times exactly match the desired snapshot probabilities, and further the status probabilities at intermediate times are interpolations of these desired probabilities (see Figure 5). In addition, the parameterization technique gives the user the freedom to tune the extent of spatial and temporal correlation (or persistence) in the wind profile. In these initial studies, these correlation-tuning parameters have been chosen so that wind deviations have significant persistence for about one hour and 100km. In the future, we expect to tune the parameters to match historical correlations in wind speeds/directions at wind farms. Since the technique for parameterizing the influence model was presented in earlier work, it is not described in detail here, see [17] for these details.

A couple of remarks about the influence modeling approach are worthwhile. First, the reader will note that the model uses a binned or discretized representation of wind. An alternate continuous-valued influence model can be envisioned [37]. However, we believe the binned approach is appropriate because of intrinsic precision limits (and limited precision needs) for the day-ahead wide-area forecasting goals of this project, and because binned models naturally permit translation to generation levels (see Section 3.2.4). The bin resolutions can be chosen at the user's discretion, so varying levels of precision are possible. Second, while our focus here has been on building a gridded model with a single resolution, the influence model permits arbitrary subregion topologies. The parameterization (model-building) technique also can be extended to more general topologies. A multi-resolution gridded model is currently under development.

Case Study

The influence model builder has been implemented for the described case study. Specifically, an influence model for wind speeds has been developed for the highlighted area of interest, corresponding to the Columbia River gorge region which has a high density of wind generation. Noting that the wind speeds during the period of interest are low (less than 12 miles/hour), a coarse binning model has been used in this initial study. Specifically, two wind-speed-based bins are assumed, one corresponding to wind speeds below 6 miles/hour (which permits no wind generation for most turbines) and the other to wind speeds between 6 mi/hr and 12 mi/hr (which is above the cut-in speed but only yields a low level of wind generation). The SREF forecast was used to build the influence model simulator, as discussed above. The implemented influence model achieved 12-fold multiplication in the temporal resolution (to provide forecasts every 15 minutes) and a 2-fold multiplication in the spatial resolution (yielding 20kmx20km grid squares) compared to the ensemble forecast.

3.2.3 Influence Model Simulator

The built influence models for wind and cloudiness characteristics can be used to simulate or produce a large number of wind/cloudiness futures over the day ahead. The software tool that does this is referred to as the influence model simulator. The influence model simulator is discussed here, with a particular focus on wind-future simulation (see [17] for a discussion of cloud-cover simulation for solar-generation prediction).

Since the wind influence model is a stochastic automaton network model, each simulation of the model yields a different future or profile. Specifically, the influence model's update rule (see Section 3.2.2) is applied over the modeled time horizon (the day ahead), for the built model. Simulating the model in this way produces a specific wind bin profile at the specified temporal and spatial resolution. By repeating the simulation many times, a large number of independent profiles or futures is obtained. These futures are each different, but their aggregate statistics match the designed statistics of the built influence model (including local status probabilities and correlations), and hence also match the ensemble-forecast statistics at snapshot times.

The influence model update rule permits fast simulation, only requiring computation of a linear function followed by a randomization (which can be achieved by producing a uniform random variable on [0,1] and comparing it with a threshold). The simulation time scales linearly with the number of grid squares and the number of time steps simulated. For a realistic-scale model (say, 100-5000 grid squares over a full day), thousands of futures can be produced in less than a second. The special structure of the wind influence model also permits efficient statistical analysis, including characterizations of temporal and spatial correlations, as well as variability in aggregate wind characteristics.

Case Study: The influence model simulator has been implemented for the Columbia River gorge case study. Specifically, the built wind influence model has been used to produce 1000 wind futures. Snapshots of two futures are shown in Figure 6. Both futures show a trend toward increasing wind speeds, reflecting the trend in the ensemble forecast. They also show certain common spatial characteristics (for example a consistently low wind speed in a couple of the Southern grid squares), which reflect topological impacts on wind characteristics. However, the two futures show considerable variability in the wind speed profile, and also display complex spatial patterns. In Figure 6, we also map the probability of the higher-wind-speed bin at the snapshot times. The influence model simulations match these local status probabilities in aggregate, but show considerable variability and also enforce spatial and temporal correlation.

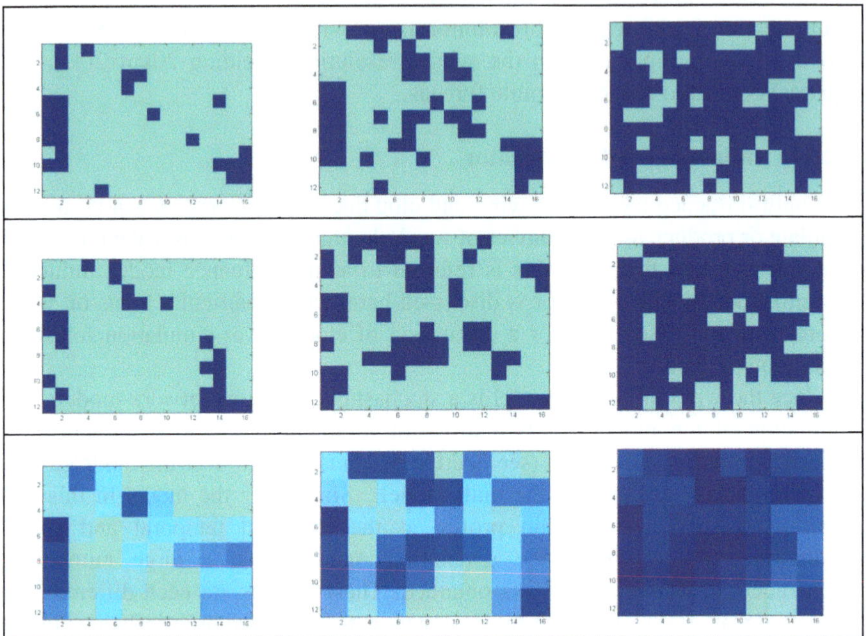

Fig. 6 Top *and middle rows:* two simulations of the influence model are shown (darker squares represent the higher-wind bin, lighter squares indicate the low-wind bin), for the Columbia River Gorge region (the boxed area on the ensemble forecast). This influence model has twice the resolution of the ensemble forecast. We notice that the influence model captures the spatiotemporal trends in wind speed in the ensemble forecast, but captures significant smaller-scale variations and correlations in wind, particularly during periods of rapidly-changing weather. *Bottom row:* the probability of the higher wind-speed bin across the region of interest is shown at snapshot times, as obtained from the ensemble forecast.

Finally, in Figure 7, we have presented some aggregate statistics of the wind-influence-model simulations (e.g., a histogram of the total number of high-wind grid squares), to illustrate the level of variability among the futures produced by the model. These statistical analyses show that the variability changes significantly with time and location, with highest variability when expected wind speeds are changing rapidly. The standard deviations in wind generation found in this way roughly match the uncertainty levels given in the literature [29, 34].

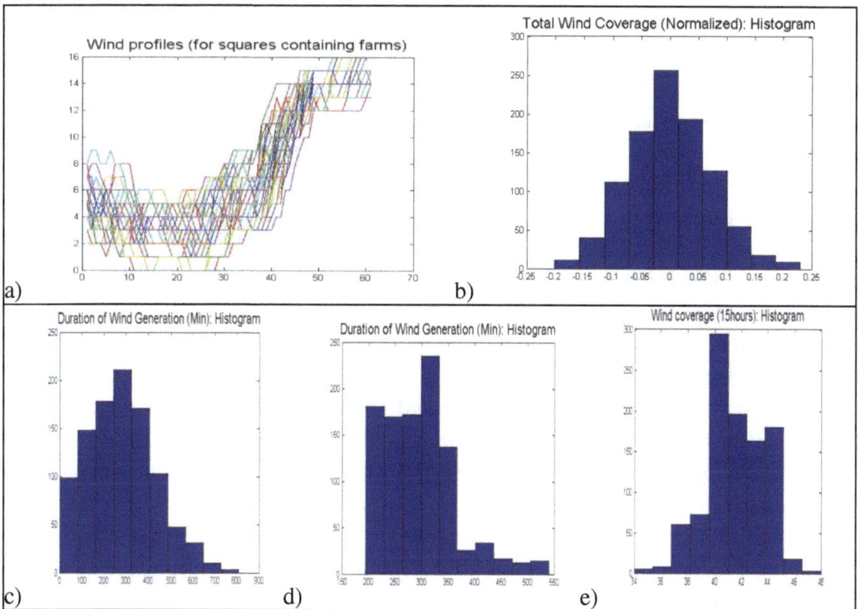

Fig. 7 Statistical characterizations of wind and wind generation profiles are obtained, using the influence-model-based simulator. a) The number of grid squares with high wind is traced, for multiple influence-model trajectories. b) A histogram of the total number of high-wind periods at all wind-farm locations is shown (normalized to the mean); the variability is about 10%, which matches with observed data. c) and d) Histograms of the time duration of wind generation (i.e., the total time such that winds are high enough to generate power) are shown, for two different wind farms. The distributions are quite different. e) The distribution of total wind coverage across the region is shown, at a snapshot time; there is high variability at times when the average wind speed is changing rapidly.

3.2.4 Wind- to Generation- Translator

The next blocks in the generation-forecasting module are responsible for translating the wind futures produced by the influence model to wind-generation futures (and analogously to translate cloud-cover futures and other environmental parameters to solar-generation levels, see [14]). There is a wide literature on modeling wind turbines and wind farms, which can be brought to bear to forecast wind generation from wind profiles [10-12]. Unit commitment requires forecasting at the resolution of wind farms across a wide area, rather than precise forecasting at a single location. Two techniques are worth reviewing. First, *binning techniques* have been used to model wind generation at the level of wind farms. In these techniques, wind bins (which may involve both speed and direction parameters) are mapped to wind generation levels for a specific wind farm, using statistical analyses of historical data. The binning approaches dovetail nicely with our solution, since the influence-model-simulator produces binned wind futures in subregions across the wide area. For a wind farm located in a particular subregion, the forecast wind bin level in this subregion can then be translated to a generation level using the binning-based model. Using maps of the wind farms' connections to the electric power grid, the wind generation at each bus during each time step can be determined. By applying this method to each influence-model-produced future, many wind-generation futures can be obtained.

Alternately, simple physics-based models for a wind turbine can be used for generation forecasting. The simplest models approximate a turbine's wind generation as a cubic function of the wind speed, between a lower cutoff speed and the turbine's rated wind speed; the model's parameters depend on the type of wind turbine being modeled. To translate the influence model futures via this model, we again determine the wind bin level at the wind-farm location of interest from the influence model. This bin level for the wind farm then is converted to a single wind speed : either the median wind speed in the bin may be used, or the speed may be randomized within the bin (with the motivation that very small scale variations in speed cannot be captured and may be modeled as uncertain). The physics-based model can be used to determine the wind generation for each turbine in the farm, and hence total farm-level generation can be determined. The remaining procedure for obtaining wind-generation futures is the same as for the first approach. We note that much more intricate models for wind turbines and farms are available, that account for wind-direction effects, capture wake effects and topographical variations, etc. However, noting that wide-area forecasts are needed, these simple approaches are approaches.

Both approaches for wind- to generation- translation described above require some data on wind farms and turbines. For both approaches, the locations of wind farms to be modeled must be known. For the binning approaches, historical data on wind speeds and generation for each farm is also needed, so that binning-based models can be constructed for each farm. These models also must be updated if generation capacity is added to a farm, and when new farms are brought online. For the physics-model-based approach, the number of wind generators of various types must be known for each wind farm. ISOs and TSOs typically have available to this information (see e.g. a discussion of wind farm data for the Texas grid [35]).

3.2.5 Representative-Future Selector

The final block in the generation forecasting module is tasked with selecting a few representative wind- and solar- generation futures, from the large set of simulated futures. Representative futures are sought both for use in unit scheduling and to provide operators with concrete illustrations of wind profiles for decision making. More specifically, our motivation for selecting a subset of futures as representative ones is three-fold: 1) to provide operators with an indication of the range of weather outcomes and consequent wind/solar generation profile that may occur on a given day, 2) reduce the computation needed for the unit-commitment problem (albeit in a somewhat limited way, see Section 4 for a discussion), and 3) facilitate performance evaluation of the unit-scheduling design by identifying typical test cases for weather outcomes.

Broadly, the purpose of the representative-future-selection block is to 1) choose a sparse set of futures that span the range of wind/solar-generation outcomes for the day of interest and 2) assign likelihoods to these typical futures. Several approaches have been developed for selecting representative samples of a random variable or random process. In this project, a technique for sample selection known as the *Probabilistic Collocation Method (PCM)* [20] is considered, which draws on a numerical-integration method known as Gaussian quadrature. This method allows selection of futures according to one or more selected performance measures (specifically, to span the range of possibilities for these performance measures). Thus, the algorithm automatically selects futures that are ordered with respect to the performance measures. If appropriate performance measures are used, the method can be used to distinguish futures which will require significantly different commitment and dispatch plans, as is needed for stochastic unit commitment. Since PCM has been developed in previous work, technical details are omitted. For this project, a Matlab software implementation of the method has been developed as part of the generation-forecasting module.

One key challenge in using the proposed approach is to choose appropriate performance measures to select representative futures. To be useful, performance measures must be able to distinguish weather/generation profiles which will require significantly different schedule profiles. In this first effort, we use as the metric the total wind generation, which should be strongly indicative of conventional generation requirements and hence these units' schedule. More broadly, we anticipate relying on operators' experience to choose performance metrics. In addition, historical data can be used to regress possible performance metrics against schedule profiles, to determine how predictive the measures are of the schedules.

Case Study

The representative-future selector has been implemented for the Columbia River gorge case study, using the total wind generation as the metric for selection. Specifically, the analysis considers generation from six large wind farms in the Columbia River Gorge area, which are connected to two buses in an example power-system model (see Section 4.3). This wind-farm example is constructed, but the locations and sizes of the farms are similar to those of actual farms in the Gorge area. For each wind future produced by the influence model, the wind-power generation at the two buses of interest is computed, using the simple physics-based model for wind-to-power translation for a wind turbine. PCM has been applied to obtain five representative futures (very low power, low power, medium power, high power and very high power generations) from 1000 wind power generation futures, using the total wind generation at the two buses as a metric for future selection. Figure 8 and Figure 9 show four of the representative futures (with Figure 8 showing the generation at Bus 1 and Figure 9 at Bus 2). The very low power generation future is not shown here, because the generation level is nearly zero. Both figures show increasing wind power generation trend, which is consistent with the weather forecasting.

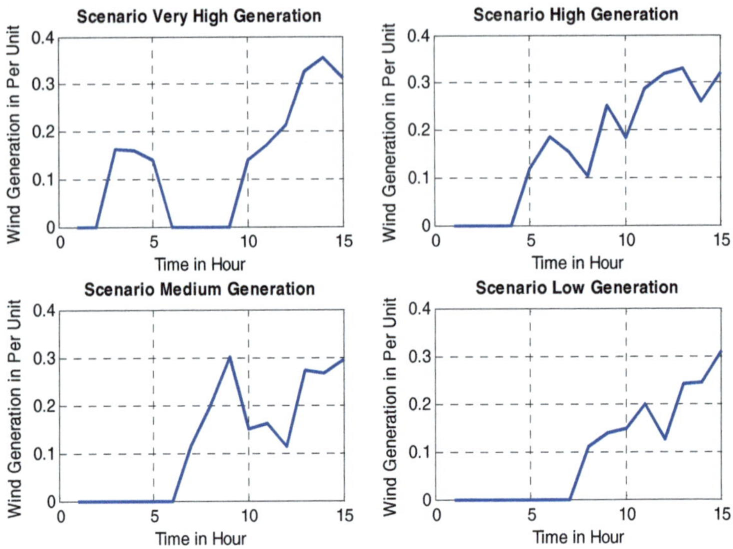

Fig. 8 Representative generation futures, Bus 6

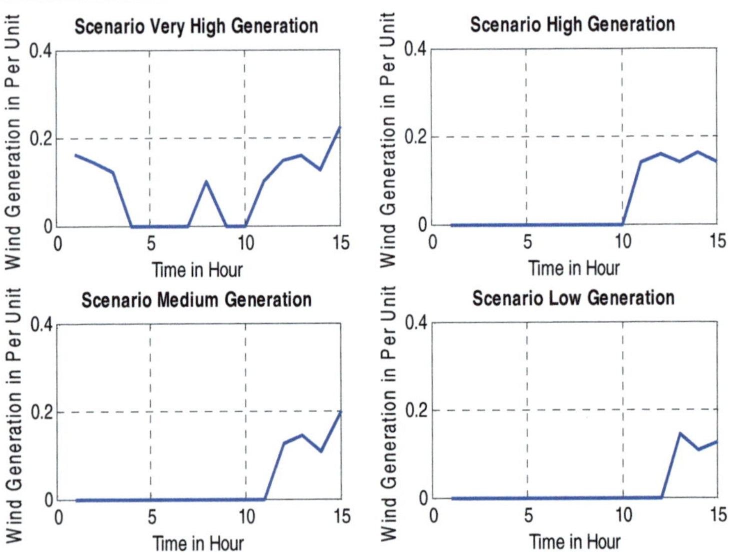

Fig. 9 Representative Futures, Bus 8

The PCM tool also assigns a probability for each scenario. These probabilities are shown in Table 1, to illustrate the likelihood of each possible representative generation future. These statistics show that wind power has more than 95% probability to be in low, medium and high generation levels. Meanwhile, the extreme cases (zero generation and very high generation) have less than 3% probability. The probability distribution of the representative futures statistically match the large ensemble of futures produced by the influence model.

Table 1 Likelihoods of the representative generation futures (scenarios)

Scenario	Probability P_i
Very High Wind Power Generation	0.0139
High Wind Power Generation	0.1314
Medium Wind Power Generation	0.4658
Low Wind Power Generation	0.3600
No Wind Power Generation	0.0289

4 Toward a Scheduling Module: Some Initial Explorations

The proposed end-to-end solution for day-ahead resource planning requires implementation of a unit commitment or scheduling algorithm, which is the focus of the scheduling module in Figure 1. This scheduling module is responsible for selecting an on/off schedule and dispatch plan for conventional generator units for the day-ahead market, which accounts for uncertain generation from the (non-dispatchable) intermittent-renewable units. Specifically, the scheduling module is tasked with using the representative wind- and solar- generation futures outputted by the generation-forecasting module, along with knowledge of the power grid and the market, to design unit schedules and dispatch levels for the conventional generators. At its essence, implementing the module requires solving a *stochastic unit commitment* problem. Our focus in building the scheduling module is to use a stochastic unit commitment algorithm that is practical for implementation in the current transmission-system operational paradigm, rather than to propose a new decision-making paradigm. Our perspective is to view stochastic unit commitment as a two-step process, first requiring an identification of critical conventional units whose schedules are dependent on the renewable units' generation futures, and second achieving an optimization of these unit's schedules and hourly economic dispatches.

The development of the scheduling module is still a work in progress, and here only some preliminary explorations are presented. Specifically, the wide literature on stochastic unit commitment algorithms is briefly reviewed, and challenges in integrating these algorithms into control-room software technologies are discussed (Section 4.1). A simple example problem is then introduced (Section 4.2), and then used to illustrate the scheduling module (the selection of critical units and the optimal scheduling of these units) in Section 4.2. The unit-commitment module is described in detail in the context of the example (Section 4.3). Finally, design results for the example problem are summarized (Section 4.4), and a performance evaluation of the model is undertaken (Section 4.5). It is important to stress that these preliminary explorations ignore many features of importance in stochastic unit commitment (e.g., security constraints), and certainly should not be interpreted as achieving a complete solution. Instead, these explorations expose subtleties in developing unit commitment plans across renewable-generation profiles, and illustrate the proposed two-step process.

4.1 Related Literature on Stochastic Unit Commitment

Unit commitment (UC) refers to the on/off scheduling as well as hourly dispatch of available generation units over a planning horizon (often, the full day ahead) to meet the time-varying electric load. Most TSOs routinely use UC software for resource planning, most commonly for dispatch in the day-ahead market. Typically, the UC plan is obtained by solving a deterministic optimization problem, to achieve a lowest-cost scheduling and dispatch of conventional generation. The growing penetration of intermittent renewables, which have significant uncertainty at a one-day look-ahead, is creating challenges to system operators to manage load/generation balance. Thus, there is a strong motivation to develop new unit commitment algorithms that allow scheduling/dispatch of conventional generation while accounting for uncertainty in renewable generation.

A number of methods for unit commitment under uncertainty have been developed in the literature (many of them focusing particularly on wind-generation integration), under the headings of stochastic unit commitment (SUC) and stochastic security constrained unit commitment (SCUC). These papers broadly focus on the problem of scheduling and dispatching generation to optimize an expected cost in the face of generation/load uncertainty, but vary significantly in 1) modeling generation/load uncertainties, 2) the cost function and specific design problem, and 3) the methods used for optimization, among other differences (30-32). Several recent works by Oren's group and others as being particularly aligned with the approach pursued here, in that they consider scheduling given multiple stochastic futures of wind generation. The study of Constinecu et al on exploiting ensemble forecasts for stochastic unit commitment [38], and the efforts of Sauer and his co-workers on uncertainty management (e.g. [39]), are also closely aligned with the research described here.

Although these stochastic SCUC models are promising tools to solve the unit commitment problem with large scale renewable energy integration, to the best of our knowledge they have not yet being used by TSOs in the control room. The perspective of this chapter is that advances in several directions are needed integration of these approaches into control-room technology. First, realistic models for wind/solar generation that leverage weather-forecast products are needed within the stochastic unit commitment solutions. Many existing studies make simpler assumptions regarding wind profiles, for example [21,28] and derivative works use Monte Carlo simulation to generate wind speed and assume the wind speed error distribution is Gaussian. Second, the unit-commitment strategies need to be tailored to permit easy implementation in the current operational paradigm. In particular, many of the stochastic unit commitment approaches assume hourly re-planning of the commitment plan per a dynamic-programming solution, but most TSOs use a binding day-ahead market and hence require a fixed optimal plan. Additionally, a practical implementation would benefit from a performance evaluation of the designed commitment plan over the possible weather futures, and simple display of plan specifics and performance characteristics. Third, stochastic unit commitment remains computationally challenging for problems of realistic scale (1000's of buses, 100's of generators), and further techniques for reducing problem complexity are needed.

4.2 An Exploratory Example

The scheduling module has been developed in the context of a small-scale example, based on the IEEE 14-bus test system. The example system is assumed to have both intermittent renewable generation and conventional generation. Specifically, buses 1, 2 and 3 are connected to conventional generators, while bus 6 and bus 8 connect wind generation in the Columbia River gorge area. Figure 10 shows the IEEE 14-bus test system. In this example, the wind generators can provide up to 40% of the total power. For the day of interest, the wind speeds are relatively low and, thus, the expected total wind generation is lower than 20%. It is assumed that the wind generation cannot be scheduled or dispatched, i.e. their generation levels are determined entirely by the wind profile. Our goal is to find the optimal commitment strategy (on/off schedule and dispatch) for the conventional generators. We approach the unit-commitment problem in two steps, first focusing on selecting critical units whose on/off profiles may significantly depend on the uncertain wind generation, and second solving the unit commitment problem using a pruned decision space.

Both the critical-unit selection and the unit commitment optimization require solution of hourly economic dispatches (ED). While use of ED is commonplace in power-system operations, it is useful to briefly introduce the specific ED problem considered here. The goal of the ED considered here is to minimize the total generation cost and real power losses, subject to transmission and operational constraints. For simplicity, a DC power flow is considered, with. the real power loss cost approximated as a penalty cost. Specifically, the objective function is

$$F = \min\left(C_{cg} + C_{wg} + \lambda P_{loss}\right)$$

Subject to:
Generator constraints

$$P_{min} \leq P_g \leq P_{max}$$

DC power flow line limits

$$-P_{ij} \leq \frac{\delta_i - \delta_j}{x_{ij}} \leq P_{ij}$$

And Real power balance

$$\sum_{i=1}^{n_g} P_{Gg,i} + \sum_{k=1}^{n_w} P_{Gw,k} = P_{load}$$

where n_g is the number of conventional generator and n_w is the number of wind (renewable) generators. The goal of the ED is to design dispatch level $P_{Gg,i}$ $(i = 1,2,\ldots,n_g)$ for the conventional generators. The conventional generator's operational cost C_{cg} is assumed to be quadratic in the power generation; the wind generator's operational cost C_{cw} is assumed to be proportional to the wind power generation P_{Gw}, which is obtained from the influence model and wind-to-generation translator. The real power loss is approximated as a linear function of the generation vector, in the standard way:

$$P_{loss} = (B^{-1}P)^T G^* (B^{-1}P) = P^T (B^{-T} G^* B^{-1}) P$$

where P is a column vector of real power generation, B is the network susceptance matrix and G is the network conductance matrix.

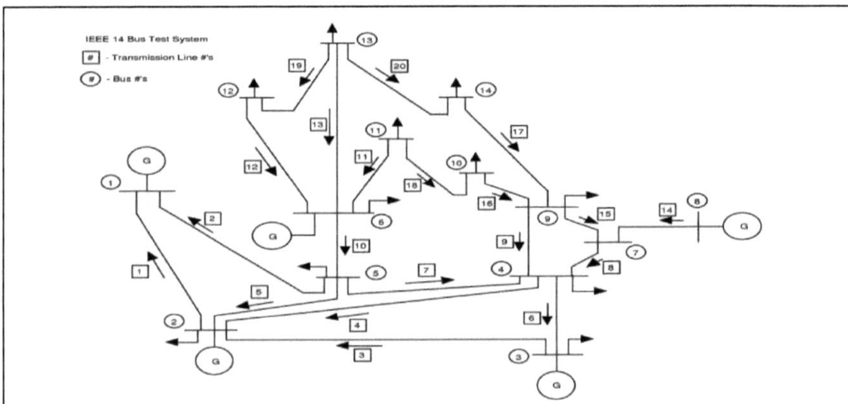

Fig. 10 IEEE 14-bus system

4.3 *Scheduling Module: Details and Simulation*

Broadly, the scheduling module has two stages. The first stage identifies critical units, and the second stage finds the optimal on/off schedule of critical units and the dispatch levels of all units (thus taking advantage of the critical-unit identifier to prune the decision space). Let us begin with brief descriptions of each stage's functionality, in the context of the example. It is worth remarking that the proposed algorithms leverage only the representative generation futures, which are expected to provide sufficient coverage of the uncertainty space to permit decision. In fact, all futures could be used for scheduling with a relatively modest increase in computational cost (roughly linear in the number of futures). This alternative can be implemented in an entirely analogous way.

4.3.1 Critical Unit Identifier

The critical unit identifier determines a small set of conventional (dispatchable) generation units whose on/off schedule and dispatch may be highly sensitive to the generation profile on the day of interest. Specifically, critical units are identified as follows. For each representative generation future (five in our case), the economic dispatch is determined for each hour on the day ahead, *assuming that all of the dispatchable units are on-line*. We notice that some units may or may not need to dispatch power, depending on the representative generation future. Units whose dispatch may or may not be zero depending on the weather future are considered critical units, since they will be difficult to schedule given the uncertainty in the generation profile. In the example, when the high wind power future is considered, we notice that some units are scheduled to produce zero power at some times based on the ED. Meanwhile, at the same times in the low wind generation case, those units are scheduled to produce some amount of power. These units are identified as critical units. Specifically, in the 14-bus example, conventional generator 2 at bus 2 is identified as a critical unit by solving the optimal dispatch problem in Section 4.2. Figure 11-1 and 11-2 illustrate the dispatch of generator 2 for the high and low wind generation futures, respectively. At each time hour, the dispatch level is determined by the ED problem solved in section 4.2. At times t=7, t=9 and t=11, generator 2 dispatches some amount of power in low wind generation case, but dispatches zero power in high wind generation case.

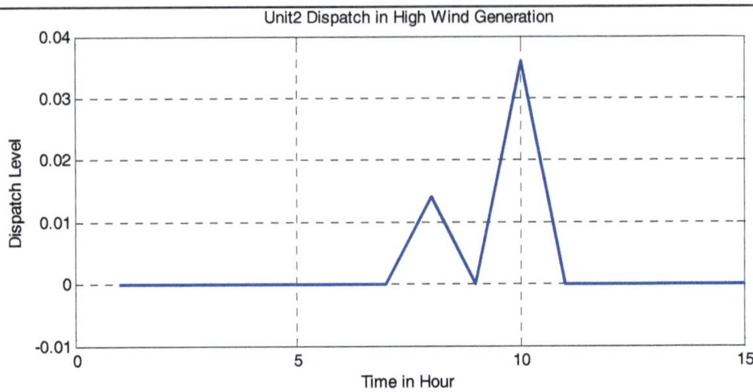

Fig. 11 -1 Unit 2 Dispatch in High Wind Generation Case

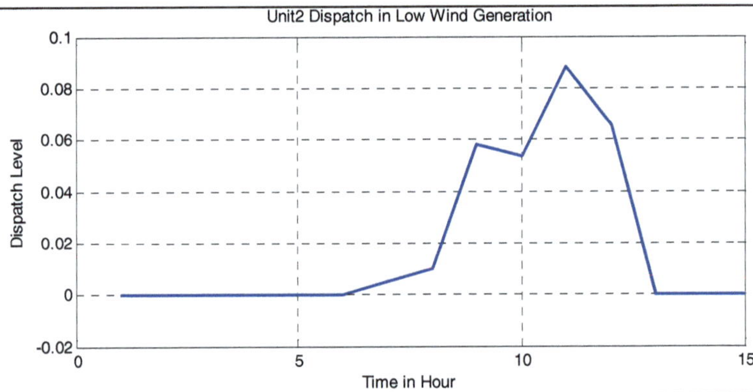

Fig. 11 -2 Unit 2 Dispatch in Low Wind Generation Case

Figure 12 and 13 demonstrate the non-critical units' dispatch levels (assuming again that all generators are on-line). Specifically, figure 12-1 and 12-2 illustrate the dispatch of generator 1 for the high and low wind generation futures respectively; and figure 13-1 and 13-2 illustrate the dispatch of generator 3 for the high and low wind generation futures respectively. No matter what the wind generation future is, the non-critical unit is assigned some nonzero dispatch level at each time hour.

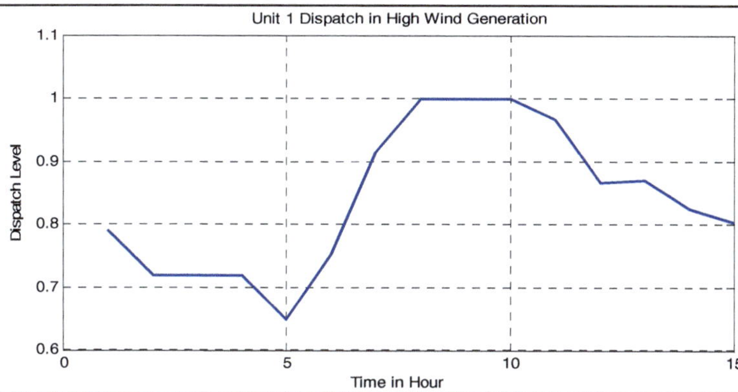

Fig. 12 -1 Unit 1 Dispatch in High Wind Generation Case

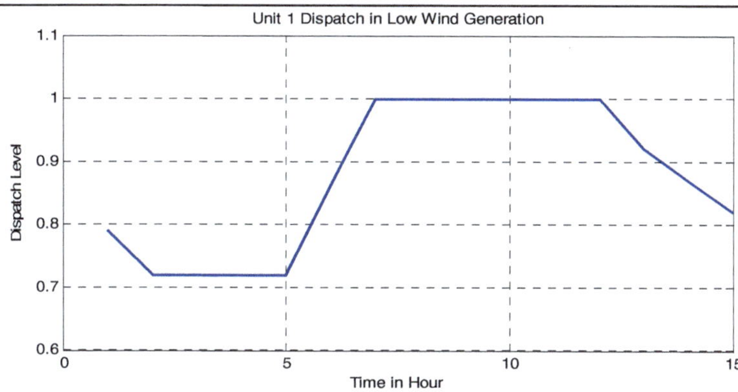

Fig. 12 -2 Unit 1 Dispatch in Low Wind Generation Case

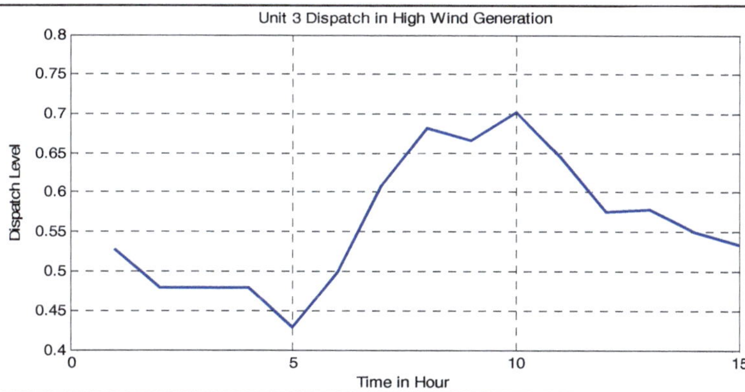

Fig. 13 -1 Unit 3 Dispatch in High Wind Generation Case

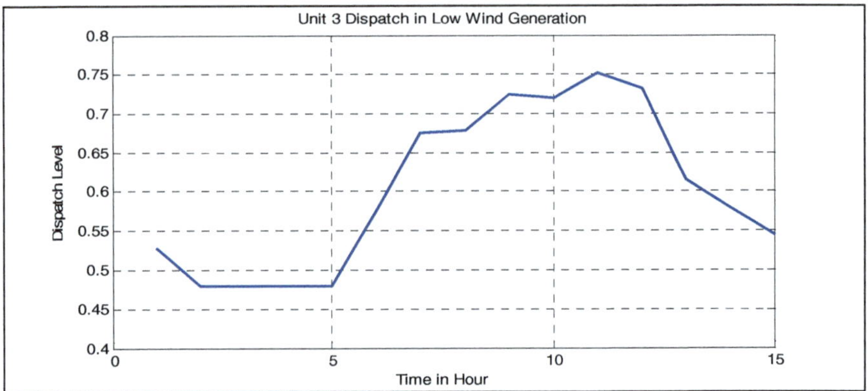

Fig. 13 -2 Unit 3 Dispatch in Low Wind Generation Case

4.3.2 Scheduling On/Off Profiles and Dispatches

The day-ahead market requires a single binding unit-commitment and dispatch schedule that performs well across weather futures. This design problem can naturally be phrased as an optimization problem, to minimize a cost or expected cost with regard to on-line/off-line schedules and dispatch level. Broadly, the optimization formulations considered can be phrased as mixed integer programming problems, which are similar in flavor to several of the stochastic unit commitment problems in the literature (although our formulation does not allow re-planning). For problems of moderate/large scale, standard optimization programs can be used to solve the unit commitment task. However, a special pruning of the design space is pursued here to reduce the high computational burden of this optimization, and to obtain insightful characterizations of uncertainty impact. Specifically, the dispatch level of all thermal units and the on/off of schedule of only the critical units are considered as design variables our formulations.

Two approaches for solving the stochastic UC have been considered. The first benchmark method uses a single mean wind power generation trajectory to design the critical units' on/off schedules and all thermal units' dispatch level. Scheduling based on mean wind profiles has been considered in the literature (see e.g. [39]), and in this sense the approach is a benchmark. The second method explicitly accounts for the wind generation uncertainty in the UC objective function, and hence seeks for a schedule and dispatch profile that minimizes the expected cost across possible weather scenarios. Let us give a mathematical description of each problem formulation.

Specifically, suppose there are n thermal units in a power system, and k of them are critical units. In method 1, the schedules of the critical units and the hourly dispatch of all units are designed to minimize the following cost:

$$f(\vec{g},\vec{x},\vec{y}) = \min(\sum_t \sum_i g_{it}C_{it} + \sum_t \sum_i y_{jt}S_{jt} + \sum_t \sum_j x_{jt}H_{jt} + \lambda \sum_t P_{loss,t})$$

Subject to the following constraints:

Power balance –

$$\sum_i g_{i,t} + E(g_{wind,t}) = P_{load,t}$$

Generation limits –

$$g_{i,min} \leq g_i \leq g_{i,max}$$

Line flow limits –

$$-g_{ij} \leq \frac{\delta_i - \delta_j}{x_{ij}} \leq g_{ij}$$

On-line time limit –

$$t_{i,online} \geq t_{i,online,min}$$

Transition costs (start-up and shut-down costs) are also modeled. They are described below after the problem formulation for method 2.

In method 2, the schedules of the critical units and the hourly dispatch of all units are designed to minimize the following cost:

$$f(\vec{g},\vec{x},\vec{y}) = \min E(\sum_t \sum_i g_{it}C_{it} + \sum_t \sum_i y_{jt}S_{jt} + \sum_t \sum_j x_{jt}H_{jt}$$
$$+ \gamma \sum_t (P_{L,t} - \sum_i g_{it} - g_{wt,m})^2 + \lambda \sum_t P_{loss,t})$$

Subject to the following constraints:

Generation limits –

$$g_{i,min} \leq g_i \leq g_{i,max}$$

Line flow limits –

$$-g_{ij} \leq \frac{\delta_i - \delta_j}{x_{ij}} \leq g_{ij}$$

On line time limit –

$$t_{i,online} \geq t_{i,online,min}$$

In both formulations, the decision variables are the following:

g_{it} is the power produced by the thermal generator i in time period t

y_{it} is 1 if critical unit starts at the beginning of period t, otherwise, 0

x_{it} is 1 if critical unit shuts at the beginning of period t, otherwise, 0

Index i represents all thermal units, and index j represents all critical units

Other parameters include:

$gw_{t,m}$, which is the total power produced by wind generator in time period t at representative scenario m.

Finally, the term $\gamma \sum_t (P_{L,t} - \sum_i g_{it} - g_{wt,m})^2$ in the second problem formulation is called the correction cost. This term requires some further discussion. Notice that, in method 2, the power balance cannot be enforced since the renewable-generation is uncertain. Instead, we model the imbalance between power generation and consumption as a correction cost, which reflects the additional cost needed to meet the power imbalance on the day-of-operations (through re-dispatch on the hourly market, use of reserves, and possibly through turning on fast-ramp units). We stress that the correction cost arises due to the uncertainty in real wind power generation. In contrast with other stochastic unit commitment efforts, we do not seek to model the re-dispatch of power at an hourly scale in detail, but propose the quadratic correction cost to encompass the family of corrective actions that may be taken. We anticipate that operators would choose the scaling constant γ based on historical costs incurred on the day of operations when there is a significant generation-load imbalance. In particular, a regression may be used to determine the dependence of the additional cost on the imbalance. We anticipate that the regression may identify a non-quadratic mapping for this correction cost: in particular, insufficient generation is likely to be more expensive than overproduction; we leave a careful analysis to future work.

Another point that requires discussion is the loss penalty cost in method 2. As discussed previously, we model the loss penalty cost as a function of all the thermal units' dispatch levels under the power balance constraints. In method 2, we do not have this power-balance constraint, which complicates computation of the loss penalty. Here, we assume the sum of all thermal units' dispatch level is close to the load consumption minus the mean renewable generation. Under this assumption, we can use the same penalty cost function in method 2 as is used in method 1. Finally, for method 2, we note that the expectation is computed across the representative renewable-generation futures.

The problem formulations described above have further been extended to include transition costs, including startup costs and shutdown costs. Startup costs involve both fixed costs C_f and variable costs C_v. Shutdown costs generally involve only fixed costs and sometimes are not significant. Generally, variable costs of start-up depend on two different shutdown states the unit is in. The two possible states are hot reserve and cold reserve. The mathematical model of the variable cost in hot reserve state is:

$$C_{vb} = C_b t f,$$

and in cold reserve state is

$$C_{vc} = C_c[1 - e^{-\frac{t}{a}}]f,$$

where t indicates the time period that the generator is in this state, and the remaining scalars are model parameters. Details are omitted.

How to Solve the Optimizations: The described optimization problems are mixed integer programming problems, and a range of tools for solving these problems can be brought to bear. We stress that the proposed methods take several steps to reduce the inherent computational complexity of the problems, including pre-selection of critical units (which reduces the number of integer variables) and abstract modeling of the correction cost. These simplifications are particularly important for optimization of an expected cost across arbitrary wind futures, since traditional simplifications of mixed integer programs often fail in this case.

For the small-scale case study described below, we have simply used an exhaustive search over the binary (on/off) variables together with the quadratic-programming tool in Matlab to find the optimum. For this small-scale example, it is worth noting that the optima obtained over the pruned design space are identical to those that would be obtained if all units' schedules were designed, and hence our critical-unit-based approach does not lead to performance degradation.

4.4 *Example Problem: Results*

Tables 3 and 4 show the dispatch levels of all units and the critical unit's on/off schedule, respectively, when the first unit-commitment method is used (i.e., the method based on expected generation).

Dispatch level:

Time	1	2	3	4	5	6	7
Gen1	0.7902	0.7188	0.7177	0.7187	0.7094	0.8486	0.9618
Gen2	0	0	0	0	0	0	0
Gen3	0.5227	0.4746	0.4738	0.4745	0.4682	0.5621	0.6385

Table 3

Time	8	9	10	11	12	13	14	15
Gen1	0.9906	1.0	1.0	1.0	1.0	0.9055	0.8568	0.8022
Gen2	0	0	0	0.0234	0	0	0	0
Gen3	0.6579	0.6915	0.7657	0.8	0.6858	0.6005	0.5676	0.5308

Table 4

Critical unit on (1) /off (0)

Time	1	2	3	4	5	6	7	8	9	10	11	12	13	14	15
Gen2	0	0	0	0	0	0	0	0	0	1	1	1	0	0	0

Meanwhile, Tables 5 and 6 show the dispatch level of all units and the critical units' on/off schedules, respectively, for the second method (i.e., minimization of the expected cost over wind futures).

Dispatch level

Time	1	2	3	4	5	6	7
Gen1	0.7751	0.7053	0.7041	0.7052	0.6961	0.8322	0.9430
Gen2	0	0	0	0	0	0	0
Gen3	0.5426	0.4927	0.4919	0.4926	0.4861	0.5834	0.6624

Table 5

Time	8	9	10	11	12	13	14	15
Gen1	0.9697	0.9745	1.0	0.9931	0.9731	0.8880	0.8408	0.7870
Gen2	0	0.0051	0.0035	0.0544	0	0	0	0
Gen3	0.6841	0.7173	0.7680	0.7819	0.7181	0.6230	0.5890	0.5508

Table 6

Critical unit on/off

Time	1	2	3	4	5	6	7	8	9	10	11	12	13	14	15
Gen2	0	0	0	0	0	0	0	0	1	1	1	0	0	0	0

4.5 *Example Problem: Evaluation*

Because the real day-ahead wind speed and power generation has large variability, operators may wish to check how an optimal UC plan performs under real weather scenario. Here, the UC plan performance across the representative scenarios is considered. We also compare the overall UC plan performance for method 1 and method 2. The objective function for performance evaluation is chosen as:

$$f = \sum_t \sum_i g_{it} C_{it} + \gamma \sum_t (P_{L,t} - \sum_i g_{it} - g_{wt,m})^2 + \lambda \sum_t P_{loss,t})$$

Tables 7 and 8 present the performance evaluation for the two methods.

Method 1 performance evaluation

Table 7

Wind Scenario	Very High	High	Medium	Low	Very Low
Probability	0.0139	0.1314	0.4658	0.3600	0.0289
Correction Cost	116.0053	3.7191	6.8265	56.3789	338.8427
Total Cost	1319.2	1206.9	1210.0	1259.6	1542.0

Method 2 performance evaluation

Table 8

Wind Scenario	Very High	High	Medium	Low	Very Low
Probability	0.0139	0.1314	0.4658	0.3600	0.0289
Correction Cost	124.3360	5.1023	9.3974	66.3734	363.8691
Total Cost	1303.1	1183.9	1188.2	1245.2	1542.7

Analysis:

The correction cost and total cost are both much higher when the UC plan is evaluated on extreme wind generation futures. Moreover, the very low case has the largest cost. This is not surprising because, for a very low wind generation scenario, more reserve power must be used which is usually expensive. One point to notice is that, even under very high wind generation scenario, the correction cost and total cost are higher. This is because the very high wind-generation case has the lowest probability, thus, it diverges more from the expected behavior of wind generation which results a higher cost in our formulation.

Comparison: Stochastic and deterministic UC

Method 2 incurs a higher correction cost but has a lower total cost compared to Method 1. This is reasonable because method 2 use an expected cost measure which accounts the uncertainty in wind generation. Thus, the solution provided by method 2 requires more flexibility in terms of correction on the day of operations, but can reduce overall cost.

5 Conclusions

Computing technologies have been used in managing the electric power grid for many years. Yet, the growing pervasiveness of cyber-systems in the control room – which include cluster- and cloud- based systems with unprecedented computational power, increasingly high-bandwidth data communications, improved visualization technologies, etc. – can provide unique opportunities for decision-making and management, which are far from fully realized. At their essence, these new cyber- capabilities allow a seamless integration of the physical-world, human, and economic aspects of power-grid management. In this sense, they are transforming the grid from a collection of disparate processes into an integrated cyber-physical system.

In this article, we have envisioned using the growing integration of cyber-technologies in the control room, to assist in the operation of power systems with high penetration of intermittent-renewables. Specifically, we have envisioned an end-to-end framework for forecasting probabilistic renewable-generation futures, and using these futures for unit-scheduling for the day-ahead market. The envisioned framework exploits new cyber- capabilities in myriad ways: it uses ensemble forecasting data in real time to inform forecasting/design, draws on stochastic network modeling tools such as the influence model, and imagines an approach to unit-commitment that distinguishes critical units to simplify computation and aid decision-makers. Here, the envisioned framework has been prototyped for a small-scale case study, using ensemble forecast data from a historical day of interest (specifically, wind forecast data for the Columbia River Gorge area of Washington/Oregon on that day). This case study, while preliminary, indicates that the framework may lead to practical new technologies for unit scheduling on the day ahead. A crucial next step is to compare the performance of the proposed methods with benchmark methods for a larger-scale example.

References

1. Ott, A.L.: Experience with PJM Market Operation, System Design, and Implementation. IEEE Transactions on Power Systems 18(2) (May 2003)
2. CAISO Smart Grid Roadmap and Architecture, Publication of the California Independent System Operator (CAISO) (December 2010)
3. NYISO Transmission and Dispatching Operations Manual (Manual 12), Publication of the New York Independent System Operator (October 2012)
4. PJM Manual 11: Energy and Ancillary Services Market Operations, Prepared by Forward Market Operations group at PJM (2013)
5. Borenstein, S.: The trouble with electricity markets (and some solutions). Working paper of the Program on Workable Energy Regulation, University of California Energy Institute (January 2001)
6. Hawkins, D., Rothleder, M.: Evolving role of wind forecasting in market operation at the CAISO. In: Proceedings of the IEEE Power Systems Conference and Exposition (October 2006)
7. Xie, L., Carvalho, P., Ferreira, L., Liu, J., Krogh, B., Popli, N., Ilic, M.: Wind integration in power systems: operational challenges and possible solutions. Proceedings of the IEEE 99(1), 214–232 (2011)

8. Smith, J.C., Milligan, M.R., DeMeo, E.A., Parsons, B.: Utility wind integration and operating impact state of the art. IEEE Transactions on Power Systems 22(3), 900–908 (2007)
9. Lange, M.: On the uncertainty of wind-power predictions – analysis of the forecast accuracy and statistical distribution of errors. Transactions of the ASME-N-Journal of Solar Energy (June 2004)
10. Wu, Y.-K., Hong, J.-S.: A literature review of wind forecasting technology in the world. In: Proceedings of IEEE PowerTech, Lausanne, Switzerland (July 2007)
11. Tuohy, A., Meibom, P., Denny, E., O'Malley, M.: Unit commitment for systems with significant wind penetrations. IEEE Transactions on Power Systems 24(2), 592–601 (2009)
12. Botterud, A., Wang, J., Monteiro, C., Miranda, V.: Wind power forecasting and electricity market operations. Proceedings of USAEE 3, 3846 (2009)
13. Orwig, K.D., et al.: Enhanced short term wind power forecasting and value to grid operations. In: Proceedings of the 11th Annual International Workshop on Large Scale Integration of Wind Power into Power Systems, Lisbon, Portugal, November 13-15 (2012)
14. Jiang, J., Roy, S.: Stochastic prediction of spatio-temporal solar-generation futures: an influence-model-based methodology, `http://www.eecs.wsu.edu/~sroy` (in preparation)
15. Asavathiratham, C., Roy, S., Verghese, G.C., Lesieutre, B.C.: The influence model. IEEE Control Systems Magazine (December 2001)
16. Roy, S., Wan, Y., Taylor, C., Wanke, C.R.: A Stochastic Network Model for Uncertain Spatiotemporal Weather Impact at the Strategic Time Horizon. In: Proceedings of 10th AIAA Aviation Technology, Integration, and Operations (ATIO) Conference, Fort Worth, TX (September 2010)
17. Xue, M., Zobell, S.M., Roy, S., Taylor, C., Wan, Y., Wanke, C.: Using stochastic, dynamic weather impact models in strategic traffic flow management. In: Proceedings of the Second Aviation, Range and Aerospace Meteorology Special Symposium on Weather-Air Traffic Management Integration, Seattle, WA (January 2011)
18. `http://nomads.ncep.noaa.gov/txt_descriptions/SREF_doc.shtml`
19. Wan, Y., Roy, S., Lesieutre, B.C.: Uncertainty evaluation through mapping identification in intensive dynamic simulations. IEEE Transactions on Systems, Man, and Cybernetics, Part A: Systems and Humans 40(5), 1094–1104 (2010)
20. Xue, M., Roy, S., Zobell, S.M., Wan, Y., Taylor, C., Wanke, C.: A Stochastic Spatiotemporal Weather-Impact Simulator: Representative Scenario Selection. In: Proceedings of the 2011 Aircraft Technology Integration and Operations Conference, Virginia Beach, VA (September 2011)
21. Barth, R., Brand, H., Meibom, P., Weber, C.: A stochastic unit-commitment model for the evaluation of the impacts of integration of large amounts of intermittent power. In: Proceedings of the 9th International Conference on Probabilistic Methods Applied to Power Systems, Stockholm, Sweden, June 11-15 (2006)
22. Takriti, S., Krasenbrink, B., Wu, L.S.-Y.: Incorporating Fuel Constraints and Electricity Spot Prices into the Stochastic Unit Commitment Problem. Operations Research 48(2), 268–280 (2000)
23. Johnson, R.B., Oren, S.S.: Equity and efficiency if unit commitment in competitive electricity markets. Utilities Policy 6(1), 9–19 (1997)
24. ERCOT – Generation, `http://www.ercot.com/gridinfo/generation/`

25. Cook, S.R., Gelman, A., Rubin, D.B.: Validation of software for Bayesian models using posterior quantiles. Journal of Computational and Graphical Statistics 15(3), 675–692 (2006)
26. Wu, T., Rothleder, M., Alaywan, Z., Papalexopoulos, A.D.: Pricing energy and ancillary services in integrated market systems by an optimal power flow. IEEE Transactions on Power Systems 19(1), 339–347 (2004)
27. Lesieutre, B.C., Oh, H., Thomas, R.J., Donde, V.: Identification of market power in large-scale electric energy markets. In: Proceedings of the 39th Hawaii International Conference on Systems Science (January 2006)
28. Meibom, P.: Stochastic Optimization Model to Study the Operational Impacts of High Wind Penetrations in Ireland. IEEE Transaction on Power Systems 26(3) (August 2011)
29. Morgan, E.C.: Probability distributions for offshore wind speeds. Energy Conversion and Management 52 (2011)
30. Wood, A.J.: Power Generation, Operation and Control. Wiley (1996)
31. Soliman, S.A.: Modern Optimization Techniques with Applications in Electric Power Systems. Springer (2011)
32. McCalley, J.D.: Lecture on Unit Commitment. Personal Collection of EE553, Iowa State University, IA (2012)
33. Drgrib (NDFD GRIB2 Decoder), `http://www.nws.noaa.gov/mdl/degrib/txtview.php?file=tkdegrib.txt&dir=base`
34. Hodge, B.: Wind Power Forecasting Error Distributions over Multiple Timescales (2011), `http://www.nrel.gov/docs/fy11osti/50614.pdf` (retrieved)
35. Generation (2014), `http://www.ercot.com/gridinfo/generation/` (retrieved)
36. European Centre for Medium-Range Weather Forecasts, `http://data-portal.ecmwf.int/data/d/interim_daily/` (retrieved)
37. Liggett, T.M.: Interacting Particle Systems. Springer, New York (1985)
38. Constantinecu, E.M., Zavala, V.M., Rocklin, M., Lee, S., Anitescu, M.: A computational framework for uncertainty quantification and stochastic optimization in unit commitment with wind power generation. IEEE Transactions on Power Systems 26(1), 431–441 (2011)
39. Ruiz, P.A., Philbrick, C.R., Zak, E., Cheung, K.W., Sauer, P.W.: Uncertainty management in the unit commitment problem. IEEE Transactions on Power Systems 24(2), 642–651 (2009)
40. Basu, S., Choudhury, T., Clarkson, B., Pentland, A.: Learning human interactions with the influence model. In: Neural Information Processing Systems (2001)
41. Khaitan, S.K., McCalley, J.D.: Cyber physical system approach for design of power grids: A survey. In: 2013 IEEE Power and Energy Society (PES). IEEE (2013)
42. Khaitan, S.K., McCalley, J.D.: Design techniques and applications of cyberphysical systems: a survey. IEEE Systems Journal (99), 1–16 (2014)

Cyber Security of Smart Grid Communications: Risk Analysis and Experimental Testing

Giovanna Dondossola and Roberta Terruggia

Abstract. The book chapter deals with the cyber security evaluation of active distribution grids characterized by a high level penetration of renewable Distributed Energy Resources (DER). This evolution of the energy infrastructure introduces significant changes in the control and communication functions needed for meeting the technical, security and quality requirements during the grid operation. The risk analysis and treatment of fully controllable smart grid energy infrastructures require effective evaluation tools and scalable security measures. The analysis focuses on a Voltage Control function in medium voltage grids addressing voltage stability of the power grid when a consistent amount of distributed renewable sources are connected. For this reason the chapter analyses the most relevant security scenarios of an ICT (Information and Communication Technology) architecture implementing this control application. The risk level resulting from the analysis are linked to security requirements and standard measures whose deployment in real scale infrastructures requires the security testing of application architectures. The chapter presents an experimental environment for the security testing and evaluation of voltage control communications. This includes the test bed set up, the test cases and the evaluation framework to be used for measuring the attack effects on substation-DER communications and verifying the mitigation capability of standard security measures.

1 Introduction

The evolution of the energy markets all over the world is imposing a significant enhancement in the control and operation infrastructures of electrical distribution grids. The new infrastructures of the energy grids are characterized by more complex system topologies, where high, medium and small size generators, loads and storage devices are connected at the different voltage levels of the electrical

Giovanna Dondossola · Roberta Terruggia
Ricerca Sistema Energetico RSE SpA, Via Rubattino 54 20134 Milan, Italy

S.K. Khaitan, J.D. McCalley, and C.-C. Liu (eds.), *Cyber Physical Systems Approach to Smart Electric Power Grid*, Power Systems, DOI: 10.1007/978-3-662-45928-7_7

networks. The SCADA (Supervisory Control And Data Acquisition), automation, control and protection systems currently deployed in bulk generation plants, transmission and distribution substations and their field devices have to be enhanced with new control functionalities needed for the technical and economic optimization of the grid operation. The underlying ICT architectures of smart grids are necessarily based on heterogeneous systems and third party telecommunication services allowing to economically and efficiently support the communication needs of multiple actors. In this power grid technological evolution the role of ICT is becoming increasingly relevant and the power system community is now aware that the security and efficiency of the power delivery depend on the resilience of the intrinsically vulnerable electronic technologies. The need of managing the ICT risks mainly motivates the high priority given to the cyber security aspects by all smart grid research roadmaps.

The methods and technologies developed for securing mass IT applications and telecommunication services are a starting ground, but new solutions specific for the power environment are needed. For this reason the state of the art of smart grid security is progressing towards the development of standard methodologies and measures declined in the context of the smart grid functions. The needs perceived from the smart grid stakeholders fall in the areas of risk assessment, security requirement definition and measure evaluation. The objective of the chapter is to exemplify, through a representative application of the power grid evolution, the correlations between the risk analysis and the evaluation of the security requirements.

The structure of the chapter is as follows: Section 2 provides an overview of related works in cyber security of smart grids; Section 3 introduces the key elements of the Voltage Control use case ICT architecture. Section 4 presents the benchmark grid and the security scenarios addressed in the analysis. Section 5 describes the use of a qualitative approach for the risk analysis of the Voltage Control scenarios. Section 6 identifies the security requirements and technical standards that need to be implemented considering the risk levels analysis. Section 7 focuses on the architecture and key features of the experimental environment implemented taking into account the Voltage Control scenarios and the related security measures. Finally Section 8 provides some highlights about future research directions.

2 Related Works

In the last five years several European and International initiatives related to the standardization of the energy grid technologies ([1],[2],[3],[4]) stressed the importance of cyber security in the smart grid context.

A first significant reference about the security of smart grid architectures and communication interfaces is the NIST report [5].

Two years later the European Smart Grid Coordination Group issued the First Set of Standards report [6] mapping the information, communication and security standards over the Smart Grid Architecture Model (SGAM), the Use Cases report

[7] about the use case approach, and the Smart Grid Information Security (SGIS) report [8] suggesting, among other things, a qualitative method for the risk analysis of the use cases.

Also the academic research moves towards these themes for example with the development of specific solutions as the CySeMoL (Cyber Security Modeling Language) tool for evaluating the security of SCADA architectures [9]. Within the Cigré working group D2.31 a first exercise on the application of CySeMoL for the estimation of attack probabilities to the Voltage Control architecture has been published [10].

In [11],[12] a Petri's net tool is used as an alternative approach of modeling the effects of cyber attacks to a SCADA architecture, in combination with a power flow simulator estimating the impact of some attack scenarios on the power infrastructure. More detailed attack models may be developed by means of the ADVISE tool [13], recently used for studying the attack probabilities to energy management systems in the customer domain [14].

A survey of the cyber physical system approach for design of power grids is presented in [23] and [24].

The communication security of smart grids is also addressed by several European projects. In the SmartC2Net project the use case methodology has been adopted for the description of the control functions and the communication requirements of a set of reference use cases and an integrated architecture view has been presented [15]. This chapter provides an insight of the communication security of the SmartC2Net Voltage Control use case. The SGIS method has been applied to this use case and the NIST security requirements have been linked with its most critical information assets. A subset of the smart grid standards are related to the tests described in this chapter, including standards for data communication with DERs [16], [17], towards control centers [18] and cyber security [19].

3 Voltage Control - ICT Architecture

The evolution from the fixed balance power grid to a dynamic balance smart grid involves to extend the control functionalities for targeting new balancing scenarios. The Voltage Control function and its related communications become important aspects because the connection of DERs to medium voltage grids can influence the status of the whole power grid affecting the capacity of the DSO (Distribution System Operator) to comply with the contracted terms with the TSO (Transmission System Operator) and directly the quality of service of their neighbor grids. This difficulty not only could be transferred into charges to the DSO, but it may also impact on the TSO operation because the scheduled voltages at grid nodes could not be observed and voltage stability problems cannot be managed properly. In order to maintain stable voltages in the distribution grids the Voltage Control (VC) function is introduced. The main functionality is to monitor the grid status from field measurements and to compute optimized set points for DERs, flexible loads and power equipment deployed in HV/MV substations. Figure 1 shows the input/output schema of the VC function.

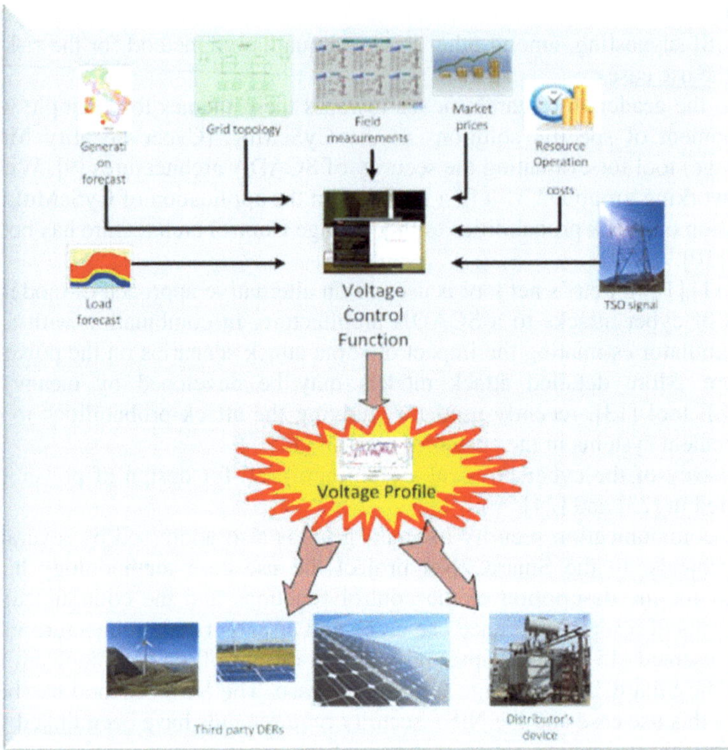

Fig. 1 Voltage Control - function

Since the DER may be outside the control of the utility and the optimization algorithm requires inputs from actors external to the DSO, the resulting overall architecture span over a multi-domain space interconnecting a variety of ICT entities and network segments. Figure 2 introduces the actors/sub-functions of the VC use case and shows how they can be mapped over the Function layer of the Smart Grid Architecture Model (SGAM) [8]. The actors of the use case are placed into the Transmission, Distribution and DER domains in the horizontal axis. The zones in the vertical axis vary from the Market zone of the Aggregator to the Field zone of the control functions of the OLTC (On Load Tap Changer), Capacitor bank, DER and Flexible Load. In the middle we have the Generation and Load Forecast functions placed in the cell Enterprise zone/Distribution domain. The EMS (Energy Management System) and DMS (Distribution Management System) control functions are in the Operation zone hosting all the active grid operation functions. The Substation Automation System (SAS) and the Medium Voltage Grid Controller (MVGC) functions are located in the Station zone.

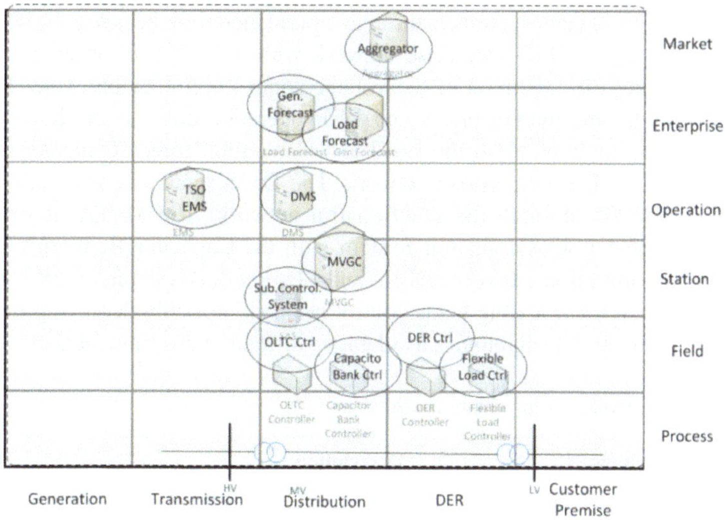

Fig. 2 Voltage Control - SGAM mapping

In order to analyze the communication and security aspects of this use case, we need to highlight the main interactions between the actors involved. The main control and communication components are presented in Figure 3. The VC function is performed by the MVGC on a node of a HV/MV substation control network. In order to compute an optimized voltage profile the algorithm involves communications through components inside the DSO area, but also exchanges of information with systems outside the DSO domain.

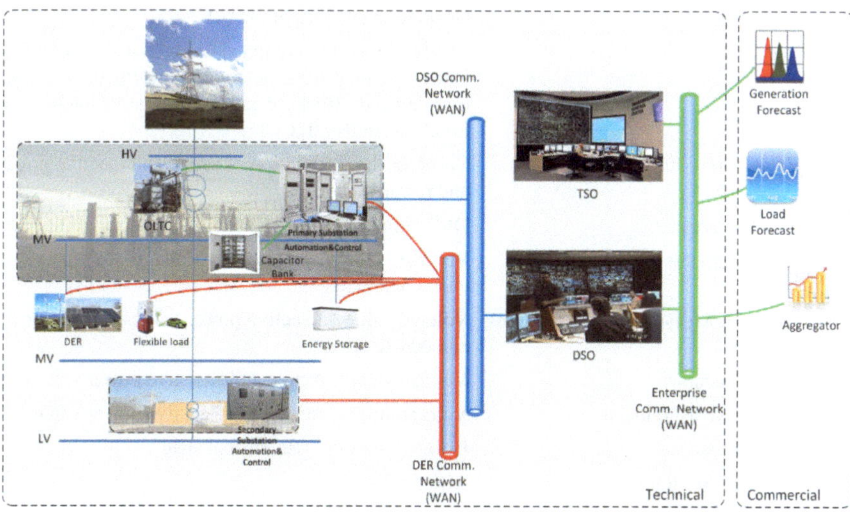

Fig. 3 Voltage Control - architecture

The TSO control center interacts through a permanent link between the TSO control network and the DSO enterprise network with the DMS in order to send the signals triggering the execution of the voltage control optimization cycle. The Aggregator provides the market prices and DER operation costs to the DMS via the DSO enterprise network. Also the Load and Generation forecast interact with the DMS through the DSO enterprise network. The DMS sends /receives information to/from the MVGC through the DSO control network. The MVGC is connected through the Substation Automation System with the Capacitor Bank and with the OLTC in the substation network. DERs and Flexible loads communicate with the MVGC via the DER /Flexible loads control network, possibly deploying heterogeneous communication technologies available in different geographical areas.

Table 1 Voltage Control - information assets

Information Exchanged	Description
Grid Topologies	Information regarding the characteristics of the grid elements (substations, loads, generators and lines). Configuration changes of the controlled grid (grid topology reconfigurations, new DER/load installations)
Weather Forecasts	Weather forecast, weather data
TSO Signals	Signals influencing the execution of the voltage control algorithm (e.g. changing optimization criteria or overriding commands): Voltage setting, Reactive Power setting, Automatic Voltage Regulator inclusion/exclusion
Generation Forecasts	Active power production plan on an hour base for a time horizon of 36 hours (36 values of active power). Generation coefficient $0<C<1$
Load Forecasts	The future load is predicted on the basis of reference loads (seasonal patterns), stochastic fluctuations, active demand effects, weather forecast, calendar day. Load coefficient $0<C<1$
Energy/Ancillary costs	Costs for the modulation of active and reactive power and reward schemes
Load/DER Features	DER Nominal Power, Capability, Controllability
OLTC Measurements and States	Voltage values, Automatic Voltage Regulator included/excluded
Capacitor Bank Measurements and States	Voltage values, Reactive power values, Capacitor included/excluded
DER Measurements	Voltage values, Active and Reactive power values
Flexible Load Measurements	Voltage values, Active and Reactive power values
Grid State Estimations	Estimation of the grid current state
Capacitor Bank Set Points	ΔQ +/-; ΔV +/-
OLTC Set points	ΔV +/-
DER Set points	ΔP +/-; ΔQ +/-
Flexible Load Set points	ΔP +/-; ΔQ +/-

Table 1 and Table 2 report the list of the basic VC information assets and the main steps of the control loop, respectively. The full template reporting a step by step analysis of the VC use case control loop is available in [15].

Table 2 Voltage Control - control steps

Step Name	Primary Actor	Triggering Event	Pre-Condition	Post-Condition
Generation Forecast Estimation	Generation forecast	Periodically	New info available	New generation forecast available
Information acquisition	DMS	Periodically / Asynchronous	TSO signal or new info	Info integrated in the data base
Forward of Forecast data	DMS	Periodically /Asynchronous	DMS receives new data	MVGC obtains input for the control algorithm
Grid measurement dispatch	Third party DER / Distributor's device	Periodically	Field dispatches new measurements	MVGC obtains new measurements
Forward of grid monitoring data	MVGC	Periodically	SAS has new SCADA and DER monitoring data	DMS receives new monitoring data
Execution of control voltage algorithm	MVGC	Values out of range	The state is not acceptable	Computation of new setpoints
Set Setpoints	SAS / MVGC	New setpoint	New setpoints computed	Devices change their settings

4 Benchmark Grid and Security Scenarios

The architecture details of the VC use case represent a key starting point in order to study the risk levels, but a real grid scale has to be set to analyze the overall system exposure to cyber threats and their global impact on the whole infrastructure. The definition of a benchmark grid for the cyber risk analysis is given in Table 3.

According to the Italian territorial configuration, the geographical area of the benchmark grid covers 19 regions served by thousands of primary substations controlled by 29 centers. As for the RES penetration a realistic 2020 scenario [20] installing 40GW of renewables in the Italian medium voltage grids is used in the analysis. The 2020 scenario will require the extension of the grid through the installation of new substations: the estimated number of substations per center is shown in Figure 4.

Table 3 Benchmark grid - cyber risk analysis

Parameter	Description
Area	Geographical extension of the area covered by the grid service: multination, nation, region, province, city
DER penetration	Total amount of Power from Renewable Energy Sources (RES)
Regulation	Applicable regulations
DER size	Installed DER capacity
Grid size	Installed grid capacity
Grid Topology	# HV/MV substations
	# MV loads
	# MV/LV substations
	# generators
	# storage devices
	# MV lines
Telecontrol Network Topology	# Control centers
	# substation links per center
	# of DER links per substation
Population density	# of people in the area

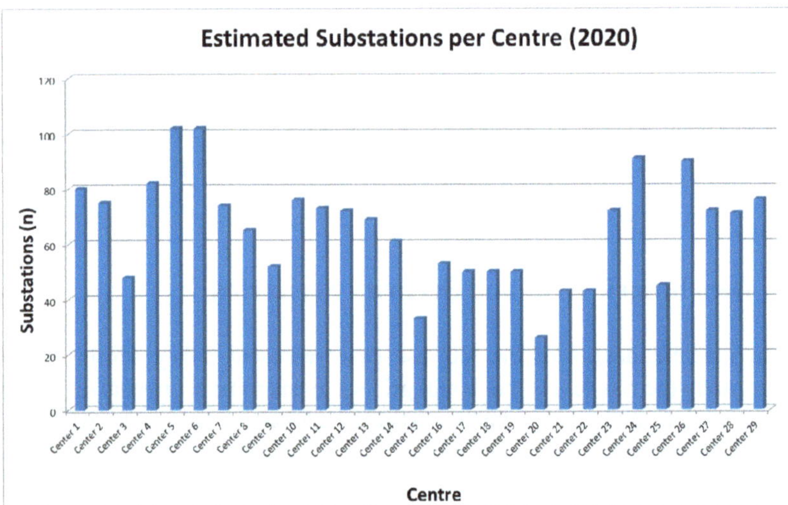

Fig. 4 Benchmark grid - telecontrol topology

According to the Italian grid code and the related connection rules [21], the size of renewable generators that have to be mandatory connected to the medium voltages falls within the power range of [0.2, 6]MW. Depending upon climate conditions in the Italian regions, the targeted amount of renewable power varies according to the estimated distribution in Figure 5.

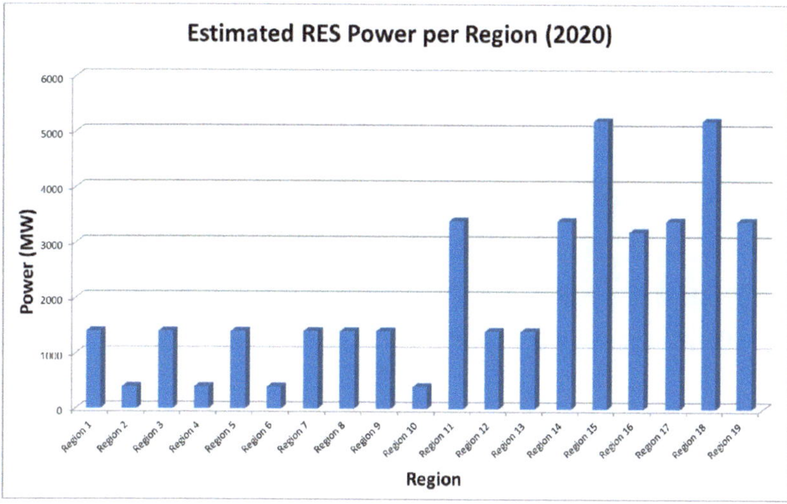

Fig. 5 Benchmark grid – regional RES distribution

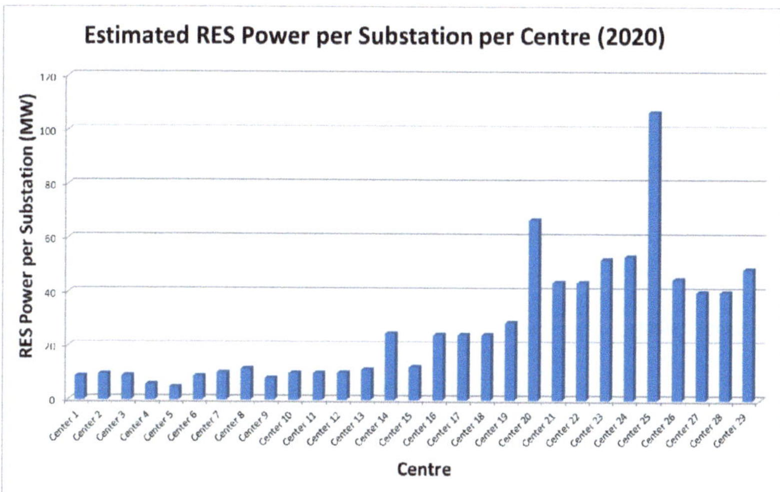

Fig. 6 Benchmark grid - RES Distribution at Substation Level

The relative distribution of population per center, calculated by currently registered population in the area, is shown in Figure 7.

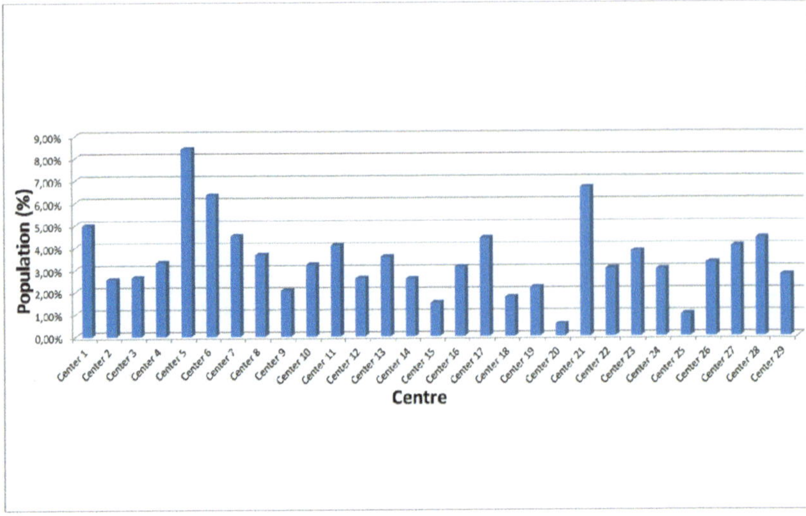

Fig. 7 Benchmark grid - population distribution

The effect of attacks to the telecontrol network of the benchmark grid depends upon the security (i.e. integrity, availability, confidentiality and non-repudiation) scenarios in the scope of the analysis, where a given security scenario is characterized by the parameters in Table 4.

Table 4 Security scenarios

Parameter	Description	Voltage Control Scenario
Attack Target	Network interface targeted by the attack	DER interfaces, substation2DER interfaces, substation2center interfaces, center2substation interfaces
Attack effect	Loss of messages (availability); insertion of fake messages (integrity)	loss of inputs to the VC algorithm, loss of output set points fake inputs to the VC algorithm, fake output set points, faked monitoring data
Attack extension	# network interfaces under attack	# DER networks # substation networks # center networks
Data frequency	Periodic / Asynchronous	periodic and asynchronous VC inputs/outputs

By instantiating the security space on the specific VC network topology and information assets, our use case security space (Figure 8) covers the security scenarios reported in the third column of Table 4.

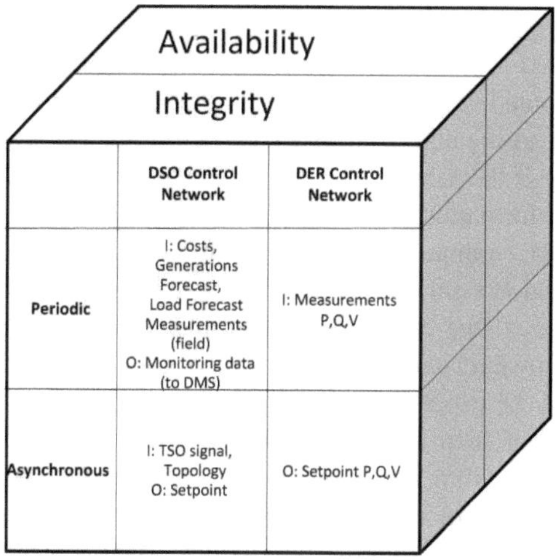

Fig. 8 Voltage Control - security space

5 Risk Analysis - A Qualitative Approach

The risk analysis of the Voltage Control ICT architecture is based on the SGIS working group of the Smart Grid Coordination Group by CEN-CENELEC-ETSI in charge of the European Mandate E/490 on smart grid standardization, and uses this method to derive qualitative Security Levels of voltage control information assets. According to the SGIS risk analysis process [8] the evaluation of the risk levels of a given smart grid use case goes through the application of the impact and threat likelihood analysis to the scenarios of the use case information assets in the security space. A risk level for each information asset/security scenario can be obtained combining the related impact and likelihood levels.

5.1 Impact Analysis

The impact of attacks is evaluated through the five-scale impact matrix in Figure 9 defining the levels of operational, financial and additional risks. From the application of the SGIS impact levels to the benchmark grid, the operational Risk Impact Levels can be assigned to the information assets/security scenarios of the VC use case. Let's evaluate the operational risks starting from the "Energy Supply" risk category (leftmost column in Figure 9). The focus is on the extreme case analysis, i.e. on those regional grids with maximum DER penetration (i.e. regions 15 and 18 in Figure 5), highest power demand and integrity scenarios introducing fake messages causing loss of loads, generators disconnections or substation trips. The loss of energy supply varies with the attack target and the damaged information assets. In the case of substation2DER interface attacks, where the setpoint information is compromised or the DER measurements are perturbed, the loss may be up to

100MW (yellow circle in the picture). The worst case considers that more than one DER connected to a specific substation is out of control (the sent setpoints differ from the computed setpoints or the measurements regarding DER status are not the real ones and so the algorithm is based on wrong values (see Figure 8). More serious is the case if the substation2center interface is attacked: in this case the entire substation (information flow) domain may be compromised and the impact may be up to 1 GW (orange circle) because the information impacting on the substation capacity, and not only specific DER capacities, is perturbed or missing. The criticality increases if the center2substation interface is under attack: in this case the information flows related to a wider grid area may be compromised and several substations may be tripped, amounting an impact value up to 6GW (red circle). As for the impact of such attack effects on the registered population, the voltage control use case falls into the medium level, while the impact on critical infrastructures may be high or critical, depending on the presence of essential or national infrastructures in the sub-regions under attack. In order to estimate these impact levels we have considered the extreme case achieving the values presented in Figure 9.

RISK IMPACT LEVELS	OPERATIONAL (availability)				LEGAL		HUMAN	REPUTATION	FINANCIAL
	Energy supply (Watt)	Energy flow (Watt/hour)	Population	Infrastructures	Data protection	other laws & regulations			
HIGHLY CRITICAL	regional grids from 10GW	from 10 GW/h	from 50% population in a country or from 25% in several countries	international critical infrastructures affected	not defined	company closure or collateral disruptions	direct and collateral deaths in several countries	permanent loss of trust affecting all corporation	Thirth party affected
CRITICAL	national grids from 1 GW to 10GW	from 1 GW/h to 10GW/h	from 25% to 50% population size affected	national critical infrastructures affected	not defined	temporary disruption of activities	direct and collateral deaths in a country	permanent loss of trust in a country	>=50% EBITDA
HIGH	city grids from 100MW to 1GW	from 100MW/h to 1GW/h	from 10% to 25% population size affected	essential infrastructures affected	unauthorized disclosure or modification of sensitive data	prison	direct deaths in a country	temporary loss of trust in a country	<50% EBITDA
MEDIUM	neighborhood grids from 10MW to 100MW	from 10MW/h to 100MW/h	from 2% to 10% population size affected	complimentary infrastructures affected	unauthorized disclosure or modification of personal data	fines	seriously injured or discapacity	temporary and local loss or trust	<33% EBITDA
LOW	home or building networks under 10 MW	under 10MW/h	under 2% population size affected in a country	no complimentary infrastructures	no personal nor sensitive data involved	warnings	minor accidents	short time & scope (warnings)	<1% EBITDA

MEASUREMENT CATEGORIES

Fig. 9 Voltage Control - SGIS Impact Levels

5.2 Threat Analysis

The likelihood of threat/attack occurrences represents the other key indicator to be estimated in order to compute the risk level. The level of likelihood is evaluated for every information asset considering parameters such as threat sources/actors, their motivations and capabilities to achieve an attack effect through compromise methods and in presence of essential security counter-measures.

The threat source represents the entity (person or organization) that wants to break the security barriers for obtaining benefits of some type. Examples of threat sources are listed in Table 5.

Table 5 Threat sources

Disaffected or dishonest employees
Foreign Intelligence Services
Amateur or professional hackers
Virus and other malware writers
Vandals
Thieves
Terrorists
Investigative journalists
Commercial competitors (i.e. industrial espionage)
Political pressure groups/activists
Organized criminal groups

Considering their different levels of capability (from formidable to very little) and priority (from focused (very high) to indifferent (very low)) it is possible to realize an identikit of the possible threat sources. In order to reach his/her scope, the threat source "uses" a threat actor that materially performs the attack. Threat actors are entities potentially having capability, opportunity and motivation to attack an asset. The different capabilities of threat actors can be used in order to delineate the possible threat actor profiles. In some cases the threat source and the threat actor could coincide and be the same entity.

Table 6 Threat Actors

Threat Group	Profile
System and Service User	Privileged User Normal User Service Consumer Shared Service Subscriber
Actors with business or network connection with the assets	Information Exchange Partner Service Provider
Actors indirectly connected to an asset through directly connected actors	
Actors having access to hardware and software before the asset commissions or are those that are responsible for implementation, configuration or management of the asset	Supplier Handler
Actors having physical access to the asset	Privileged User Normal User Bystander Person Within Range Physical Intruder

The threat actor is relative to an asset. A threat actor with particular privileges respect to an asset might not have the same privileges, and so not be able to attack also the other assets. The threat actors can be grouped considering the relationship with respect to a specific asset and similar applicable compromise methods. The threat actors have authorized logical access to the different assets and any service they provide. Table 6 includes sample groups of threat actors and associated profiles.

Coming back to the Voltage Control use case, possible threat sources should be identified by correlating investigative data. For now we assume that they may be employees, industrial espionage agents, vandals, cyber hackers, viruses and worms, thieves and terrorists. Both the identification of threat actors and the evaluation of their threat capabilities to compromise the information assets may be driven by the analysis and management of roles in the control application. By focusing on the DSO domain of the VC use case, we may have several user/service roles for grid operation and ICT maintenance that could become threat actors. Examples of possible user roles are: local power operator, remote power operator, normal ICT user, ICT administrator, ICT security administrator. Examples of possible service roles are: DER controller, MVGC, SAS and DMS. Each role defines a trust level and it is used to take authorization decision. For this reason an authentication mechanism is associated to each role. The access control matrix assigns to each data type for each (user or service) role specific rights (read, write, update, delete). In the VC use case, for example, only MVGC has the right to write set point to DER. Furthermore an important aspect to take into account is the number of users/services for each role.

A further step of the threat analysis considers architecture characteristics such as types of services running on the components, technologies and implementation aspects. In the VC use case the key components are the control IEDs (Intelligent Electronic Devices), the servers and the routers at different DSO subdomains such as ICT maintenance center, control center or substation. For each of them it is necessary to consider the configuration parameters of software layers/modules, for example the operating systems and protocols used for implementing the communications. In the VC use case we suppose that the servers run a UNIX based Operating System. DER-substation and intra-substation communication uses the standard IEC 61850 over the MMS protocol and for substation-center information flows the IEC 60870-5-104 standard protocol is used. They both are connection-based flows supported by the TCP/IP reliability mechanisms. The VC use case might exploit heterogeneous network technologies. The center-substation links usually deploy IP based wired networks, whereas the substation-DER links might use wired as well wireless networks depending on the geographical coverage of the technology.

Besides these "structural" aspects the knowledge about the control loop behavior, as reported in Table 2, is essential for building effective attack processes whose actual effectiveness also depends on the data frequency. For example the success of DoS (Denial of Service) attacks, such as flooding, buffer overflows and resource exhaustion will be higher on periodic information flows (e.g. measurements and monitoring data) than on asynchronous information flows (e.g. setpoints).

By grouping the VC use case information assets and attack scenarios considering similarity in their parameters, we identify three main categories of assets according to the attack target interfaces and five most relevant attacker profiles.

By applying the SGIS five scale likelihood levels in [8], the analysis described above identifies for the VC use case the threat levels presented in Figure 10.

	substation2DER	substation2centre	centre2substation
Dishonest employee (Admin)	Very High	Very High	Very High
Dishonest employee (normal user)	High	Medium	Medium
Vandal	Very High	High	Low
Hacker	Very High	High	Medium
Terrorist	Medium	Very High	Very High

Fig. 10 Voltage Control - Likelihood Levels

5.3 *Risk Levels*

Figure 11 represents a numerical approach for the calculation of risk levels proposed by SGIS, where the qualitative values of impact and likelihood are summed.

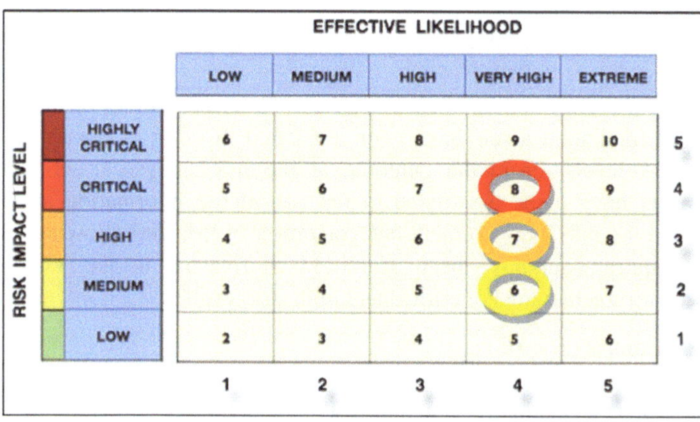

Fig. 11 Voltage Control - risk calculus

These numerical values are mapped through the matrix in Figure 12 where the risk (security) levels are identified.

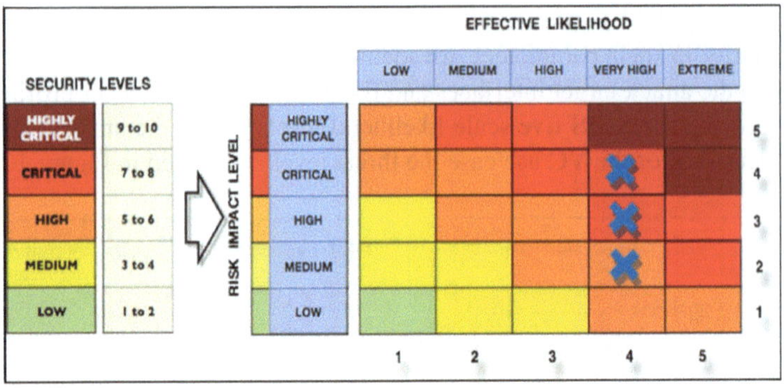

Fig. 12 Voltage Control - Risk Levels

Combing the VC impact levels (Figure 9) with the likelihood levels (Figure 10) by means of the SGIS risk matrix, the High and Critical risk levels are identified for the VC use case, depending on the information assets/security scenarios under consideration. To be noticed that the combination of the impact with the likelihood analysis has increased the need of security protection of substation-DER communications (from a medium impact level to an high risk).

Qualitative approaches as the SGIS toolkit provide only a rough estimation of the risk value for each assets and scenario. In order to obtain more precise evaluations the application of quantitative risk assessment methods is envisaged.

6 From Risk Levels to Security Standards

Considering the information assets and scenarios related to the VC use case, the impact and likelihood levels have been evaluated in order to obtain the corresponding risk levels. From the outcome of the risk analysis a set of security requirements have to be associated to the considered information assets. With reference to the NIST requirement categorization in [5], the following groups of security requirements have been identified as relevant to the VC use case assets/scenarios achieving the critical and high risk levels:

- Access Control (SG.AC)
- Identification and Authentication (SG.IA)
- Smart Grid Information System and Communication Protection (SG.SC)
- Smart Grid Information System and Information Integrity (SG.SI)
- Cryptography and Key Management.

In order to meet the Voltage Control use case security requirements, the list of security measures from technical standards in Table 7 can be selected. To be noticed that the maturity level of the selected standards varies from available international standard to work in progress.

Table 7 Voltage Control - security standards

Standard Type	Standard Reference
Communication protocol security standards	IEC 62351 Parts 3/4/5/6
Network security standards	IEC 62351 Part10
Role-based access control	IEC 62351Part 8
Key and certification management	IEC 62351Part 9
XML security	IEC 62351Part 11
Enabling standard IT security protocols	TLS
	IPSEC
	SNMP
	https
	ssh

Figure 13 depicts where the different parts of IEC 62351 have to be applied according to the communication protocols of the VC use case. Depending on the risk levels of the related information assets, more or less costly implementations of the security measures, i.e. for the key management and the grid/network monitoring, will be deployed.

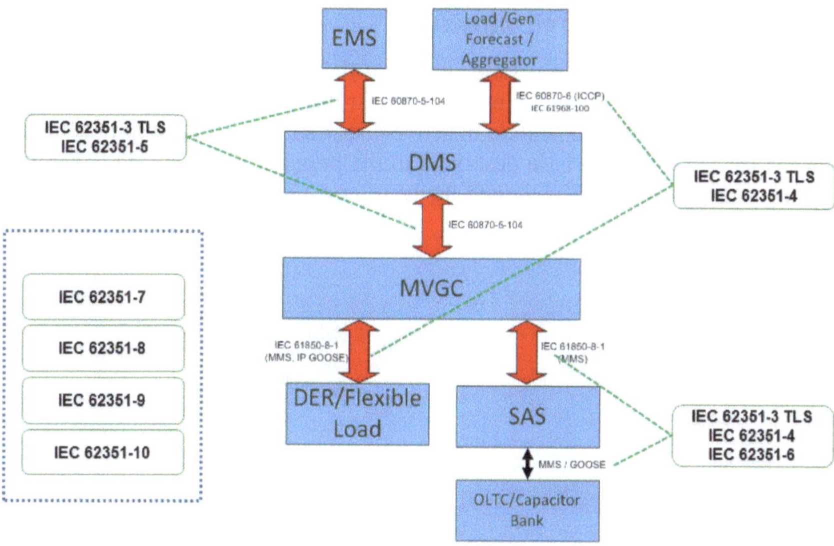

Fig. 13 Voltage Control - mapping of IEC 62351 parts

Figure 14 summarizes two examples of the overall security analysis process considering a couple of the assets identified during the analysis.

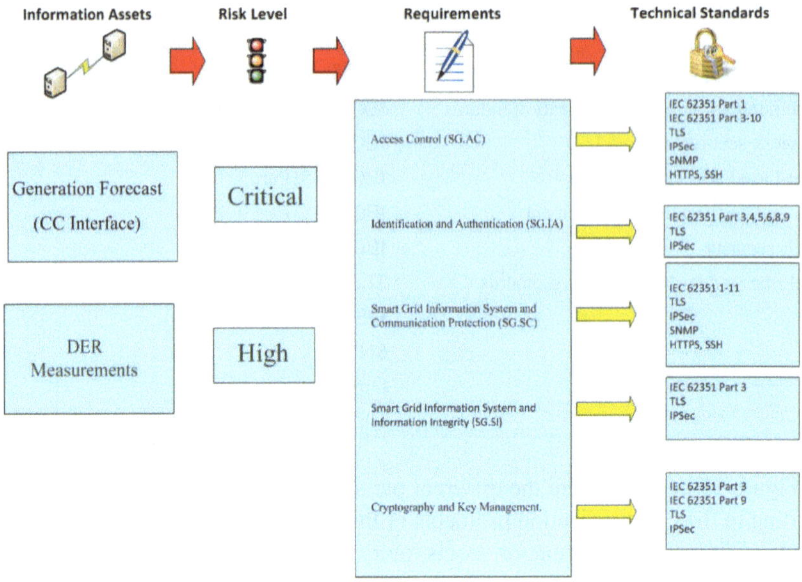

Fig. 14 Schema of the approach

7 Experimental Environment

In order to collect precise measurements about the deployment of security measures in control applications an experimental test bed is implemented focusing on the Voltage Control communication in active distribution grids. The test bed architecture is based on the use case described in the previous sections and covers the components and networks highlighted by the red oval in Figure 15.

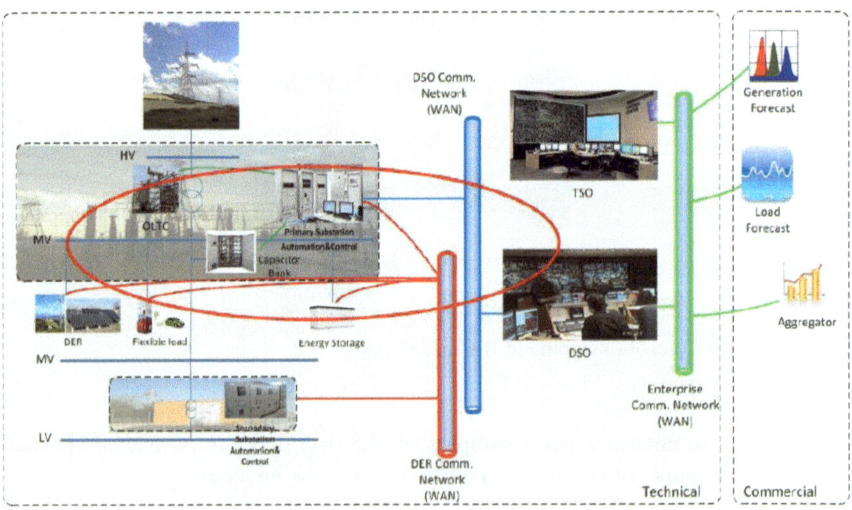

Fig. 15 Test bed – use case coverage

A schematic view of the test bed ICT architecture is presented in Figure 16, whose components and networks are described in Table 8.

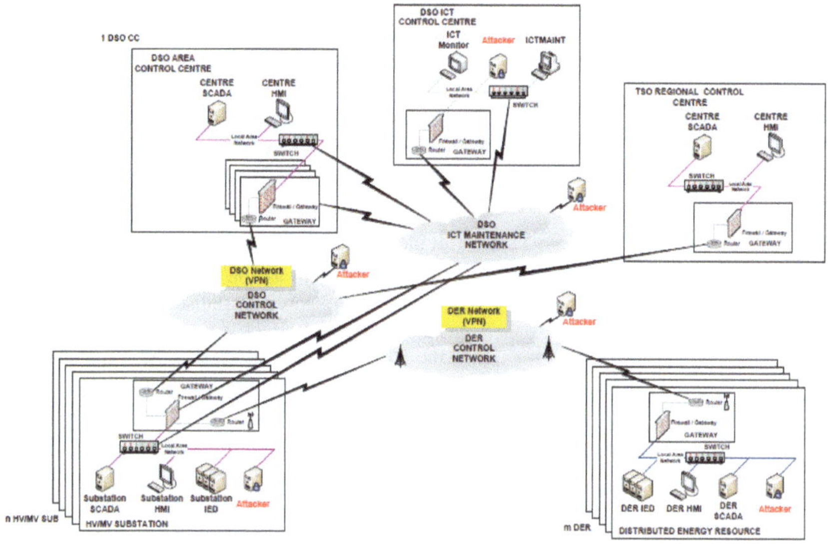

Fig. 16 Test bed – architecture

From the outcome of the VC risk/security analysis described in the previous section (risk levels, security requirements and technical standards), the test bed has given priority to the implementation of the Part 3 of the IEC 62351 security standard for the substation/DER communications based on the IEC 61850/MMS protocol. Part 3 is dedicated to describe the TLS (Transport Layer Security) implementation aspects that may be included in power system information exchanges in order to preserve the integrity and the authentication of the messages. A further priority is given to the integration in the test bed of the functions addressed by the new edition of the Part 7, currently still under development. Part 7 is related to the network and system management performed through the identification of specific data objects used to monitor and control end systems and networks. The SNMP (Simple Network Management Protocol) protocol is used in the test bed for the implementation of the monitoring data objects relevant for the VC security scenarios.

In order to measure some security key performance indicators several test runs may be performed collecting experimental data of VC communications. Table 9 describes the types of tests to be executed in order to verify the communication behavior during different operating/security/attack conditions.

Table 8 Test bed - components and networks

Component/Network	Description
DSO Control Center	It remotely controls a partition of the distribution grid. Each DSO CC interacts with different HV/MV substations where the MVC function is executed
HV/MV Substation	It includes automation, communication, SCADA and Operator HMI functions. Each substation may control different DER sites
TSO Center	It supervises critical regions of a transmission grid
DER site	It includes large DER connected to MV grids
ICT maintenance control center	It remotely controls the ICT components of DSO networks. Collects data statistics related to network monitoring and attack successfulness measuring the effects of cyber attacks to the communications involved in grid operation and maintenance
Attacker	It performs malicious actions. It may be placed inside the DSO ICT control center, substations, DER sites and corresponding control networks
DSO control network	It connects the DSO control center with the HV/MV substations. It uses a dedicated service on a shared, possibly third party, infrastructure. The protocol IEC 60870-5-104 is used for these communication flows
DER control network	It connects each DSO substation with multiple third party DER sites located in different geographical areas possibly deploying heterogeneous wired/wireless communication technologies. The communication uses the MMS profile of the IEC 61850 standard
ICT maintenance network	It is used for the configuration and management of the control and communication devices deployed in the DSO control center and HV/MV substations
Local Area Network	Each site deploys its own Local Area Network for the interactions among the local components

Table 9 Tests

Test Case	Description
Normal	Tests verifying the VC communications with essential security measures in absence of ICT faults/attacks
Secured	Tests verifying the VC communications deploying enhanced security measures in absence of ICT faults/attacks
Attack	Tests verifying the VC communications with varied degrees of security measures in presence of ICT attacks

The test cases may be applied to different information flows involving the Control and Monitoring of the power grid and of the ICT network. More in specific, we address the control center –Substation communications, the Substation – DER communications and the ICT communications.
Considering the attack scenarios described in the risk analysis section the following attack processes are experimented:

- DoS Attacks to DER (gateways). The traffic between DER and Voltage Controller is perturbed and some DER measurements are not able to reach the Voltage Controller.
- DoS Attacks to Substation (gateways). The traffic between the Voltage Controller and the DMS is perturbed; some DER and SCADA measurements are not able to reach the DMS.
- Fake DER setpoints. Either an (additional) fake setpoint is sent to DER, or a legal setpoint is intercepted and modified with wrong set point values
- Fake TSO signals. A fake TSO signal is sent to the Voltage Controller.

For each test case a set of tests may be performed and the results compared. Quantitative requirements related to network measurements may be verified as latency, bandwidth and packet loss. Grid related requirements may also be evaluated such as # of DER affected by the attack, # of Substation, amount of power delivered and power quality.

7.1 Test Analysis: Normal Test Case versus Secured Test Case

In this subsection an example of test performed and results obtained are presented. They address the DER – Primary Substation communications for the exchange of the DER measurements and setpoints. In our tests we assume that the DER emits the measurements periodically every 2 seconds and the MVGC sends the setpoints every 30 seconds (this information flow is mostly sent in asynchronous mode, but in order to obtain comparable results in this test we consider it as a periodic one).

We performed normal and secured tests composed by different runs: in the secured tests we protected the communication by the use of the TLS protocol. The evaluated communication measures are presented in Table 10.

In Table 11 we compare the results obtained in the two test cases. The overhead brought by TLS on the different metrics can be seen in the table: the results show that the inclusion of the TLS causes the increase of the time for each single communication phase. In the Handshake Time we have an extra time of 0.03137 sec for the TLS handshake. We can conclude that the total time for the initial handshake and session phases is 0.141333 seconds without TLS and 0.176704 including TLS security which means an overhead of 0.035371 seconds corresponding to an increment of 25% of the total time. Also the measurement and setpoint communications are perturbed by the introduction of the TLS, but not in a critical way.

Table 10 Evaluation measures

Measures	Description
Handshake Time	Time interval needed to create the connection at different stack levels
RTT (Round Trip Time) -Measurements	Time interval between the output of a Measurement and the reception of the corresponding TCP ack by the DER
RTT-Setpoint	Time interval between the output of a setpoint request and the reception of the corresponding TCP ack by the MVGC
Inter-Measurements Time	Time interval between each two consecutive Measurements
Inter-Setpoint Time	Time interval between each two consecutive setpoints

Table 11 Test results

Test Case	Metrics (time in seconds)				
	Handshake Time	*Inter-Measurements Time*	*RTT-Measurements*	*Inter-Setpoint Time*	*RTT-Setpoint*
Normal	0.141333	2.0105	0.0000981	30.0637	0.00111
Secured	0.176704	2.0105	0.0000992	31.0588	0.00117

8 Conclusions and Future Work

In the research context about smart grid cyber security the chapter addressed the perceived need of tools and measures mitigating the risks originated by intrinsically vulnerable ICT infrastructures. In order to estimate the SGIS impact and likelihood levels the chapter includes a study of the Voltage Control use case detailed ICT architecture as well as benchmark grid data and attack scenarios. Through their application to the use case, the key steps of the security analysis process have been performed to illustrate the parameters and the outcome of the risk analysis and their links with the security requirements and ongoing standards. The value of security testing of control scenarios is emphasized by detailing the test performed using a Voltage Control experimental architecture.

The results obtained by the experimental activity will be used as inputs for more comprehensive analysis based on simulation and analytic modeling [22]. The experimental measures will allow to test the accuracy of the models and the model based evaluations will calculate the key performance indicators scaling the addressed scenarios up to the benchmark grid.

Acknowledgments. The research leading to these results has received funding from both the EU 7th Framework Program (FP7/20072013) under grant agreement no 318023 for the SmartC2Net project, and by the Ministry of Economic Development with the Research Fund for the Italian Electrical System for the project "Distributed Generation and Active Networks".

References

[1] IEC Smart Grid Standardization RoadMap. SMB Smart Grid Strategic Group SG3, Edition 1.0 (2010)

[2] EPRI Smart Grid Resource Center (August 2010), `http://www.smartgrid.epri.com/`

[3] Smart Grid Mandate M/490 EN, Standardization Mandate to European Standardisation Organisations (ESOs) to support European Smart Grid deployment (March 2011), `http://ec.europa.eu/energy/gas_electricity/smartgrids/doc/2011_03_01_mandate_m490_en.pdf`

[4] European Technology Platform for the Electricity Networks of the future (2012), `http://www.smartgrids.eu/`

[5] National Institute of Standards and Technologies, The Smart Grid Interoperability Panel Cyber Security Working Group NISTIR 7628 "Guidelines for Smart Grid Cyber Security" (2010)

[6] CEN-CENELEC-ETSI Smart Grid Coordination Group SGCG/M490/B_Smart Grid Report First set of standards Version 2.0 (November 16, 2012)

[7] CEN-CENELEC-ETSI SGCG/M490/E_Smart Grid Use Case Management Process — Use Case Collection, Management, Repository, Analysis and Harmonization (2012)

[8] CEN-CENELEC-ETSI Smart Grid Coordination Group. Smart Grid Information Security (November 2012)

[9] Sommestad, T.: A Framework and theory for cyber security assessment. PhD Thesis in Industrial Information and Control Systems, Royal Institute of Technology, Stockholm (November 2012)

[10] Ekstedt, M., Korman, M., Terruggia, R., Dondossola, G.: Application of a cyber security assessment framework to smart grid architectures. Paper D2-01_11 in the Proceedings of the Cigré Study Committee D2 Information Systems and Telecommunication, 2013 Colloquium, Mysore – Karnataka, India, November 13-15 (2013)

[11] Ten, C.-W., Hong, J., Liu, C.C.: Anomaly Detection for Cybersecurity of the Substations. IEEE Trans. Smart Grid (2011)

[12] Ten, C.-W., Govindarasu, M., Liu, C.C.: Cybersecurity for Critical Infrastructures: Attack and Defense Modeling. IEEE Trans. Systems, Man, and Cybernetics – Part A: Systems and Humans, 853–865 (July 2010)

[13] LeMay, E., Ford, M.D., Keefe, K., Sanders, W.H., Muehrcke, C.: Model-based Security Metrics Using ADversary VIew Security Evaluation (ADVISE). In: Proceedings of the 2011 Eighth International Conference on Quantitative Evaluation of SysTems (QEST 2011), pp. 191–200. IEEE Computer Society, Washington, DC (2011)

[14] Hägerling, C., Kurtz, F., Wietfeld, C., Iacono, D., Daidone, A., Giandomenico, F.: Security Risk Analysis and Evaluation of Integrating Customer Energy Management Systems into Smart Distribution Grids. In: CIRED Workshop 2014 (June 2014)
[15] SmartC2Net European Project, Deliverable D1.1. SmartC2Net Use Cases, Preliminary Architecture and Business Drivers (September 2013), http://www.smartc2net.eu
[16] International Standard IEC 61850-7-420 ed1.0. Communication networks and systems for power utility automation - Part 7-420: Basic communication structure - Distributed energy resources logical nodes, Technical Specification (2009)
[17] International Standard IEC 61850-8-1. Communication networks and systems in substations - Part 8-1: Specific Communication Service Mapping (SCSM) - Mappings to MMS (ISO 9506-1 and ISO 9506-2) and to ISO/IEC 8802-3. International Standard, 2nd edn. (June 2011)
[18] International Standard IEC 60870-5. Telecontrol equipment and systems - Part 5-104: Transmission protocols – Network access for IEC 60870-5-101 using standard transport profiles. International Standard, 2nd edn., Reference Number IEC 60870-5-104(E) (June 2006)
[19] International Standard IEC 62351. Power System Management and associated information exchange - Data and Communication Security – Parts 1-11
[20] Petroni, P.: Smart Grids Operation, automation and protection issues. In: Cired 2012, Lisbon, May 29-30 (2012)
[21] Comitato Elettrotecnico Italiano Norm CEI 0-16. Reference technical rules for the connection of active and passive consumers to the HV and MV electrical networks of distribution Company (2013)
[22] SmartC2Net European Project, Deliverable D5.1. Methodologies Synthesis (September 2013), http://www.smartc2net.eu
[23] Khaitan, S., McCalley, J.: Design Techniques and Applications of Cyber Physical Systems: A Survey. IEEE Systems Journal PP, 1–16 (2014)
[24] Khaitan, S., McCalley, J.: Cyber Physical System Approach for Design of Power Grids: A Survey. In: IEEE PES GM 2013, Vancouver, BC, July 21-25, pp. 1–5 (2013)

Acronym	**Definition**
DER	Distributed Energy Resource
DG	Distributed Generation
DMS	Distribution Management System
DoS	Denial of Service
DSO	Distribution System Operator
EMG	Energy Management Gateway
HV	High Voltage
ICT	Information and Communication Technology
IED	Intelligent Electronic Device
IP	Internet Protocol
LAN	Local Area Network
LV	Low Voltage
MIM	Man In the Middle
MMS	Manufacturing Message Specification

MV	Medium Voltage
MVGC	Medium Voltage Grid Controller
OLTC	On Load Tap Changer
P	Active power
Q	Reactive power
RES	Renewable Energy Sources
SAS	Substation Automation System
SCADA	Supervisory Control And Data Acquisition
SGAM	Smart Grid Architecture Model
SGIS	Smart Grid Information Security
SNMP	Simple Network Management Protocol
TLS	Transport Layer Security
TSO	Transmission System Operator
V	Voltage
VC	Voltage Control
WAN	Wide Area Network

Reliable and Scalable Communication for the Power Grid*

Christopher Zimmer and Frank Mueller

Abstract. Future smart power grids require constant data availability for actuation of control decisions. The job of ensuring the timely arrival of data falls onto the network that connects these intelligent devices. This network needs to be fault tolerant. When nodes, devices or communication links fail along a default route of a message from A to B, the underlying hardware and software layers should ensure that this message will actually be delivered as long as alternative routes exist. Existence and discovery of multi-route pathways is essential in ensuring delivery of critical data.

In this work, we present methods of developing network topologies of smart devices that enable multi-route discovery in an intelligent power grid. This is accomplished through the utilization of software overlays that (1) maintain a digital structure for the physical network and (2) identify new routes in the case of faults. The resulting cyber network structure is scalable, reliable and inexpensive to build by extending existing infrastructure.

1 Introduction

Today's critical infrastructure often governs control decisions of intelligent devices that can have a significant impact on human life, the environment and the economy. Ensuring that the appropriate data is available is crucial for making informed decisions. Such considerations are becoming increasingly important in cyber-physical systems (CPS) that combine computational decision making on the cyber side with

Christopher Zimmer · Frank Mueller
North Carolina State University, Raleigh, NC 27695-8206
e-mail: mueller@cs.ncsu.edu

* This work was supported in part by NSF grants 1329780, 1239246, 0812121 and U.S. Army Research Office (ARO) grant W911NF-08-1-0105 managed by NCSU Secure Open Systems Initiative (SOSI). This is an extended version of a prior conference paper [22].

S.K. Khaitan, J.D. McCalley, and C.-C. Liu (eds.), *Cyber Physical Systems Approach to Smart Electric Power Grid*, Power Systems, DOI: 10.1007/978-3-662-45928-7_8

physical control on the device side, let it be the power grid, medical devices or automotive subsystems.

Conventional embedded systems and a CPS differ in that the later governs physical devices through embedded control in a *networked environment* and has a direct impact on people who rely on such devices. Failure of network equipment in a CPS environment may have a number of impacts, such as

- faulty decisions regarding device malfunctioning,
- incorrect actuation (decisions) due to lack of data,
- system reconfigurations/restart, or
- severe performance degradation with missed deadlines.

In smart (power) grids that rely on commodity communication infrastructure, as one example, these types of failures are expensive and cause inefficiencies. Decisions are being made that affect the real world based on data passed within the network of smart grids. This impact on the real world makes it necessary to improve communication within the smart grid to ensure that the *correct* decisions are made in a *timely* manner .

Assuming correct device behavior, the timeliness requirement falls onto the network that connects these intelligent devices. Failures may occur in today's CPS because of a lack of flexibility in routing decisions. Routing decisions are an important part of networking. Commodity networking equipment often relies on static routing techniques within networks. When there is a failure along a static route, any messages sent along that route will time out and result in communication failure. In these scenarios, many systems will assume points along this route to be out of service. This does not have to be the case.

Networks can be designed to be fault tolerant. When nodes, devices or communication links fail along a default route of a message from A to B, the underlying hardware and software layers should ensure that this message will actually be delivered as long as alternative routes exist. Networks of devices can be configured to contain multiple pathways to connect clusters of nodes in a redundant manner. One can ensure delivery of critical data via different network routes, i.e., multi-route pathways need to exist. A network, upon discovery of a faulty route, then needs to be able to utilize an alternate route.

In conventional networks, the main objective is to maximize throughput. Commodity network equipment is designed to provide high levels of throughput. This design choice runs counter to the needs of an intelligent distributed network required for next-generation CPS infrastructure. For example, in a power grid, the guarantee that a message is delivered is more important than high rates of throughput. Sample tasks the power grid must perform, such as distributed load balancing [4], substantiate this need. Thus, system components need to collaborate intelligently upon a component network failure to accommodate sustained communication needs at all times.

Network failures may conventionally result in message delivery failure. This can be avoided through smart routing technologies that can bypass faulty equipment in modern network topologies. However, such fault tolerance is only feasible in situations where the faulty equipment does not constitute a single point of communication failure. Therefore, it is important to maintain redundant pathways through networks. Another problem with smart routing technology is that in current topologies routers are sparely distributed as their cost is significantly higher than that of switches. Since switches lack routing capabilities, this severely limits the ability of CPS devices to sustain network failures through automatic re-routing over alternate paths.

Contributions: This work develops novel methods for designing network topologies of smart devices that enable multi-route discovery in an intelligent power grid. This is accomplished through the utilization of software overlays that (1) maintain a digital structure for the physical network and (2) identify new routes in the case of faults. To this end, we first present a method of utilizing software network overlays to enable the discovery of additional communication pathways throughout a network. Using abstracted network information, the system is able to react in case of faults and generate new routes through the network in a manner that is transparent to the user by providing a software overlay middleware. In this network, any single node in the system can act as a message-passing agent to dynamically route messages within the network. This paradigm enables us to use inexpensive network devices abundantly within the network and ensure a resilient communication infrastructure at the same time.

The primary aim of this study is to determine appropriate topologies in which to structures the various devices used in Distributed Grid Intelligence (DGI) to insure that the Intelligent Energy Management (IEM) nodes are able to make optimal decisions. By formally creating a network topology in this system we are better able to guarantee critical services that would be delegated through efficient communication of the IEM nodes.

This work further presents a visualization capability to monitor connection states and pathways through the network aimed at helping external entities to understand the states of the network.

Our vision is that the application of this approach in an intelligent power grid will enable IEM and IFM devices to make automated, decentralized decisions and to maintain state of lower-level devices.

2 Related Work

Wauters et al. [20] survey network overlays for computer networks and assess their suitability for smart grids. They identify reliability and cost metrics for different topologies, which are more costly than our work and deliver comparable reliability.

GridStat [18] enables the allocation of node specific redundant pathways for high-level power-grid networks. Our work is orthogonal to GridStat as it is designed for low-level micro grids with switches. We also enable dynamic arbitration over many redundant pathways, which allows for a generic application of redundancy that is resilient to link faults.

Software overlays have also been utilized in the High Performance Computing (HPC) domain, e.g., by Varma et al. [19]. That work focused on structural compression and reduction of data over a radix tree irrespective of physical topology while COMIG focuses on making physical tree topologies resilient via crosslinks.

Commodity tools such as Cacti [3] or Ganglia [14] provide graphical monitors of networked systems and components for conventional computing ensembles. Our distributed live monitoring (DLM) work (Section 8) differs in that it is able to provide the visualization for non-IP devices such as those used in a Zig-Bee platform that are only MAC-addressable. Our DLM interfaces with IP-addressable nodes to detect any MAC-based devices connected to it and displays their current status.

Berthier et al. [7] discuss the impact of cyber network topology on state estimation for power grids using contingency analysis based on an ad-hoc exploration of this topology. Yan et al. [21] model the vulnerability of power network topologies to cascading failures from the security angle. Our work focuses on cyber network connectivity to sustain or even prevent power outages while most of these prior works focus on network security.

Motivated by early considerations about the power grid [5], Nguyen et al. [15] and Huang et al. [10] assess the impact of clustering due to partitioning for power grids by developing a 1-to-1 and an k-n model, respectively, (with k control nodes and n power nodes) and empirically simulate the partitioning characteristics of different topologies. Other models include multiple-to-multiple [16] and regular allocation [16] assumptions. Our topology is an example of a regular model, but our work differs in that we consider how to retrofit existing cyber networks to create a more robust topology. We also consider an abstract tree overlay, which may or may not match the physical cyber network topology.

This work extends our prior publication [22] by the following contributions: It contains a more detailed motivation of the problem, additional background information, clarifications of its relation to micro grids, more details of micro grid assumptions, cost considerations in retrofitting power systems with additional cyber network paths, more detailed explanations of our technical approach, a refined lower-cost placement of crosslinks plus an algorithm, more comparisons to related work, and a discussion of future work on cyber security specific to CPS.

3 Micro Grids and Renewables

The power grid is currently undergoing a significant transition from a centralized architecture centered around large capacity generation resources of power plants to a future distributed architecture where large generation sources are complemented by many small ones due to renewable generation sources, such as photovoltaic and wind. To address the transitional challenges of our power infrastructure, the NSF Engineering Research Center (ERC) for Future Renewable Electric Energy Delivery and Management (FREEDM), a multi-institutional project, investigates the cyber-infrastructure of micro grids harboring renewable generation sources [2].

In the FREEDM system, power management of green energy is provided in a highly distributed and scalable manner. The system has to ensure that Intelligent Energy Management (IEM) and Intelligent Fault Management (IFM) devices have the appropriate data to make control decisions for micro grids and with respect to micro grid connectivity to an upstream utility power grid. This is described in more detail later.

At a grander scale, FREEDM is contributing to methods to overcome the looming energy crisis. The objective is to reduce reliance on fossil fuels that are increasingly scarce, reduce reliance on non-renewable sources of energy, and create a system that can reduce the world's CO_2 emissions to combat climate change. To overcome these challenges, the FREEDM center is developing a revolutionary power grid with the following characteristics:

- It supports distributed intelligent control mechanisms;
- it enables plug-and-play of power resource and storage devices;
- it provides stability and reliability of power delivery;
- it improves energy efficiency; and
- it combines scalable and secure communications.

Today's power grids will ultimately transition to become a system with the FREEDM characteristics. Today's systems generally utilize dated technologies and are unable to provide high levels of fault tolerance as they operate under centralized control structures.

A high-level view of the FREEDM infrastructure design is depicted in Figure 1. The goal is to create an Internet for power that supports the incorporation of a variety of power sources and storage devices to operate in a plug-and-play manner. This includes incorporating a variety of green power generation mechanisms, such as photo-voltaic, wind, and hydro-power. In the proposed system, consumers can generate their own power and sell it back to the utility. Such micro grids feature plug-in hybrid electric vehicles (PHEVs), local wind turbines, and other consumer-level load and generation sources.

There are a number of challenges associated with a power grid with distributed generation sources in terms of CPS design in general and for FREEDM specifically. One of the interesting challenges is its Distributed Grid Intelligence (DGI). The goal of DGI is to facilitate the departure from centralized power control in favor of distributed control with multiple control objectives. DGI is developing two types of systems to be utilized in the power grid. The first is the intelligent energy management (IEM) system. IEMs are responsible for enabling the power grid to make distributed decisions, i.e., load balancing and system control. The second type is is the intelligent fault manager (IFM). IFMs are responsible for working with IEMs to detect faults and to make islanding decisions. Islanding refers to temporal isolation of a micro grid where selective loads are still served by local power generation capabilities.

Figure 1 shows the topology of the DGI system and its interface with the IFM and IEM nodes. DGI within the FREEDM system features a communication network through the Reliable Secure Communications (RSC) layer. RSC is investigating ways of integrating a complete communication system into the intelligent

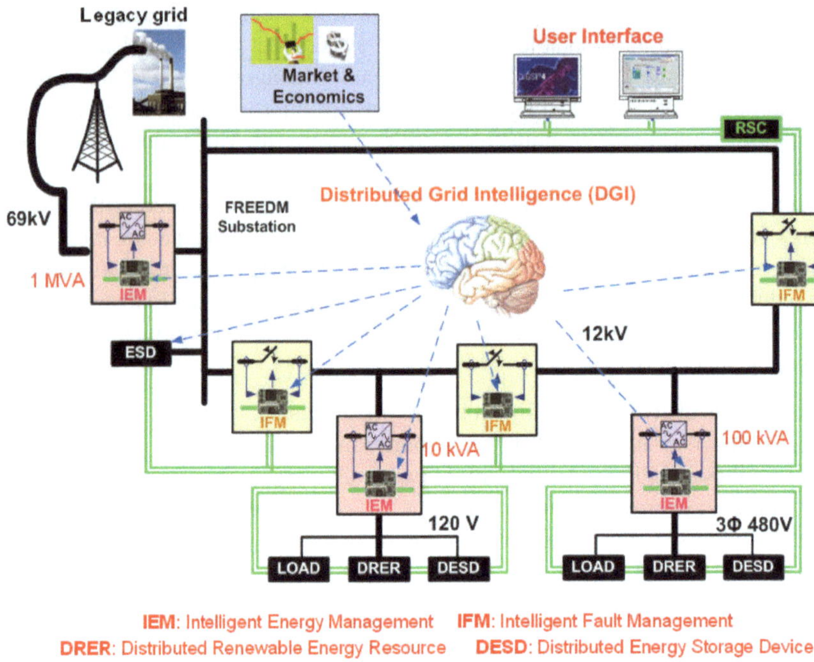

Fig. 1 FREEDM System: Locally generated power is fed into the power grid via solid state transformers; such decentralized generation requires a new level of Distributed Grid Intelligence (DGI) to monitor fault and manage energy in order to ensure sustained reliability compared to centralized power generation.

power grid. The network is composed of several different network types to support the scope of devices in the project. The current design of the RSC network is a hybrid network that combines a wireless in-home design with a wired external design. The current model of the in-home design is that of several wireless ZigBee devices that communicate through a StarGate concentrator in the home. This is important for many reasons. First, this allows the IEMs in the system to have a more accurate assessment of the current load of a house. Based on the information, future loads are predicted. Second, this allows the intelligent power grid to shut off non-essential devices in a home during critical times. Third, this enables power generating homes to sell back energy to the power grid in times of excess supply.

4 Distributed Control for the Power Grid

In micro grids with a FREEDM design, CPS control needs to ensure system reliability in the sense of sustained power supply to costumers. To this end, this work focuses on the cyber-networking side of providing fault tolerance for micro grids.

Micro grids are a significant deviation from modern power grids. Today's power grids, particularly their hierarchical control below the substation level, serve as the closest analog for us to envision improvements for the overall power architecture. This work will inherently benefit and eventually change the overall power grid through efforts to develop techniques that improve the operation of micro grids.

The common design of today's power grids follows a centralized command and control structure, *i.e.*, most notably Supervisory Control and Data Acquisition (SCADA) systems relying on human monitors for decision making. SCADA systems provide the mechanism for identifying faults. However, they represent a single point of failure within today's power grid. Even when SCADA systems are running within specified parameters, catastrophic faults can occur.

The most severe faults are cascading failures, which occur when some initial (power) nodes in the physical power system fail at first. Upon failure, their power load is passed to another local node. When this occurs, it can overload the node that received the shifted load. When it fails, its load is passed on in a transitive manner, which may result in cascading effects. Conventional power grids provide little protection against cascading failures. As demand for power increases and the complexity of the power grid increases due to micro grids, the ability of the power grid to support the increased load may result in more frequent blackouts.

The overloads described above are common in power grids and were responsible for historic blackouts, such as in 2003 [1]. In August of 2003, a blackout occurred that affected 45 million people in the US and 10 million people in Canada. Several estimates say the cost of this blackout exceeded $6 billion dollars. The original cause of the power failure occurred due to overgrown vegetation that struck power lines. After the power lines failed, a series of cascading failures occurred resulting in an 80% loss of power in the Northeastern US. The SCADA system within the region of the original blackout failed to detect the faults due to a race condition in its software that caused the centralized system to fail. As a result of the failure of the SCADA system, human monitors failed to be alerted to the problems for over an hour. The missing alerts from the SCADA system caused the human monitors to disregard a phone call that would have pointed them to the cascading failures. Due to these failures, 256 power plants went offline that day.

The problems of cascades like to 2003 blackout might have been averted had a smart grid been in place. In essence, the centralized nature of the SCADA controller creates a single point of failure. This design resulted in a system failure at an inopportune time that led to the Northeastern blackout. Due to the distributed nature of decision making components in a smart grid, it would take many more faults before these systems were to fail.

Trends to decentralize control, such as distributed controllers to locally isolate faults, load power factor corrections and voltage regulators for generation sources, help conventional power systems to reduce the threat of cascading faults. However, we conjecture that with the projected wide-spread deployment of micro grids, additional safeguards are required.

The FREEDM model features IEM nodes and IFM nodes that are distributed throughout the micro grid. In the event of a failure, the IFM nodes would disconnect the breaker when detecting failures and notify the IEM nodes. If a local controlling IEM node that governs high-level decisions of other nodes fails, the remaining IEMs can distribute the load to accommodate the loss. When cascades occur, the IEMs can identify the fault and circumvent further cascades. They can resort to islanding to isolate the micro grid from either incoming or outgoing faults.[1] This may result in internal partial or complete failure but it would stop further damage to the remaining components. If the failure originated from outside, the micro grid would be isolated from the cascade. Secondly, the IEMs can redistribute the load through immediate control of the transformers used within the network. Zig-Bee nodes inside homes may provide feedback to the IEMs so that IEMs could shut off unnecessary loads to reduce power flow.

The challenge here is to devise a mechanism to support distributed control, a key component of which is a strong communications network. In a conventional network design, routers and switches present a single point of failure, just as in a SCADA network, due to the static nature of routing protocols. If a switch were to fail in a critical location within the network, even if redundant pathways existed, many commodity networks could not exploit such pathways as switches generally do not provide dynamic routing capabilities. In the above example, this could lead to misinformation being disseminated throughout the network. If these types of failures existed at the time of faults within the power network, it would be difficult to circumvent faults or operate effectively in spite of their presence. To improve upon this paradigm, the communications network must be one of the most robust components of the system. In a robust communication network, failure results in the reorganization of communication pathways. This allows messages to still be transmitted to all operating nodes in the network by routing around failed components. Thus, CPS systems in general and the power grid in particular become more resilient.

Let us briefly describe how such distributed control with fault-tolerant networks generalizes to other CPS infrastructure. There is a large scope of applications, especially within critical infrastructure, that could potentially move away from centralized SCADA systems. Researchers are rolling out components in communities to monitor underground water pipes. These devices monitor the flow on the pipe to maintain pressure as well as monitor for slow leaks. The current scope of these devices is to record this data so that it can be collected. But as this technology improves, it will provide real-time feedback to utilities to help them quickly identify failures [17]. Another example is oil refineries that currently use SCADA systems to collect the data complemented by humans monitoring the system. This exposes SCADA to human mistakes with potentially severe consequences. Distributed control shifts responsibility from human operators and creates a decentralized system that reacts to faults in a more robust manner. Fault tolerant communication infrastructure would increase reliability in both of these scenarios.

[1] Incoming faults are predicted cascades that originate outside of the micro grid with a potential to destabilize local power balance (or even damage local power devices) while outgoing ones originate inside the micro grid with a potential to destabilize immediately surrounding power balance, below a substation or beyond (or even damage power devices in this realm).

5 A Resilient Network Model

The objective of this work is to provide resilience in CPS control specifically for the power grid. We consider existing physical infrastructure in terms of its topology, propose a cost-effective topological extension and abstractly analyze its resilience characteristics. To this end, we assume a model that is agnostic to the type of faults affecting the network. In other words, our approach works equally well with, *e.g.*, node failures and link failures, where a node is an IEM/IFM node in the FREEDM model or any other compute node in a power grid control system. The detection of such faults is orthogonal to this work and could be accomplished by timeout-based monitoring, such as in our prototype, assuming fail-stop fault behavior. Fail-stop behavior refers to failures where a compute device stops working altogether, i.e., other devices will not receive any response at all when querying the device. We do not consider Byzantine failures where devices may provide incorrect responses, either because their security has been compromised or due to partial hardware/software failures that produce incorrect results. Any loss of communication in our model is mitigated by attempting to find a route through the network that will bypass the point of failure and still deliver the message using a different route, albeit with potentially different (higher/lower) latency than before and potentially with a reconfiguration delay due to timeouts.

As a starting point, we assume a tree topology as our network topology. The tree topology is a good fit for modern power grids that are hierarchically designed, and power line corridors owned by power utilities often already harbor cyber network lines linking control and monitoring devices along the hierarchical structure.

Resilience can be improved by utilizing redundant physical network paths. Assuming that sufficient paths exist, software overlays may be utilized to improve network resilience, an idea first described by Anderson *et al.* [6]. Their work presented the basis for a resilient overlay network (RON) by partitioning distributed nodes that may contain a different topological perspective than the external, physical network topology. Their work assumed nodes to potentially be geographically scattered across the Internet, i.e., their topology assumed inherent physical link capability for multi-path routing.

Our work differs in that we assume a proprietary network topology that may lack multi-path routes in its current design due to the strictly hierarchical structure of the power grid. Nonetheless, we utilize a similar partitioning for the routing of messages. In contrast to RON, our overlays focus on much smaller local area networks (LAN) to facilitate fault-tolerant communication in a micro grid setting. As such, it complements switches found in LANs due to their low cost with advanced fault-tolerant routing capabilities otherwise only available for expensive routers. When multi path routes are already available in wide area networks (WAN) at higher levels of the power grid, our method can equally be applied, but our focus in this work is on the local side.

The physical network is assumed to be implemented as a tree network topology by default, which is consistent with today's power (and corresponding proprietary cyber network) infrastructure. Communication in this model follows that of a typical network in which messages are sent from switch to switch. The standard communication links in this model are referred to as uplinks. As shown in Figure 2, uplinks are the vertical communication lines that create the tree structure. These vertical lines represent today's physical infrastructure.

Our method complements vertical connections with a set of horizontal links placed at various points throughout the network, effectively creating a cross-linked tree, as depicted in Figure 2 (discussed in detail later). These links, designated as crosslinks, are only intended to be used in fault scenarios if we assume that initial communication follows the legacy, vertical links. During link outages, *e.g.*, when nodes start incurring timeouts for sending messages, the nodes incurring the timeouts will transparently morph routing from the path given by the physical switch network to specially designated crosslinks between lateral nodes. These crosslinks facilitate the delivery of messages upon partial link/node failures.

Notice that crosslinks are still periodically monitored by sending heartbeat messages between both endpoints, but these heartbeat messages do not carry any power communication payloads. Such monitoring ensures that crosslink status is known in case of link/node failures to facilitate the discovery of an alternate routing path.

The distinction of vertical and horizontal links here is mainly to illustrate the additional investment required in physical infrastructure. And when new leaf-level (consumer) infrastructure is installed, initial paths are established over vertical links. During 24/7 operation, however, distinction between horizontal/vertical links becomes impertinent as the main objective is connectivity. While latencies due to number of hops may differ, worst-case response times are calculated based on upper bounds considering the longest path though our topology.

Let us next outline the operational model of routing messages in a software overlay over the given cross-linked tree. Each node in the tree contains a prioritized list of nodes containing crosslinks within a predefined radius r relative to their physical location. By enumerating these lists and passing messages via crosslinks, dynamic routes are created throughout the network.

The details on how to implement this model over existing network devices, such as routers and switches, are provided in Section 6.1. Before considering implementation variants, let us first analyze the resilience characteristics of the cross-linked tree, i.e., we are deriving a probabilistic model to study graph partitioning as an indication for disconnected sub-networks (in the cyber sense).

The partitioning/isolation property signifies the likelihood of a communication (cyber) network outage in power grids. Such disconnects in general may result in power outages or reduced capacity islanding due to lack of coordination within the sub-network of the power grid, as detailed next. While the legacy grid continues to supply power during a cyber outage, power efficiency may degrade in the absence of micro-grid control. In the event of simultaneous failure of connectivity to the legacy grid (*e.g.*, when physical power lines and communication lines are clipped simultaneously), outages are unavoidable. In micro grids, in contrast, cyber isolation still allows islanding if a micro grid has generation capabilities, but only for a selected subset of quintessential loads while all other devices remain without power. Hence, cyber isolation serves as a basis to quantify the reliability of the overall system and that of individual nodes.

Our objective is to keep the probability for such partitioning (the likelihood of isolation and potential outages) low while controlling the cost of software overlays due to retrofitted cross links. Let us assume that a given single unit may fail with a probability of p. For simplicity, we assume an equal probability for node and link failures here.

Let us express the cross-linked tree as an abstract graph $G = (V, E)$ of vertices V and edges E, where the former combines nodes and switches while the latter represents network links. The height of the graph is denoted as h. In the graph, vertical tree edges T are distinguished from crosslinks C, such that

$$E = T \cup C \wedge T \cap C = \phi.$$

Let $v = |V|$ and $e = |E|$ be the number of vertices and edges in G where

$$v = 2^h - 1 \text{ and } e = 2^h + 2^{h-2} - 3 \quad (1)$$

Consider the (transformed) graph representing an overlay tree depicted in Figure 2 with a height of $h = 4$. G has a total of 15 nodes and 17 links for the example in Figure 2, as given by the above equations. Notice that smaller trees are irregular with respect to crosslinks, as discussed below, i.e., the selected height provides the smallest example of an otherwise scalable generalization.

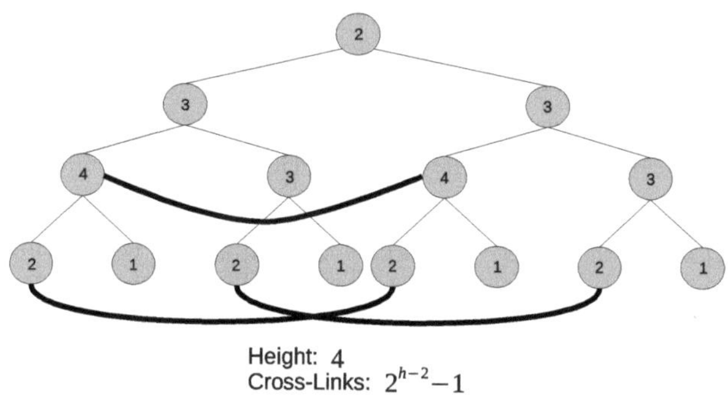

Fig. 2 Sample Overlay Tree: Virtual links overlay the physical cyber network topology with bi-directional connections that form a (in this case) binary tree of height four plus crosslinks (bold links) to increase the reliability. The core tree reflects common physical cyber network topologies close to homes (i.e., the generation sources for micro-grids) while crosslinks represent proposed, additional connections.

We then derive a probabilistic model based on graph analysis and combinatorial theory. Our overlay graphs have a number of unique properties that we utilize: Any vertex has a degree of $d \in \{1, \ldots, 4\}$ (number of edges), including crosslinks. In Figure 2, crosslinks are depicted in bold, and vertex degrees as the label per vertex. At each level $l > 2$, crosslinks are created to connect every other node at the respective level. Hence, the total number of crosslinks is $2^{h-2} - 1$. This guarantees a uniform distribution of crosslinks that remains proportional to the growth of the overall tree. More significantly, we intend to show that graph connectivity is preserved with at least the same probability as the tree grows, which provides a stability invariant. The consistency and resilience of our model under scaling of power grids to large sizes can thus be exposed via reasoning over strong guarantees for stability and, implicitly, resilience to failure.

Stability is derived in terms of resilience to graph partitioning. Depending on a component location in the graph, partitioning may only occur under a certain combination of simultaneous failures. It suffices to consider single, double and triple failures of units in this model: single link (L), single node (N), double link (LL), double node and single link plus single nodes (LN) failures etc. based on the independent per unit failure probability p. Using combinatorics, the number of failures that results in graph partitioning (isolation) of at least one vertex (node) can be enumerated per class (see Table 1). The table shows the unique failures (omitting identical pairs and isolation of lower degree nodes since units are unordered in their enumeration). For instance, a single-link leaf becomes isolated when its parent or its link fail. A dual-link leaf can be isolated when both its links (1 case), a link and a node on opposite sides (2 cases) or 2 nodes fail (1 case), where multi-partitioning is only counted once. For triple-link nodes at level $h-1$, two cases each exist with unique partitioning. Any other vertex cannot be isolated by just a dual failure. It would require triple unit failure for degree 3 nodes above level $h-1$, such as LLL (12 cases), and so on. Due to the low degree of vertices in cross-linked tree (by construction), this covers all cases for larger partitions as well.

Table 1 Enumeration of Isolation Scenarios

# Nodes	**Degree**	case 1	case 2	case 3	case 4
2^{h-2}	1	1 L	1 N		
$2^{h-2}+1$	2	1 LL	2 LN	1 NN	
2^{h-3}, l=h-1	3	2 LL	2 LN	2 NN	
2^{h-3}, o/w	3	12 LLL	12 LLN	12 LNN	4 NNN
2^{h-3}	4	4 LLL	12 LLN	12 LNN	4 NNN

Notice that multi-unit failures are counted only once by ensuring that only (a) nodes on independent paths (without common vertices) and (b) links on edge-independent paths (without common edges) are counted. The former is also captured by the minimum vertex cut while the latter represents the minimum edge cut (see Menger's theorem [8]). All unique cuts need to be counted once, and higher degree cuts subsumed by lower degree cuts can be omitted. However, non-omission only increases the overall partitioning probability insignificantly since higher-degree cuts are significantly less likely than lower ones. Some lower cuts are included at higher degrees in Table 1 to simplify the problem.

Example: Consider a failed link between the root node and its left child in Figure 2. A second simultaneous link failure between this child and its 3rd level left child would be included in the upper link (single) failure consideration, but also counted separately in our approach to provides a closed formula. This formula still presents a sound upper bound approximation that is tight as argued next.

The systematic structure of our overlay graph construction ensures that the number of these cuts remains constant as the height increases, which is significant for the stability argument in terms of resilience. The approach is thus sufficient to characterize network stability by absence of partitioning. The overall partitioning (isolation) probability P can then be approximated (by omitting any additive constants) as follows, where each term corresponds to the respective entry in Table 1:

$$\begin{aligned}
P \approx\ & \frac{2^{h-2}}{e}p + \frac{2^{h-2}}{v}p \\
& + \frac{2^{h-2}}{e^2}p^2 + 2\frac{2^{h-2}}{ev}p^2 + \frac{2^{h-2}}{v^2}p^2 \\
& + 2\frac{2^{h-3}}{e^2}p^2 + 2\frac{2^{h-3}}{ev}p^2 + 2\frac{2^{h-3}}{v^2}p^2 \\
& + 12\frac{2^{h-3}}{e^3}p^3 + 12\frac{2^{h-3}}{e^2v}p^3 + 12\frac{2^{h-3}}{ev^2}p^3 + 4\frac{2^{h-3}}{v^3}p^3 \\
& + 4\frac{2^{h-3}}{e^3}p^3 + 12\frac{2^{h-3}}{e^2v}p^3 + 12\frac{2^{h-3}}{ev^2}p^3 + 4\frac{2^{h-3}}{v^3}p^3
\end{aligned}$$

The overall partitioning probability has multiple implications. First, the probability of graph connectivity remains constant since denominator and numerator grow at the same rate since $2^h - 1 = 2^{logv} - 1 = v$ (see Equation 1). This indicates that graph connectivity remains *stable* regardless of tree height. Second, for a large number of nodes, partitioning only depends on the probability p for single node/link failure, *i.e.*, our overlay is *scalable*.

Using numerical approximation, these properties become obvious by another simplification step based on the fact that $v \approx e \approx 2^h$ (due to Equation 1):

$$P \approx \frac{1}{2}p + \frac{7}{4v}p^2 + \frac{9}{v^2}p^3 \text{ and } \lim_{h \to \infty} P = \frac{1}{2}p \tag{2}$$

Notice that h is going to grow significantly to accommodate an explosion of devices with a wide-spread deployment of micro grids with IEMs/IFMs.

Moreover, we conjecture that fewer crosslinks would actually suffice as along as they were growing at a rate of at least $O((log_2 n)^2/n)$ total crosslinks for any cross-linked graph, such that first-order failures (single link/node) increase by only a constant factor. Such a refinement is subject to future work. But it may have practical value as a lower constant implies a potential for proportional cost savings when retrofitting trees with crosslinks within the physical infrastructure.

The placement of crosslinks poses another interesting aspect. In the analysis, an equal distribution of connections across a level is assumed. An algorithm for systematic crosslink placement can be constructed by "alternating" their node source/sink such that the subtree rooted in node v with the lowest aggregate cross connectivity distance $c(v)$ is connected to its equivalent neighbor at a distance l linear to the respective height, where $c(v)$ is defined as the sum of the shortest paths from nodes of a subtree (rooted in v) to the nearest crosslink. Let us outline two such algorithms with different growth rates.

Algorithm: At each level at height $h, h \geq 3$, $2h - 4$ crosslinks are created, where $l = 2h - 3$ and $c(v) = min_{v_c \in V}(|v - v_c|)$, i.e., c is the minimum distance to the nearest cross node v_c in G. The number of crosslinks grows proportionally to n, i.e., it is upper bounded by $2h^2 = 2(log_2 n)^2$, which is less than $O((log_2 n)^2/n)$ total crosslinks (for $h \geq 3$).

The algorithm is configurable in terms of the number of crosslinks. For example, a lower number of just $h - 2$ crosslinks could be established per level for the same l and c, subject to the same upper bound. However, any lower rate of crosslinks per level, e.g., $log_2 h$, would violate the bound and result in insufficient alternate routes to ensure stability and, hence, scalability under sustained resilience.

Notice that c can be calculated by depth-first-search (DFS) in linear time. An even more efficient algorithm to calculate can be incorporated into the systematic construction of G using the following steps. (1) The initial c of root is zero. (2) c is then inherited from a parent to the children and incremented by one upon creating a child node. (3) Upon construction of a crosslink, a subtree of height $log(h)$ has to be updated for each node receiving a new crosslink, where such a node is a leaf in the respective subtree. The resulting complexity is only $O(log(h)^2)$ instead of linear DFS complexity.

6 Realistic Network Overlay Designs

Cross-linked trees provide a theoretical basis and a means to reason about reliability. In the following, different design options for mapping and embedding these cross-linked trees into existing networks and with conventional network devices will be developed. The main objective is to provide resilience while keeping the cost of retrofitting networks low. This is mostly a consideration in terms of the required crosslinks as the overlay protocols can be established in software.

6.1 COMIG: A Communication Overlay for MIcro Grids

The first approach considers devices organized into software partitions that are calculated locally based on their IP address. Partitions are created as a side effect of subnet masks. Each partition is assumed to be locally connected to a switch. These partitions are then grouped together in clusters of a certain static size. The combined group of clusters and partitions are interconnected with horizontal crosslinks and vertical uplinks. Figure 3 depicts example of COMIG.

Our software overlay network represents a tree-based topology utilizing vertical uplinks. These uplinks serve as the default routing path for general message communication in the absence of failures. Figure 4 depicts the vertical uplinks and shows the resulting tree formed by them. Uplinks are necessary to provide inter-cluster communication. They constitute the network backbone of COMIG. To increase fault tolerance, horizontal crosslinks are introduced. Figure 4 depicts these

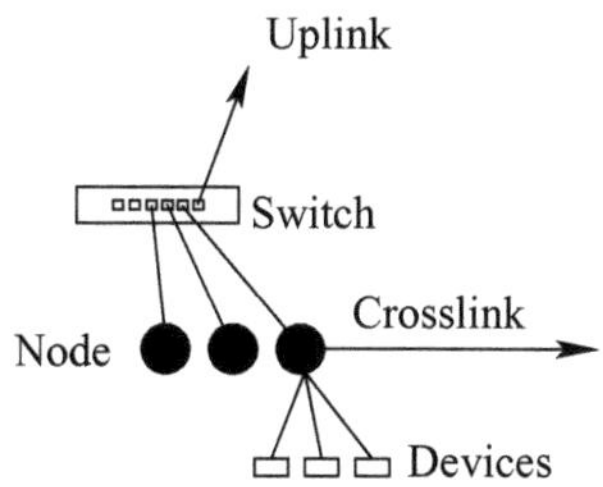

Fig. 3 Device Cluster: Nodes are connected to switches with crosslinks from a node to another tree trunk

crosslinks, which serve as secondary paths through the network, activated by the overlay network protocol upon primary path failure.

A COMIG overlay is an abstract software overlay that fits arbitrary intelligent power grids. Most importantly, it provides redundant communication pathways and the potential to connect the network in alternate ways in case of faults in the system via its software middleware layer. This capability is crucial for allowing intelligent nodes in the system to coordinate the actions of system control tasks and to maintain appropriate state.

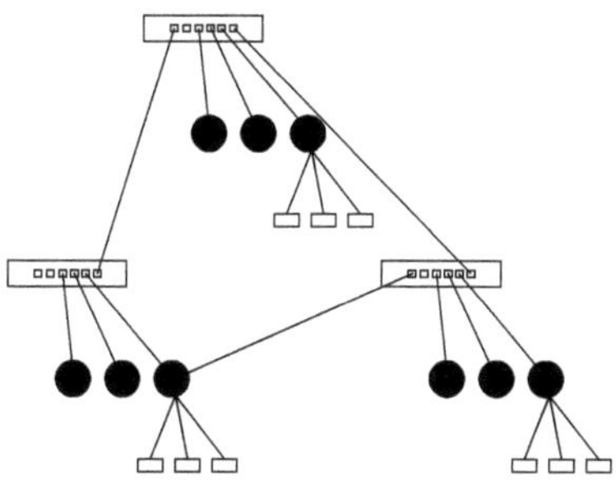

Fig. 4 Cluster Tree: Uplinks connect switches while crosslinks connect a switch to a node on a second network port

Communication pathways are primarily used through the switching interface composed of uplinks, as depicted in Figure 5(a). COMIG differs from a regular network in the composition of a series of intelligently placed crosslinks that can be implemented as node-to-switch or node-to-node links. The abstract network will enter into a reorganization mode upon loss of an uplink connection.

A message timeout is utilized an indicator for link failure and results in reorganization. The reorganization mode explores alternate routes in the network based on meta-information describing the characteristics of the network. The collection of this meta-information occurs as follows. Nodes can derive partition information from their network overlay data, e.g., to determine its neighbors on a switch and the partitions above and below it in a tree. A node in reorganization mode can communicate with its neighbors to determine the location of crosslinks. It can further determine if this is a node failure on the receiving end or a link failure along the switching path by utilizing the crosslinks. Figure 5(b) depicts the utilization of a crosslink in an attempt to resend a previously failed message. If the failure was in the switch link, then a switch link is indicated as a failure type by the nature of the response from the receiving node.

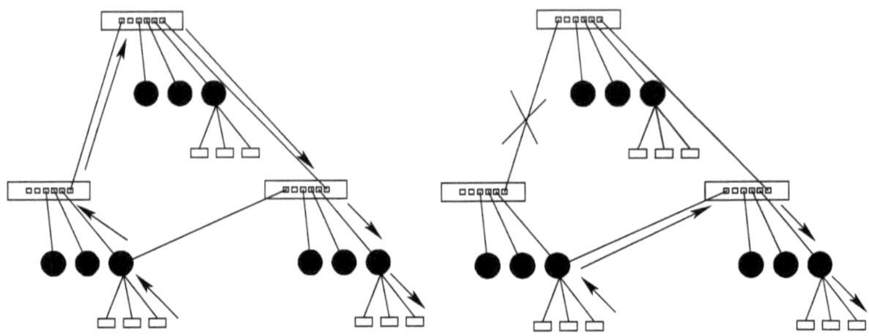

Fig. 5 Message Pathways: Without faults, messages travel along the up/down links of the tree; upon failure of an uplink, crosslinks dynamically re-route messages horizontally to another switch before traveling down (or up) to the destination node.

Overall, COMIG provides essential functionality to an intelligent power grid utilizing a distributed network. In case of wide area faults in the power grid, it aids distributed grid intelligence of the micro grid by ensuring reliability through reorganization.

6.2 SWOMIG: A SWitch Overlay for Micro Grids

COMIG has one pitfall: It relies on an overlay structure imposed on the network even at times when faults do not exist, which adds performance overhead. In general, overlays impose a trade-off of performance for fault tolerance that may initially not always be satisfactory but can be refined in terms of minor design changes.

A power grid may require only very moderate bandwidth while the conventional Internet may have higher bandwidth requirements due to consumer needs for streaming services. Nonetheless, certain levels must be maintained to insure timely decisions can still be made in power grids. This observation motivates our second design termed SWOMIG, a switch-based overlay for micro grids. While both COMIG and SWOMIG can operate agnostically of the underlying physical network structure, SWOMIG allows for static communication pathways to be used at times when faults are absent. In contrast, COMIG forces communication over abstracted routes with overhead even in the absence of failures.

Cross-linked trees are also the basis of SWOMIG's design with crosslinks that are disjoint from uplinks representing the static route of the network. This is easily accomplished through default routing configuration tools. In SWOMIG, during normal operation, the network utilizes the static pathways. This provides high throughput. In the presence of a fault, i.e., when a message timeout occurs, an overlay is imposed, but only on the node that experiences a timeout. Each node maintains a list of surrounding nodes' crosslinks. In the presence of a fault, a node determines if an alternate route exists to transmit the messages by traversing the list of crosslinks.

An example of a commodity configuration that can self-organize via discovery of alternate routes is depicted in Figure 6. In the first part of the figure, the commodity network is using the default pathways to enable communication. The remaining portion of the figure explores possible reorganizations using crosslinks. We primarily consider link failure in this example because the communication network may parallel the physical transmission corridors of the power grid. In such a case, simultaneous power and cyber network cuts may occur due to external (physical) intervention, such as fallen trees or unintentional construction-related line cuts, just to name a few.

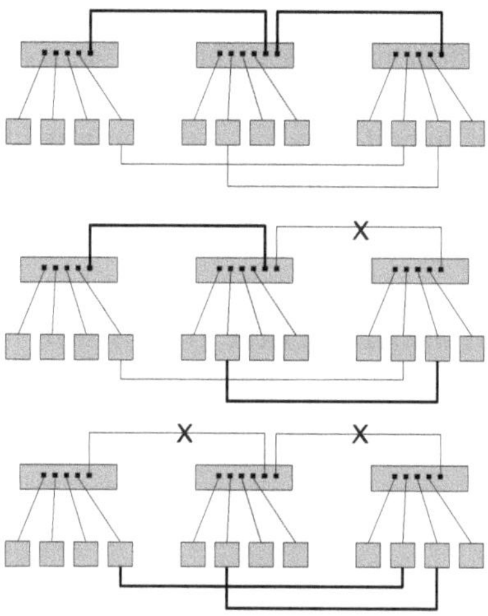

Fig. 6 Device Cluster: Switches are directly connected to one another on the default path; upon failure of links on the default path, traffic is re-routed via crosslinks that connect two nodes with one another across distant switches

Components of devices themselves tend to be physically protected in hardened enclosures to provide protection from weather and other environmental stress, but may be subject to electrical faults. Links are more exposed to the elements and may experience disconnects when cut. (Recall that the analytical model considers node and link failures to have equal probabilities of failure. If these probabilities were to significantly differ, a refined model is needed, which is subject of future work.) These lines running parallel to the power lines represent the default path. Crosslinks can be implemented using commodity Internet connections or specialized lines connecting nodes such as IEM/IFM nodes in micro grids, optionally not using above-ground cabling to further protect them. In this example, link failure is used as a cause of fault in the network. Another scenario is device failure, where a switch or a node fails. In the case of switch failure, if subnet partitioning were exploited to provide localization information, this information could help isolate the fault location. In this case, local communication among nodes on a switch would fail within the network. Nonetheless, this failure can be confirmed by a representative in the network with a crosslink as an alternate path to replay a timed-out message.

6.3 Discussion

In practice, the cost of crosslinks depends on the deployment methods. Wired connections between end-points of micro-grid generation sources (homes) may require trenches across corridors not owned by power companies. A more viable method may be wireless connections, which could be realized by short-range Wifi (802.11) or medium range Wimax technology. Wired is more robust and thought to be more secure but the low cost of wireless and ease of installation may be more realistic. If we assume a hybrid deployment of (existing) wired connections along the tree links combined with wireless crosslinks, the default connectivity (the uplink in the tree) would be more reliable than crosslinks. Thus, crosslinks become natural backups for broken uplinks, where the latter can be repaired while the former ensure sustained connectivity. This hybrid approach seems to provide a good cost-reliability trade-off. Inclusion of different reliability levels for uplinks vs. crosslinks is subject to future modeling.

7 Implementation and Interface Definition

We have designed and implemented a unified message passing API that facilitates coordination between nodes ranging from the large and sophisticated IEM nodes to small ZigBee devices. An API is important for the development of applications for a complex system of software, such as a power grid. Using this API, we can guarantee a common messaging-passing standard that will be utilized ubiquitously within micro grids (and possibly above). This API has currently been deployed by other software teams within the FREEDM project and is being used to create applications for load balancing and power management as a proof-of-concept. We have also found the API beneficial for creating and testing the implementation of our overlays.

We design the API to loosely resemble that of Active Messages [9] as implemented in Tiny OS [12]. In particular, messages are non-blocking and asynchronous. This design choice allows less sophisticated devices that simply use a MAC-based designation to be incorporated into the network. Such low-end devices can then be accounted for by more sophisticated nodes. In this message passing API, a device or node registers a message type to receive a message handler. The handler is then used in sending and receiving messages. The current API provides a number of constructs for basic communication using point-to-point messages. These include

- non-blocking sends and receives,
- conditioned waiting and signaling, as well as
- handle generation.

Non-blocking network abstractions facilitate resilience in network overlays when faults are considered at arbitrary rates and when timeouts are utilized for fault detection in a distributed network. This allows devices within the network to send messages without waiting for acknowledgments before proceeding with other work. The same approach is applied to receives to avoid a need for actively monitoring a queue. In a non-blocking approach, a received message is handled by the network API. When a new message is received, the application is able to use it right away or defer it until a later time. From this asynchronous API, one can thus create blocking semantics of a layered blocking API if desired. This is done using the conditioned wait and condition signaling methods supported by our API. This allows a running process on a device to send a message and then blocks. Once a corresponding message is received, the same process is woken up again.

We utilized the Mace distributed prototyping language [11] to implement this API. Mace is a C++ abstraction that enables the low-level network details to be abstracted from the programmer while leaving significant amounts of flexibility in the message-handling abilities and supporting timeout-based fault detection, which is central to our fault tolerance network overlay approach. The basic prototype of our proposed system on top of Mace comprises a universal basis for message passing over our API in the FREEDM infrastructure.

8 Network Overlay Monitoring

Locating faults in a conventional power grid can present a challenge. Current fault localization practices often require the operating utility to field phone calls that allow it to determine a rough location of where the fault may have occurred. Using such a rough estimation approach can increase the duration of the power failure. Smart grids have the capability to remedy this situation. The distributed nature of control within a smart grid allows agents to detect faults much faster than complaint triangulation. When these faults are detected, the discovering nodes can report the node failure. The identification of the failed (cyber) node should increase the accuracy of locating the fault in the (power) network. We are developing a distributed live monitoring (DLM) tool that identifies failed power devices among other features to aid in fault localization.

The ability to understand the structure and status of the network is imperative, particularly when the network is distributed in nature. A truly global status may often be difficult to obtain due to the distributed nature of the network. To aid the maintainers of the system in identifying problems and correcting them, our DLM tool provides a real-time view of the state of the networked devices in the system and the dynamic routing through our software overlays. This system provides information regarding a node's current running status as well as a topological layout of the network, both derived from data provided to the operational model.

Nodes can communicate with an interface server in our first prototype, which features a centralized single server. The server provides a graphical representation of the status of the nodes and current messages in flight. The projected design of the

DLM presents a fully detailed representation of the underlying LAN. This enables one to monitor system activity, to detect failed components, to observe alternate routing activity, and to sustain partial functionality in the presence of partitioning / islanding of micro grids. As such, one can determine which links have failed and, more specifically, which nodes have failed. Working nodes report the status of successful and failed communication attempts within the network. This information is relayed to and concentrated at the server to allow visualization.

The DLM is used to display the communication paths of three separate devices as depicted in Figure 7. The DLM can be provided with information detailing the locations of software links between nodes to create a graphical representation of the network. The network structure is being coded as a tree network resembling the shape of the network utilized in our Mace prototype in this figure.

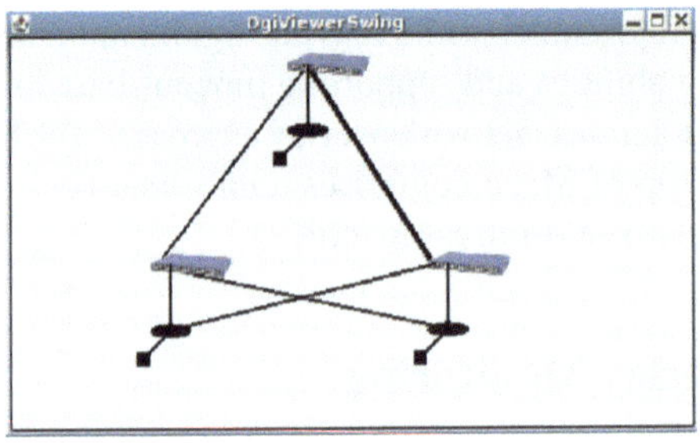

Fig. 7 Distributed Live Monitoring Swing Window

The DLM server and its node components (runtime support / daemons) were developed using the Java Swing graphics packages. We utilize a network socket API to connect with our distributed prototype written in C++. The Mace distributed program library provides the distributed message abstraction and inherently supports fault detection [11]. Our implementation supports a variety of services, such as the ability to

- monitor and set connection status,
- define links and partitions,
- visualize paths and messages, as well as
- topologically arrange output in a hierarchical manner.

The current implementation assumes a cross-linked topology but the principle design supports other topologies for visualization as well. The prototype of the message-passing system is instrumented with calls that relay messages during each critical step in the program communication path. At each step, a command message is sent to the visualizer that renders the information on screen. Failed nodes may not be able to report their current status to the DLM. The status of a node is updated should another node detect link/node failures via timeouts.

9 Future Work

Current limitations of this work in SWOMIG make it very difficult to compose multiple overlays into a single path. A single path would be beneficial in situations where multiple failures occur. Multiple composed overlays could be used to facilitate a single path throughout the network, combining all of the nodes. The problem here is that current localization data is limited to determining neighbors on switches. This provides very little information relating one group position to another. In such a system, to compose two overlays blindly would require a recursive enumeration of crosslinks in the network to identify a single path. This exhaustive type of search is inappropriate in terms of its latency within a network and would not scale well. An alternate approach to this that we will be investigating in the future is to not only provide crosslink information to the nodes in the network. One would also provide localization information that can be used by switches to identify locations of their partition in relation to other partitions surrounding them with a constant set radius. We can then explore statistical means of deriving the best trade-off of composing multiple paths to improve fault detection with a higher radius.

Another future direction of research focuses on scalability by utilizing crosslink offload. In some ad-hoc networks, studies suggest that, after scaling to a certain size, considerable processor time is spent passing other nodes' messages around, which makes it difficult to make computational progress anymore [13]. This type of fault, similar to life lock, could easily affect our current model if the physical network structure had a single crosslink connecting two halves of the network, which could overwhelm the single connecting node. To overcome this challenge, we will study models that include both priority and overhead evaluations. A priority evaluation allows the use of global priority values to be assigned to nodes. In negotiating the use of a crosslink with another node at times of failure, the priority value ensures that important nodes can communicate at the cost of the lower-level nodes that may be shut out. This would also have to account for a load metric that crosslink nodes maintain to insure that their own tasks can be accomplished — unless, of course, the crosslink node is a low priority node, in which case it would have a mode change to serve only as a message-passing link for other high-priority nodes.

10 Conclusion

Today's increased prevalence of intelligent devices in critical infrastructure imposes a need for fault tolerant communication. Automated decisions actuated by devices of such infrastructure can have a direct impact on the environment and human life. In an intelligent energy grid, this may include decisions on supplying power to critical devices, such as medical life support systems, while shutting down power to non-critical devices of a hospital.

This work contributes a fault tolerant communication mechanism for micro grids at low cost and high scalability. Such a provision enables IEM and IFM nodes to communicate, even in the event of multiple link failures. The first step to accomplish this is through introducing increased but intelligently distributed redundancy in the links of the network. We introduced a framework of middleware components that utilize software overlays to support fault-tolerant communication. The network overlay proves to be resilient by exploiting redundancy through utilization of alternate communication paths at the software level, and it is shown to provide stability in terms of sustained resilience as networks are scaled up in future micro grids.

The work further allows cheap switching equipment to be deployed as a means to complement legacy hierarchical network topologies. Since switches lack dynamic routing capabilities, our middleware realizes re-routing using the software overlay. This is significantly less costly than deployment of routers instead of switches. Our development of low-overhead route detection algorithms to assist in the presence of single and multiple link failures constitutes the key contribution to provide such fault tolerance in a transparent manner to other control software. Our middleware layer provides the means for higher-level distributed grid intelligence (DGI), such as providing hierarchical control schemes within this software overlay architecture. Thus, the vision of sustainable, scalable and reliable decentralized energy management on the software side in the FREEDM system and for other CPS domains in general is provided by our software middleware architecture for fault tolerant network overlays.

References

1. NERC final report, `http://www.nerc.com/docs/docs/blackout/ch5.pdf`
2. North carolina state university freedm project, `http://www.freedm.ncsu.edu`
3. Cacti: The complete rrdtool-based graphing solution (2005), `http://www.cacti.net`
4. Akella, R., Meng, F., Ditch, D., McMillin, B., Crow, M.: Distributed power balancing for the freedm system. In: Proceedings of the 2010 Annual FREEDM Conference (2010)
5. Albert, R., Albert, I., Nakarado, G.: Structural vulnerability of the north american power grid. Physical Review E 69(2) (2004), doi:10.1103/PhysRevE.69.025103
6. Andersen, D., Balakrishnan, H., Kaashoek, M.F., Morris, R.: The case for resilient overlay networks. In: Proceedings of the 8th Annual Workshop on Hot Topics in Operating Systems HotOSVIII, pp. 152–157 (2001)
7. Berthier, R., Bobba, R., Davis, M., Rogers, K., Zonouz, S.: State estimation and contingency analysis of the power grid in a cyber-adversarial environment. In: NIST Workshop on Cybersecurity for Cyber-Physical Systems (2012)
8. Bondy, J.A.: Graph Theory With Applications. Elsevier Science Ltd. (1976)
9. von Eicken, T., Culler, D.E., Goldstein, S.C., Schauser, K.E.: Active messages: a mechanism for integrated communication and computation. In: International Symposium on Computer Architecture, pp. 256–266 (1992)
10. Huang, Z., Wang, C., Nayak, A., Stojmenovic, I.: Small cluster in cyber physical systems: Network topology, interdependence and cascading failures. IEEE Transactions on Parallel and Distributed Systems PP(99), 1 (2014)
11. Killian, C., Anderson, J., Braud, R., Jhala, R., Vahdat, A.: Mace: language support for building distributed systems. In: ACM SIGPLAN Conference on Programming Language Design and Implementation, pp. 179–188 (2007)

12. Levis, P., Madden, S., Polastre, J., Szewczyk, R., Whitehouse, K., Woo, A., Gay, D., Hill, J., Welsh, M., Brewer, E., Culler, D.: Tinyos: An operating system for sensor networks, pp. 115–148 (2005), http://dx.doi.org/10.1007/3-540-27139-2_7, doi:10.1007/3-540-27139-2_7
13. Li, J., Blake, C., De Couto, D.S., Lee, H.I., Morris, R.: Capacity of ad hoc wireless networks. In: MobiCom 2001: Proceedings of the 7th Annual International Conference on Mobile Computing and Networking, pp. 61–69. ACM, New York (2001), http://doi.acm.org/10.1145/381677.381684
14. Massie, M.L., Chun, B.N., Culler, D.E.: The ganglia distributed monitoring system: Design, implementation and experience. Parallel Computing 30, 2004 (2003)
15. Nguyen, D., Shen, Y., Thai, M.: Detecting critical nodes in interdependent power networks for vulnerability assessment. IEEE Transactions on Smart Grid 4(1), 151–159 (2013), doi:10.1109/TSG.2012.2229398
16. Shao, J., Buldyrev, S.V., Havlin, S., Stanley, H.E.: Cascade of failures in coupled network systems with multiple support-dependent relations. CoRR abs/1011.0234 (2010), http://dblp.uni-trier.de/db/journals/corr/corr1011.html#abs-1011-0234
17. Stoianov, I., Nachman, L., Madden, S., Tokmouline, T.: Pipeneta wireless sensor network for pipeline monitoring. In: IPSN 2007: Proceedings of the 6th International Conference on Information Processing in Sensor Networks, pp. 264–273. ACM, New York (2007), http://doi.acm.org/10.1145/1236360.1236396
18. Tomsovic, K., Bakken, D., Venkatasubramanian, V., Bose, A.: Designing the next generation of real-time control, communication, and computations for large power systems. Proceedings of the IEEE 93(5), 965–979 (2005)
19. Varma, J., Wang, C., Mueller, F., Engelmann, C., Scott, S.L.: Scalable, fault-tolerant membership for MPI tasks on hpc systems. In: International Conference on Supercomputing, pp. 219–228 (2006)
20. Wauters, T., De Turck, F., Develder, C.: Overlay networks for smart grids. IEEE Smart Grid Research: Communications, 1–27 (2013)
21. Yan, J., He, H., Sun, Y.: Integrated security analysis on cascading failure in complex networks. IEEE Transactions on Information Forensics and Security 9(3), 451–463 (2014), doi:10.1109/TIFS.2014.2299404
22. Zimmer, C., Mueller, F.: The freedm architecture of fault tolerant network routing through software overlays. In: FREEDM Conference (2009)

Biologically Inspired Hierarchical Cyber-Physical Multi-agent Distributed Control Framework for Sustainable Smart Grids

Jin Wei and Deepa Kundur

Abstract. It is well known that information will play an important role in enhancing emerging power system operation. However, questions naturally arise as to when the increased data-dependence may be considered excessive. Two practical considerations emerge: 1) communications and computational overhead, in which redundant and irrelevant information acquisition and use results in heavy computational burden with limited performance return, and 2) increasing risks of cyber attack whereby indiscriminate cyber-dependence and -connectivity increases attack scope and impact. In this chapter, we present a hierarchical cyber-physical framework of power system operation based on flocking theory in the context of the smart grid stability problem. We study strategies to harness an appropriate degree of cyber technology by effectively leveraging physical couplings. Our formulation enables the identification of large-scale distributed control strategies for robust power grid operation. Furthermore, our formulation also enables a novel witness-based cyber-physical protocol whereby physical coherence is leveraged to probe and identify phasor measurement unit data corruption and estimate the true information values for attack mitigation.

1 Introduction

1.1 Background

The National Academy of Engineering hails the electric power grid as the 20th century's innovation most beneficial to civilization [25]. The electric power grid

Jin Wei
Department of Electrical and Computer Engineering, The University of Akron, Akron, OH 44325, USA
e-mail: `jwei1@uakron.edu`

Deepa Kundur
Department of Electrical and Computer Engineering, University of Toronto, The Edward S. Rogers Sr., Toronto, ON M5S 2E4, Canada
e-mail: `dkundur@comm.utoronto.ca`

S.K. Khaitan, J.D. McCalley, and C.-C. Liu (eds.), *Cyber Physical Systems Approach to Smart Electric Power Grid*, Power Systems, DOI: 10.1007/978-3-662-45928-7_9

started in 1896, based in part on Nikola Tesla's design published in 1888 [58]. It is the fundamental infrastructure of modern society. Transportation, communications, finance, and other critical infrastructures are dependent upon its secure, reliable electricity supplies for energy and control. The term "electric power" is the rate at which electrical energy is transferred by an electric circuit to produce useful work involving heat, light, motion, sound, information technology processes, and chemical changes. Energy is a quantity that measures the ability of a physical system to produce change on another physical system. Changes are produced when the energy is transferred from one system to another through (1) physical/thermodynamical work, (2) heat and/or (3) mass transfer. Electricity is an energy carrier. Although energy is not naturally available in the form of electricity nor is electricity directly used to produce change, its conversion to and from electricity enables the transmission of power from generation to consumption over a complex interconnected grid. The term grid in the context of power systems has traditionally been used to represent the network of electrical components used to supply, transmit and consume electric power. This term can refer to the complete or a suitable subset of electricity generation, transmission, and distribution infrastructure [48, 74, 77]. Popular grid topologies in North America are radial and mesh while loop topologies are predominant in Europe.

In recent years, electricity demand is changing and growing very fast. For example, the devices and infrastructures needed to operate the fundamental communication network, data centers, and storage alone add more than 2500 Megawatt hours (MWh) of demand globally per year that did not exist five years ago. In 2012, the average monthly electricity consumption for a U.S. residential utility customer was 903 kWhs [6]. It is expected that the world's electricity demand will be triple by 2050. The increasing electricity demand causes electric transmission congestion and atypical power flows threaten to overwhelm the power grids which face many challenges that they were not designed and engineered to handle. Because modern infrastructure systems are so highly interconnected, a change in conditions at any one location can have immediate impacts over a wide area, and the effect of a local disturbance even can be magnified as it propagates through a network. Large-scale cascade failures can occur almost instantaneously and with consequences in remote regions or seemingly unrelated businesses. On the North American power grid, for example, transmission lines link all electricity generation and distribution on the continent. Wide-area outages in the late 1990s and summer 2003 underscore the grids vulnerability to cascading effects [11, 103]. Furthermore, with the increasing energy demand, the modern power grid is growing into a complex network with numerous interconnected regional grids, owned and operated by power corporations at all levels and scales. The complex interests, operations, and management among different power corporations often complicate cross-region transmission tasks and sometimes result in an inefficient or poorly-coordinated power delivery. The deregulation of the energy industry necessitates high granularity of informational, financial and physical transactions to assure adequate power system operation in a competitive electricity market. However, the traditional grid has not kept pace with these modern challenges [44]. Moreover, mitigating climate change requires large-scale incorporation of renewable sources into the energy mix. The International Energy Agency predicts that hydro power will remain the major source of renewable energy for the next two decades, followed by wind and solar. The challenges of integrating these renewable energy sources into the electrical system are different for each technology but the system of the future must accommodate them all. Therefore, achieving high levels of renewables will require the systems to be more flexible, responsive and intelligent, which is substantially different from the existing grids [5]. Therefore, the existing grids are under pressure to deliver the growing demand for power, as well as provide a stable and sustainable supply of electricity. These complex challenges are driving the evolution of Smart Grids, which are considered as the next-generation electric power grids.

1.1.1 Smart Grid Visions

A smart grid can be described as the result achieved by integrating advanced control and communication technologies with the traditional power grid. Because of this integration, in a smart grid, there are both bidirectional information flow and bidirectional physical power flow. One of the key components is improved (human) operator interface and decision support. There is not yet an internationally unified definition of a smart grid. The North American Electric Reliability Corporation (NERC) defines the smart grid as the integration and application of real-time monitoring, advanced sensing, communications, analytics, and control, enabling the dynamic flow of both energy and information to accommodate existing and new forms of supply, delivery, and use in a secure, reliable, and efficient electric power system, from generation source to end-user [2].

The marriage of information technology with traditional power grids enables the smart grids exhibit advanced functionalities. For example, by broadly deploying advanced sensors on critical components, a smart grid is able to visualize the power system in real-time. By upgrading the control and protection techniques, a smart grid is able to more effectively utilize the grids' capacity. A smart grid is able to be situationally-aware and self-healing via wide-scale deployment of power electronic devices such as power electronic circuit breakers and Flexible AC Transmission Systems (FACTS). Furthermore, the integrated communication networks in the smart grids enhance consumer-centricity such that the power delivery system is expanded by using Supervisory Control and Data Acquisition (SCADA) systems and other wide-area monitoring techniques, electricity services are improved by developing the home automation systems and enabling the real-time charging and billing information.

The smart grids' advanced functionalities facilitate their goals on delivering high efficiency from technical, environmental, and economic perspectives. Technically, the smart grids intend to protect physical and information assets from man-made and natural threats, develop self-healing delivery infrastructure, and ensure extremely reliable delivery of "digital-grade" power to increasing numbers of end-users. From the environmental prospective, the smart grids target to reduce carbon footprint by accommodating renewable and traditional energy sources. Economically, the smart grids enhance consumer-centricity and propose affordable maintenance in order to stay globally competitive.

Besides the definition of smart grid provided by NERC, there are various alternative views of smart grids suggested by different organizations. For instance, in Electric Power Research Institute's (EPRI's) viewpoint, the objective of the smart grid is the convergence of greater consumer choice and rapid advances in communications, computing and electronic industries [4, 45]. The U.S. Department of Energy (DOE) denotes operating principles of the smart grid where open but secure system architecture, communication techniques and standards are used to provide value and choice to consumers [50, 111]. The smart grid criteria defined by the multinational corporation ABB includes adaptive, predictive, integrated, interactive between customers and markets, optimized to maximize reliability, availability, efficiency and economic performance, and secure from attack and naturally occurring disruptions [60]. Overall, although there is no definition of the smart grid that prevails, all the smart grid visions agree on the general theme that the smart grid aims to improve functionality of power delivery system with use of advanced technology which are both cyber and physical.

1.1.2 Security Challenges and Fundamental Questions

While the extensive integration of cyber technology with the power system significantly improves reliability and efficiency, it also introduces additional risk from cyber attacks. The security of a system is as strong as its weakest link. Thus, the high complexity of the smart grid cause the system weakness to become aggravated and result in previously unknown emergent properties. The increased connectivity provides external access to the system weakness, which in turn can lead to compromise and infection of components. Furthermore, the tight collaboration of cyber technology and the power grid enables the attackers to increase the capabilities to exploit the system weakness. The interaction of these three components creates a host of unfamiliar vulnerabilities stemming from cyber intrusion and corruption potentially leading to devastating physical effects. For example, the first-ever control system malware called Stuxnet was found in July 2010. This malware, targeting vulnerable SCADA systems, shows that attackers have the ability to develop this type of cyber-physical attacks [40, 113]. From a technical perspective there is increased opportunity for cyber attack because of the greater dependence on intelligent electronic devices, communications and advanced metering amongst other intelligent systems. Such cyber infrastructure typically employs standardized information technologies that may have documented vulnerabilities. Coupled with increased economic motivations for attack that stem, in part, from privatization of the energy industry, cyber security of the smart grid represents a timely research and engineering problem.

Furthermore, enhancing the smart grid security is also important for protecting the public from terrorism, vandalistic hackers, disgruntled insiders of the electric power industry and cascading failures from the loss of other critical infrastructures. The associated attacks on availability can result in damaging instability such as blackouts and brownouts. Moreover, securing a smart grid makes business sense.

Protection of cyber devices is necessary to establish compliance to cyber security requirements to be able to compete in the electricity marketplace. Security also represents a means to reduce or divert technical liability and assure revenue by discouraging competitor component cloning.

Numerous reports are appearing which acknowledge current security concerns of the smart grid [28,59,66,83,87,121]. Some guidelines have also been published by government agencies and other authoritative organizations, such as NISTIR 7628 Guidelines for Smart Grid Cyber Security developed by the National Institute of Standards and Technology (NIST) [52], the document Roadmap to Achieve Energy Delivery System Cyber Security released by the DOE [3], and Critical Infrastructure Protection (CIP) standards proposed by the NERC [1]. These reports and guidelines raise three fundamental research and development questions for improving the smart grid security: (1) What are the electrical system impacts of a cyber attack? (2) How should security resources be prioritized for the greatest advantage? (3) Is the additional information available through advanced cyber infrastructure worth the increased security risk? Moreover, two main concerns on cyber attacks are specified by the reports and guidelines: (1) the possibility of attacks on information accuracy such as the false data injection attacks, and (2) the possibility of attacks on timely data delivery such as denial of information access on the SCADA control system.

1.2 Prior Art

Recently, smart grid researchers have been trying to develop potential solutions for the fundamental questions to enhance the smart grid security. It has been realized that security vulnerability analysis for the smart grid is able to aid in answering those questions. Cyber attacks on the smart grid, commonly classified as either outsider or insider, can occur within devices or along the communications paths of the cyber infrastructure. To address outsider attacks, in which an opponent has no specialized security information such as secret keys, mechanisms for authentication, access control, data integrity, confidentiality and non-repudiation suitable for smart grid infrastructure are being developed [8, 12, 14, 16, 18, 19, 24, 29, 31, 35, 38, 41–43, 47, 53, 55–57, 61, 62, 67–71, 86, 90, 91, 97, 98, 100, 104, 107, 112, 114, 118, 122]. Essentially, cryptographic primitives are applied to make such attacks either practically impossible or detectable thus alerting appropriate parties of an attack. The problem of insider attacks, in contrast, involves a trusted but corrupted entity such as a smart meter that has full access to secret keying information; here, the corrupted entity can apply numerous attacks such as falsification or delaying of data and go undetected possibly for some time or until, for example, a power delivery disruption occurs. Typically, it is difficult to immediately identify the exact source of a cyber attack and mechanisms such as islanding can be applied to isolate the corrupted components from causing large-scale disruption [10].

Research focused on cyber security often takes an information-centric perspective in which data protection is of paramount importance [23]. For smart grid applications where consumer-centricity is emphasized, efficient and safe power delivery services are a more significant concern to stakeholders than the health of the support-data used to control it. It is possible that investment in cyber security that leads to improvements in information technology has only negligible advantage for the power system [68]. It is therefore important to focus on assessing the impacts of cyber attacks on the electricity network to identify possible new vulnerabilities, develop countermeasures and prioritize mitigation investment. Initial research into cyber security of power systems focused solely on the cyber infrastructure [8, 12, 14, 16, 18, 19, 24, 29, 31, 35, 38, 41–43, 47, 53, 55–57, 61, 62, 67–71, 86, 90, 91, 97, 98, 100, 104, 107, 112, 114, 118, 122]. It is true that protection of the data better facilitates a safer electrical grid. However, because of the limited resources of electric power utilities, it is also necessary to understand the cost-benefit trade-offs of protection mechanisms. Proper smart grid risk analysis necessitates that vulnerability assessment take into account the physical impacts of cyber attack [32, 89]. Thus recently there has been a movement to incorporate cyber-physical information. For emerging smart grid topologies this interface commonly occurs at the sensors and actuators, such as intelligent electronic devices (IEDs), remote terminal units (RTUs), programmable logic controllers (PLCs), that are acquiring data from and using data to control electrical components [73, 82, 84, 85, 93, 94].

Recently, power system cyber security research thrusts have focused on modeling this unique cyber-to-physical bridge for a smart grid which aids in analyzing the impact of cyber attack on the power system. These techniques can be grouped into a number of classes. One class of static methodologies identify the cause-and-effect relationships within the cyber-to-physical bridge [26, 63, 64, 81, 110] to relate one or more cyber attacks to one more more physical consequences that are further analyzed using power system-specific tools. To account for the effects of time scale and timing on the overall system security, one class of empirical approaches has focused on merging well-developed simulators/emulators for the communications infrastructure, power systems, and control centers [36, 36, 38, 54, 80, 101, 106] to account for the dynamic nature of the interactions. These two forms of simulators are combined such that an attack is applied in the communication simulator that transfers data to the power systems simulator which makes decisions based on this possibly corrupt information. Typical traditional power system reliability metrics are used to assess impact of the cyber attacks. In cyber-physical leakage approaches confidentiality of the cyber network is studied by identifying how voltage and current measurements of the physical power system can be analysed for any clues about cyber protocol activity [17, 17, 51, 88, 108, 109]. Similarly, such contextual information relating cyber and physical dependencies have been exploited for intrusion detection [27, 27, 72, 105, 119, 120]. Testbed systems research addresses the exploration of practical vulnerabilities through SCADA testbed development and construction [30, 33, 46]. Much of this valuable research has proven that cyber attacks have the potential to cause significant disruptions in power delivery. However, the individual cyber and electrical simulators are often incompatible for study within a common framework. Commonly, exhaustive searches must be employed in order to understand worst-case scenarios. Attempts to provide more analytic insights into the problem for general feedback control system architectures have also been pursued [9, 20, 21, 21, 22], which focuses on how data corruption of denial of information access can affect the control of the power grid. Finally, the research in [75, 76] represented a work in progress towards the development of a comprehensive and practical framework for electric smart grid cyber attack impact analysis.

1.3 Methodology and Motivation of Biologically Flocking-Based Perspective

As illustrated in Fig. 1, in order to achieve our research objectives, we make use of the tool-sets consisting of graph theory, dynamical-system formulation, and flocking rules. A graph is defined by a collection of vertices (also called nodes) and a

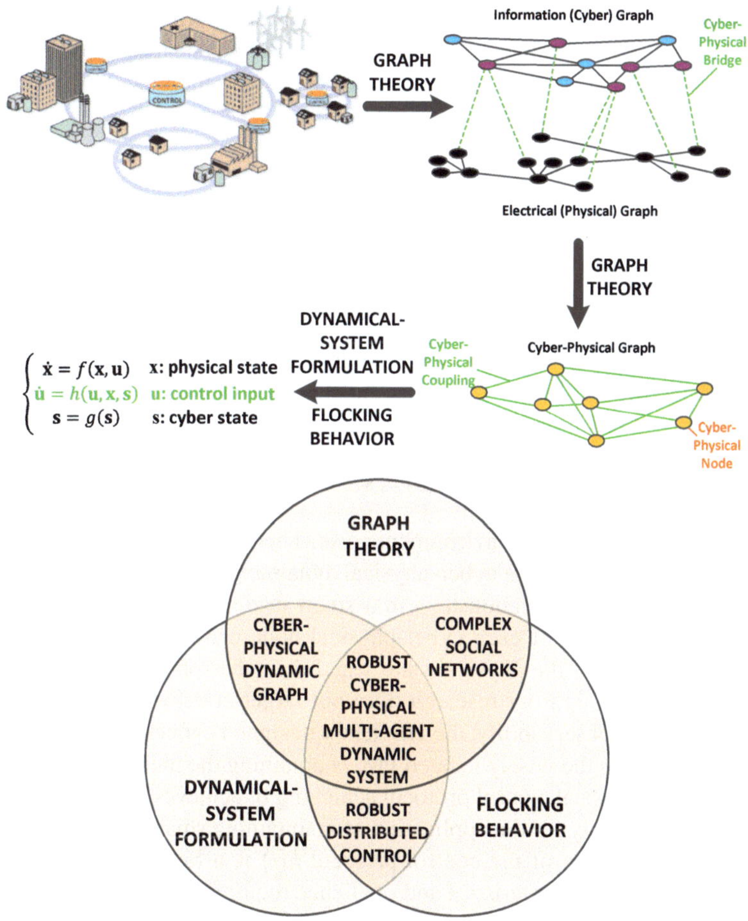

Fig. 1 Methodology overview: tool-sets consisting of graph theory, dynamical-system formulation, and flocking rules

collection of edges that connect node pairs. It is a mathematical structure that represents pairwise relationships between a set of objects. Depending the use of a graph, its edges may or may not have direction leading to directed or undirected classes of graphs, respectively. Graphs provide a convenient and compact way to describe the cyber-physical interactions and relate dependencies within a power system as witnessed by recent papers that use this tool [37, 39, 54]. However, as stated in [39], purely graph-based approaches do not sufficiently model the state changes within the physical system. Moreover, they do not effectively account for the unique characteristics of the system at various time-scales nor provide a convenient framework for modeling system physics. We assert that modeling the electrical grid is a vital component to an effective impact analysis framework.

One approach to physically modeling complex engineering interactions employs dynamical systems. A dynamical system is a mathematical formalization used to describe time-evolution of a system state, which can typically represent a vector of physical quantities. As shown in Fig. 1, $\mathbf{x}$ denotes the physical state of the system. Because of the physical characteristics of power system, the time-evolution of $\mathbf{x}$ is described by the following differential equation:

$$\dot{\mathbf{x}} = f\left(\mathbf{x}, \mathbf{u}\right),$$

where $\dot{\mathbf{x}}$ is the time-derivative of $\mathbf{x}$, $\mathbf{u}$ is the control input obtained by the cyber-physical interaction, and the function $f\left(\cdot\right)$ is determined by the power system network topology in our work.

Dynamical systems theory is motivated, in part, by ordinary differential equations and is well-suited to representing the complex physical interactions of the power grid. Furthermore, $\mathbf{s}$ in Fig. 1 represents the cyber measurement of the system and the measurement function $g\left(\cdot\right)$ in our work is formulated as follows:

$$g\left(\mathbf{s}\right) = \mathbf{s} + \mathbf{n},$$

where $\mathbf{n}$ denotes the random environment noise. Therefore, the graphs and dynamical systems tool-sets enable a cyber-physical dynamic graph representing the cyber and physical grid entity relationships in a smart grid. As shown in Fig. 1, in the graph, the state change of each cyber-physical node can be formulated by a dynamic function $f(\cdot)$ of the physical state $\mathbf{x}$ and the cyber-physical control input $\mathbf{u}$. We clarify that although our research does not target at achieving complete state controllability and observability, the efficiently designed cyber-physical integration in our work, such as the wisely located PMUs obtaining the measurement $\mathbf{s}$ and the proposed cyber-physical control protocol achieving $\mathbf{u}$, achieves sufficient controllability and observability for the application of maintaining smart grid stability.

However, the design of the control protocol $h(\cdot)$ is a big challenge due to the complex networked characteristics and resilience requirements of smart grids. Fortunately, flocking behavior in the nature sheds light on the robust distributed control design for complex systems. The collective behavior coordination and local interaction in flocks contribute to an effective solution for accomplishing the system objectives via robust distributed control and communication. Furthermore, the emergent behavior in flocks, such as obstacle avoidance, provides an essential idea to achieve the situational-awareness in real-time for smart grids.

We assert that the tool sets consisting of graph theory, dynamical-system formulation, and flocking behavior are effective for a smart grid vulnerability assessment and security design for a variety of reasons. First, effective smart grid attack analysis necessitates relating the cyber attack to physical consequences in the electricity network. A dynamical systems paradigm provides a flexible framework to model (with varying granularity and severity) the cause-effect relationships between the cyber data and the electrical grid state signals and ultimately relate them to power delivery metrics. Second, graphs enable a tighter coupling between the cyber and physical domains. For a smart grid, the cyber-to-physical connection is often represented through control signals that actuate change in the power system and the physical-to-cyber connection is typically due to the acquisition of power state sensor readings. These connections can be conveniently expressed as specifically located edges of the graphs. This way cascading failures and emergent properties from the highly coupled system can be represented. Mitigation approaches such as active control or islanding of the grid or partitioning of the core smart grid components for optimal functions, and a graph-based dynamical systems formulation can naturally portray such separation as well. Third, the flocking behavior exhibits novel and essential principles to efficiently design the security strategies for an overall system resilient to cyber and physical disruption. Last, a primary effect of including cyber attacks in traditional reliability analysis is that it increases the size of the system under study by several orders of magnitude. Our proposed mathematical formulation has the potential to keep studies tractable because our granularity of detail can be tuned and the use of dynamics can enable sophisticated behaviours without a corresponding increase in complexity.

1.4 Contributions

In this chapter, we propose a flocking-based hierarchical cyber-physical security analysis framework which incorporates cyber intelligence and control behaviors by taking a *flocking* perspective commonly used to model large-scale natural phenomenon. We assert that our framework has the following advantages. First it enables the convenient integration of cyber (communications and control) systems within dynamical models of power system physics. Second, the structure of our models conveniently enables the study of the important smart grid stability problem. Third, the models of cyber system dynamics can be employed to gain insight on effective smart grid distributed communications and control strategies for system performance and stabilization. Fourth, the analogy between the dynamics of synchronous generators and the flocking behavior in the nature enables the exploration on how information and physical couplings can be synergistically harnessed for re-stabilizing a power grid under severe attack or fault. Through analysis we assess how hierarchy and the selective use of cyber information can benefit scalability and robustness to information attack. Through a flocking-based paradigm we develop distributed control methodologies that leverage cooperation between distributed energy resources (DERs) and traditional synchronous machines to maintain transient stability in the face of severe disturbances. We also introduce and apply the notion of state-dependent hierarchy in which coherent generator clusters from disturbance are leveraged such that strong physical couplings are identified to selectively apply distributed cyber-control where necessary. Furthermore, based on the proposed hierarchical cyber-physical security analysis framework, we consider a cyber-physical viewpoint to the problem of data corruption in smart grid systems. We take the perspective that one may leverage natural physical couplings amongst power system components as telltale signs to identify information corruption and demonstrate how *cyber* corruption can be identified within the power system by taking a hierarchical cyber-physical perspective. Specifically, the *physical* coherence within the second tier of a two-tier cyber-physical structure is probed to execute a "witness"-based cyber-physical protocol to identify and mitigate cyber attack in first tier.

This chapter is organized in the following sections. In Section 2, we introduce a dynamic multi-agent system framework on cyber-physical integration modeling for the application of smart grid stability maintenance. A hierarchical control protocol design inspired by the analogy to flocking behavior is proposed in Section 3. Our proposed timely dynamic agent coherency identification for achieving hierarchy is briefly introduced in Section 4. In Section 5, we develop a witness-based verification and estimation protocol for detection and mitigation of information corruption on critical PMU data. The performance is evaluated in Section 6 and the conclusions are provided in Section 7.

2 Dynamic Multi-agent System Framework for Cyber-Physical Integration Modeling

In our research, we consider the smart grid stability from the power system (physical) perspective, which derives from standard control stability [78] and can be seriously impacted by the cyber-physical interactions in the system. In contrast to the control stability, the power system stability is defined as the ability of an electric power system, for a given initial operating condition, to regain a state of operating equilibrium after being subjected to a disturbance, with all system variables bounded so that system integrity is preserved [78, 79].

2.1 *Smart Grid Stability*

There are three types of stability are considered for power systems: rotor angle stability, frequency stability, and voltage stability. Our research focuses on improving the rotor angle stability and frequency stability of the system in the face of large system disturbance. Therefore, let $\theta_i(t)$ denote the rotor phase angle of Generator i at time t and ω_i be the normalized relative frequency of Generator i with respective to f_0 at time t. Based on the definitions and requirements of rotor angle stability and frequency stability, we are able to characterize the smart grid stability which is of interest to our research as follows:

Smart Grid Stability: a smart grid is able to achieve both phase angle cohesiveness and exponential frequency synchronization within 1 to 3 seconds following a severe disturbance:

1. Phase angle cohesiveness:

$$\left|\theta_i(t) - \theta_j(t)\right| \leq \gamma, \; for \; \forall t, \tag{1}$$

where the threshold γ is normally set as $5\pi/9$ in the realistic application as discussed in *[102];*

2. Exponential frequency synchronization:

$$\omega_i(t) \to 0, \; as \; t \to \infty. \tag{2}$$

2.2 Physical Dynamics and Interaction

According to the definition of Smart Grid Stability, the synchronous generators are the critical physical components. Therefore, modeling the physical dynamics of the synchronous generators and analyzing the interaction between them are necessary for maintaining smart grid stability. We describe the *physical* system by abstracting the information on the physical coupling between these critical components. We employ the well-known interconnected swing equations to describe rotor dynamics [78] of the Kron-reduced [15] power system as detailed by Dörfler and Bullo [34] to give the following dynamical representation for each agent:

$$M_i \dot{\omega}_i = -D_i \omega_i + P_{m,i} - |E_i|^2 G_{ii} - \sum_{j=1}^{N} P_{ij} \sin\left(\theta_i - \theta_j + \varphi_{ij}\right) \tag{3}$$

where $i \in \{1, 2, ..., N\}$ represents the generator index, θ_i denotes the rotor phase angle measured with respect to a rotating frame reference at frequency $f_0 = 60$ Hz, $\omega_i = \dot{\theta}_i$ is the normalized relative frequency, $M_i > 0$ and $D_i > 0$ represent the generator inertia and the damping parameters, respectively, and E_i, $P_{m,i}$ and G_{ii} are the internal voltage, mechanical power input and equivalent shunt conductance of Generator i, respectively. $P_{ij} = |E_i||E_j||Y_{ij}|$ and $\varphi_{ij} = \arctan\left(G_{ij}/B_{ij}\right)$ where Y_{ij}, G_{ij} and B_{ij} are the Kron-reduced equivalent admittance, conductance and susceptance, respectively, between Generators i and j.

2.3 Hierarchical Cyber-Physical Integration Framework

Based on the achieved dynamic graph providing an abstract representation of the power system, we are able to design a cyber-physical integrated framework in which the cyber and phsyical systems work synergistically such that the bidirectional cyber information and power flows are efficiently used to enhance system resilience.

As stated in [103], the essential characteristics of a smart grid include: 1) situational awareness in real time, 2) energy storage used and controlled to support system goals, 3) distributed control and protection integrated with other functional units. According to these characteristics, we model the cyber-physical integration in the smart grid with a two-tier hierarchical multi-agent framework shown in Fig. 2.

Each agent consists of both cyber and physical elements: (1) a dynamic node representing a physical power system element, in this case a generator, (2) a phasor measurement unit (PMU) that acquires generator phase angle and frequency data from the dynamic node, and (3) a local cyber-controller that computes a control signal that is applied to the agent's generator using PMU data. Each agent's frequency, phase angle, and coherency characteristics are those of its generator. The PMU and local controller are both considered to be cyber elements due to their data acquisition, communication and computation tasks. The physical coherency between active agents is timely achieved by using our real-time dynamic coherency identification method which will introduced in Section 6. The agents with high physical coherency are considered to form a *cluster* and one agent within the cluster (typically with highest generator inertia) is selected as the *lead* agent.

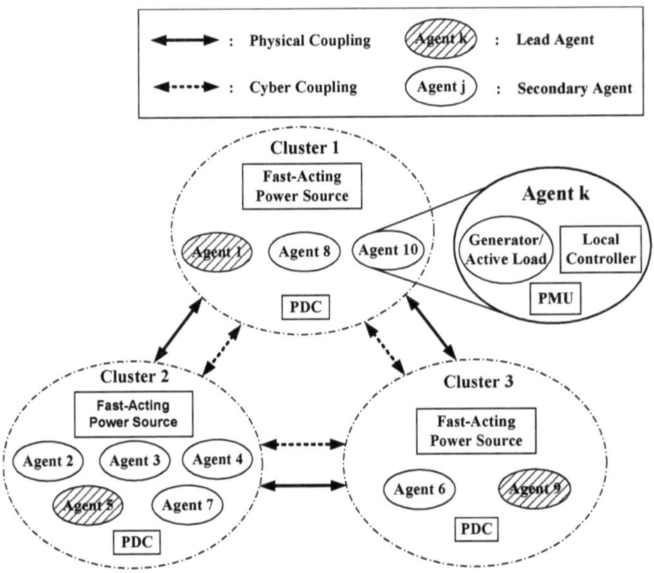

Fig. 2 Proposed two-tier hierarchical cyber-physical integrated multi-agent framework

We illustrate the implementation of the hierarchical control framework for the well-known New England 39-bus system in Fig. 3. Here, we assume there are three clusters and the lead agent of each cluster is denoted with a shaded (green) generator. Effective PMU information (cyber) and power (physical) flows are presented as dashed and solid arrows, respectively. To further delineate the tiered nature of communications, red, blue and magenta dashed arrows represent tiered communications from lowest to highest level. Therefore, only the lead agent's PMU and local cyber-control are activated for overall cluster regulation and the phasor data concentrator (PDC) in each cluster is implemented to guarantee synchronization of the data information flows amongst lead agents. Therefore, this enables a state-dependent system hierarchy whereby inter-cluster interactions are cyber-physical (tier-1) and intra-cluster synergies are physical (tier-2). Since our focus is on smart grid stability problem, the objective of the local controller is to achieve generator phase angle cohesiveness and exponential frequency synchronization in the face of cyber-physical disturbance. As such, the local controllers may require fast-acting External Energy Storages (EESs) in order to achieve their objectives as shown in Fig. 2. These storages in practice may include battery storage devices, flywheels, renewable energy sources, and other types of massive energy storage [7,49], and may be separate from each agent.

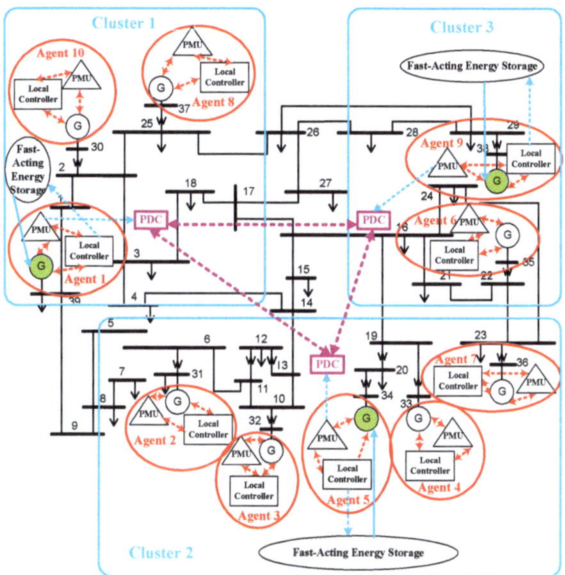

Fig. 3 Hierarchical cyber-physical control for New England 39-bus system

2.4 *Cyber-Physical Interaction*

We have introduced the concept of hierarchical cyber-physical integration framework for smart grids by modeling the system as a hierarchical multi-agent system. In this section, we continue to formulate the cyber-physical interaction between the multiple agents.

2.4.1 Dynamical Description of Cyber-Physical Interaction

In this hierarchical framework, the *cyber* network (PMU data + local controllers) is integrated into this framework through controlling the fast-acting EES power absorption/injection, $P_{u,i}$, to Generator Bus i to compensate for fluctuations in demand power in the system after a severe disturbance. Letting the control signal $u_i = P_{u,i}$ and α_i be a binary number defined as follows:

$$\alpha_i = \begin{cases} 1, \text{ if the } i\text{th agent is the lead agent;} \\ 0, \text{ otherwise,} \end{cases} \tag{4}$$

we can formulate the dynamics of our cyber-physical integrated framework as follows:

$$M_i\dot{\omega}_i = -D_i\omega_i + P_{m,i} - E_i^2 G_{ii} - \sum_{j=1, j\neq i}^{N} \underbrace{P_{ij}}_{\text{phys } \S} \sin\left(\theta_i - \theta_j + \underbrace{\varphi_{ij}}_{\text{phys } \S}\right) + \underbrace{\alpha_i u_i}_{\text{cyber } \S}. \tag{5}$$

where u_i is the control signal for the ith agent computed from PMU data (θ_j, ω_j for $j \in \{1,2,...,N\}$). The control can be interpreted as power injection for $P_{u,i} > 0$ or absorption for $P_{u,i} < 0$ at the corresponding generator buses from the fast-acting external power sources. Thus, it represents a cyber-to-physical bridge whereby computation of u_i is converted to active power flow. Similarly, a physical-to-cyber bridge exists at the measurement devices in which physical phase angle and frequency are converted to PMU data. Thus, the dynamics of Eq. (5) represents both cyber and physical interactions. Physical inter-agent couplings (denoted phys §) are characterized by parameters P_{ij} and φ_{ij} and cyber couplings (cyber §) through u_i. For normal operation $u_i = 0$. However, when a disturbance strikes, u_i will excite the system to re-achieve (smart grid) stability.

We design the control signal u_i under two assumptions. First, we assume that, in the face of severe disturbance, $u_i = P_{u,i}$ changes much faster than the mechanical power input $P_{m,i}$ for each agent and the time span to recover smart grid stability is short; thus we treat $P_{m,i}$ as a constant during the procedure of maintaining smart grid stability. This assumption is reasonable for future smart grids where fast-response energy storage such as battery storage and flywheels will be available to inject and absorb energy for periods of brief control. Second, we assume that the problems of voltage regulation and frequency synchronization are decoupled. This enables us to consider the voltage E_i as a constant during controller excitation to re-achieve the frequency synchronization.

In order to reformulate the problem of cyber-physical control for maintaining transient stability as a task of flocking formation control, we intend to present the dynamics of each agent in our cyber-physical integrated system, which is originally formulated in Eq. (5), in the form of a double integrator model. Under these assumptions, computing derivatives of the both sides of Eq. (5), and reformulating gives:

$$\begin{cases} \dot{\theta} = \omega, \\ \mathbf{D}\dot{\omega} = -\mathbf{M}\ddot{\omega} - \mathbf{L}\omega + \alpha\dot{\mathbf{u}}. \end{cases} \tag{6}$$

where the index assignments are reordered such that Agents $i = 1, \ldots, C$ correspond to lead agents, C is the number of clusters in our hierarchical framework, $\alpha = \mathbf{diag}[\alpha_1, \ldots, \alpha_N]$, $\alpha_i = 1$ for $i \leq C$, and $\alpha_i = 0$ otherwise. $\theta = [\theta_1, \ldots, \theta_N]^T$, $\omega = [\omega_1, \ldots, \omega_N]^T$, $\mathbf{u} = [u_1, \ldots, u_N]^T$, $\mathbf{M} = \mathrm{diag}[M_1, \ldots, M_N]$, $\mathbf{D} = \mathrm{diag}[D_1, \ldots, D_N]$, and $\mathbf{L}$ is a $N \times N$ physical coupling matrix whose elements can be represented as:

$$l_{ij} = \begin{cases} \sum_{j=1, j\neq i}^{N} P_{ij} \cos(\theta_i - \theta_j + \varphi_{ij}), & \text{if } i = j; \\ -P_{ij} \cos(\theta_i - \theta_j + \varphi_{ij}), & \text{if } i \neq j, \end{cases} \tag{7}$$

2.4.2 Hierarchical Cyber-Physical Dynamics

In our hierarchical framework, the agents are grouped into the same cluster if they have high physical coherency. Since the term of Generator Coherency refers to the characteristics that the states of the coherent generators are close to each other [78], it is reasonable to assert that the deviations between the states (i.e. phase angle and normalized relative frequency) of the secondary agents and their lead agents are very small. Therefore, we propose to treat the states (θ_i, ω_i) of Secondary Agent i as "noisy" versions of those of Lead Agent k which is in its cluster and estimate (θ_i, ω_i) as follows:

$$\begin{cases} \widehat{\omega}_i = \omega_k + \triangle_i \\ \widehat{\theta}_i = \theta_k + \varepsilon_i^0 + \varsigma_i \end{cases} \tag{8}$$

where ε_i^0 denotes the phase angle difference between the ith and kth agents in the static (pre-fault) state, and $\triangle_i \sim \mathscr{U}(-a, a)$ and $\varsigma_i \sim \mathscr{U}(-b, b)$ are uniform random noises on $[-a, a]$ and $[-b, b]$, respectively, with $a \ll 1$ and $b \ll 1$.

By using Eq. (8), we are able to estimate the information of the physical coupling matrix $\mathbf{L}$ by only using the lead agents' states. To simplify, we partition $\mathbf{L}$ as follows:

$$\mathbf{L} = \begin{bmatrix} \mathbf{R}_{C\times C} & \mathbf{S}_{C\times(N-C)} \\ \mathbf{T}_{(N-C)\times C} & \mathbf{U}_{C\times C} \end{bmatrix}.$$

By using Eq. (8), we can approximate the matrix $\mathbf{S}$ with $\widehat{\mathbf{S}}$ whose element is shown as follows:

$$\widehat{\mathbf{S}}(j,k) = -P_{jk} \cos\left(\theta_j - \theta_k - \varepsilon_i^0 + \varphi_{jk}\right), \tag{9}$$

where the ith secondary agent belongs to the kth cluster. Using Eq. (9), we can approximate the matrix $\mathbf{R}$ by using $\widehat{\mathbf{R}}$ whose element is defined as follows:

$$\widehat{\mathbf{R}}(i,j) = \begin{cases} \mathbf{R}(i,j), & \text{if } i \neq j; \\ -\sum_{j=1, j\neq i}^{C} \mathbf{R}(i,j) - \sum_{j=C+1}^{N} \widehat{\mathbf{S}}(i,j), & \text{otherwise.} \end{cases} \tag{10}$$

Based on Eqs. (9), (10), (6), and (8), we achieve the hierarchical cyber-physical dynamics as follows:

1. *The lead agents (tier-1):*

$$\begin{cases} \dot{\theta}_l = \omega_l, \\ \mathbf{D}_l \dot{\omega}_l = -\mathbf{M}_l \ddot{\omega}_l - \left(\widehat{\mathbf{R}} + \widehat{\mathbf{S}}\Psi\right) \omega_l + \dot{u}_l - \widehat{\mathbf{S}}\triangle, \end{cases} \tag{11}$$

where the subscript, $\omega_l = [\omega_1, \ldots, \omega_C]^T, \theta_l = [\theta_1, \ldots, \theta_C]^T$ $\mathbf{D}_l = \text{diag}[D_1, \ldots, D_C]$, $\mathbf{M}_l = \text{diag}[M_1, \ldots, M_C]$, $\mathbf{u}_l = [u_1, \ldots, u_C]^T$, $\triangle = [\triangle_{C+1}, \ldots, \triangle_N]^T$,

$$\Psi(i,j) = \begin{cases} 1, \text{ if the } (C+i)\text{th agent is in the } j\text{th cluster;} \\ 0, \text{ otherwise.} \end{cases}$$

2. *The secondary agents (tier-2):*

$$\begin{cases} \dot{\theta}_s = \omega_s, \\ \mathbf{D}_s \dot{\omega}_s = -\mathbf{L}_s \omega_s - \mathbf{M}_s \ddot{\omega}_s, \end{cases} \tag{12}$$

where $\mathbf{L}_s$ denotes the physical coupling matrix for secondary agents, $\mathbf{M}_s = \text{diag}[M_{C+1}, \ldots, M_N]$, $\theta_s = [\theta_{C+1}, \cdots, \theta_N]^T$, and $\omega_s = [\omega_{C+1}, \cdots, \omega_N]^T$.

3 Hierarchical Control Protocol Design by Analogy to Flocking

Based on the dynamical modeling of the hierarchical cyber-physical integration framework introduced above, we design the control protocol to maintain the smart grid stability in the emergent situation by leveraging the flocking theory.

3.1 Flocking Theory and Formation Control

In a system comprised of a large number of coupled agents, flocking refers to an aggregate behavior amongst the entities to achieve a shared group objective. In [99], Reynolds introduced three heuristic rules that led to the creation of the first computer animation of flocking:

1. *Flock Centering*: agents attempt to stay close to nearby flockmates,
2. *Velocity Matching*: agents attempt to match velocity with nearby flockmates,
3. *Goal Seeking*: each agent has a desired velocity towards a specified position in global space.

Based on these three rules, Olfati-Saber [95] provided a framework for design and analysis of scalable distributed flocking algorithms using a double integrator model:

$$\begin{cases} \dot{\mathbf{q}} = \mathbf{p} \\ \dot{\mathbf{p}} = \mathbf{u}, \end{cases} \tag{13}$$

where $\mathbf{q} \in \mathbb{R}^N$ is the position vector of the flockmates, $\mathbf{p} \in \mathbb{R}^N$ denotes the velocity vector, $\mathbf{u} \in \mathbb{R}^N$ represents the control signal, and N is the size of the flock.

To achieve the objectives of flocking, the control signal $\mathbf{u}$ is comprised of three terms:

$$\mathbf{u} = -\nabla V(\mathbf{q}) - \mathbf{L} \cdot \mathbf{p} + F(\mathbf{p}, \mathbf{q}, \mathbf{p}_r, \mathbf{q}_r). \tag{14}$$

The first term is the gradient of a potential energy function $V(\mathbf{q})$ which characterizes system objectives and constraints. The second term represents a velocity consensus protocol where $\mathbf{L}$ is the Laplacian matrix associated with the flock communication graph. Finally, the third term models navigational feedback which is designed to ensure each agent tracks a reference $(\mathbf{p}_r, \mathbf{q}_r)$.

The stability of the control protocol described in Eq. (14) has been analyzed in [95] to provide the following sufficient conditions for stability: (1) $V(\mathbf{q})$ is a nonnegative continuously differentiable potential energy function that achieves the global minimum at a desired formation; (2) $\mathbf{L}$ is a standard Laplacian matrix, which is positive semidefinite and has a zero row sum [92]; (3) $F(\mathbf{p}, \mathbf{q}, \mathbf{p}_r, \mathbf{q}_r)$ is a linear combination of $(\mathbf{p} - \mathbf{p}_r)$ and $(\mathbf{q} - \mathbf{q}_r)$.

3.2 *Design by Analogy to Flocking*

Let the state of each agent be given by $(\theta_i(t), \omega_i(t))$, which is the associated generator's state. Given the self-regulation goals of the transient stability problem, we consider cyber-control between agents using *deviations* of their phase angle $\theta_i(t) - \theta_j(t)$ and frequency $\omega_i(t) - \omega_j(t)$. In doing this, we are able to recognize the analogies between the transient stability problem and that of flocking. The problem of transient stabilization becomes equivalent to that of designing the collective cyber-physical dynamics of smart grid agents to be analogous to a stable flock of birds. This is achieved through the appropriate computation of cyber dynamics u_i using PMU data, which is then converted to energy injection/absorption $P_{u,i}$ at Generator Bus i.

We design u_i as follows:

$$u_i = -B_i \omega_i + h_i(\theta, \omega), \tag{15}$$

where B_i is a cyber parameter which satisfies that $B_i \geq (100 \times D_i)$ and $h_i(\cdot) : \mathbb{R}^N \times \mathbb{R}^N \to \mathbb{R}$ is a function of the vector $\theta = \{\theta_i | i \in \mathscr{I}_C\}$ and the vector $\omega = \{\omega_i | i \in \mathscr{I}_C\}$, and $\mathscr{I}_C$ represents the index set of the lead agents. We can rewrite the second line of Eq. (11) as follows:

$$(\mathbf{D}_l + \mathbf{B})\dot{\omega}_l = -\mathbf{M}_l \ddot{\omega}_l - \left(\widehat{\mathbf{R}} + \widehat{\mathbf{S}}\Psi\right)\omega_l + \mathbf{h} - \widehat{\mathbf{S}}\triangle, \tag{16}$$

where $\mathbf{B}$ is a pre-designed $C \times C$ cyber coupling diagonal matrix with diagonal element B_i and $\mathbf{h}$ is a C-dimensional cyber control column vector with: ith element is as follows:

$$h_i = \begin{cases} \frac{d}{dt} h_i(\theta, \omega), & \text{if } i = 1, \ldots, C; \\ 0, & \text{otherwise,} \end{cases} \tag{17}$$

In practice, for the ith synchronized generator, the ratio between the inertia M_i and the damping parameter D_i satisfies $M_i/D_i \in \mathcal{O}(10)$ [102]. We therefore find that the associated perturbation parameter for Lead Agent i is $\varepsilon_i = M_i/(D_i + B_i) \in \mathcal{O}(0.1)$ representing an *overdamped* system, which enables the application of singular perturbation techniques to in Eq. (16) to study the dynamics of the lead agents over a longer time scale. Specifically, applying singular perturbation analysis and letting $\mathcal{M} = \mathbf{D}_l + \mathbf{B}$ [65] gives:

$$\begin{cases} \dot{\theta}_l = \omega_l, \\ \mathcal{M} \dot{\omega}_l = -\left(\widehat{\mathbf{R}} + \widehat{\mathbf{S}}\Psi\right) \omega_l + \mathbf{h} - \widehat{\mathbf{S}}\triangle, \end{cases} \tag{18}$$

Here, the simplification has allowed the physical notion of generator "jerk" related to $\ddot{\omega}_l$ to be eliminated from the dynamics.

Since the nonlinear dynamical system of Eq. (18) is feedback linearizable, we can define a new control vector $\widetilde{\mathbf{u}}$ and rewrite the equivalent reduced order model as:

$$\begin{cases} \dot{\theta}_l = \omega_l, \\ \mathcal{M} \dot{\omega}_l = \widetilde{\mathbf{u}} - \widehat{\mathbf{S}}\Delta. \end{cases} \tag{19}$$

Furthermore, we can represent the relationship between the original control vector $\mathbf{u}$ and the new control vector $\widetilde{\mathbf{u}}$ as:

$$\dot{\mathbf{u}} = \widetilde{\mathbf{u}} + \left(\widehat{\mathbf{R}} + \widehat{\mathbf{S}}\Psi\right) \omega_l - \mathbf{B}\dot{\omega}_l. \tag{20}$$

Equation (19) represents a double integrator system analogous to Eq. (13) known to model the standard dynamics of flockings. By setting $\widetilde{\mathbf{u}}$ to the following form we thus ensure flocking formation and hence transient stability of the power network:

$$\widetilde{\mathbf{u}} = \underbrace{-\nabla V(\theta_l)}_{\text{phase cohesiveness}} - \underbrace{\widetilde{\mathbf{L}}\omega_l + F(\omega_l, \omega_r)}_{\text{frequency synchronization}}, \tag{21}$$

where $V(\theta_l)$ represents the potential energy function to guarantee that the phase angle differences between pairs of lead agents are bounded, $\nabla V(\theta_l)$ is its associated gradient with respect to θ_l, $\widetilde{\mathbf{L}}$ is the effective Laplacian matrix that ensures frequency consensus (i.e., lead agents' frequencies converge to a common value), and $F(\cdot)$ is the navigation feedback designed to lead the frequencies to converge to the desired value ω_r; typically relative frequency is normalized such that $\omega_r = 0$.

3.2.1 Potential Energy Function

Based on the sufficient condition of Eq. (1), we consider the following potential energy for our control scheme:

$$V(\theta_l) = \frac{1}{2}\sum_{i=1}^{C}\sum_{j=1,j\neq i}^{C} \chi(\theta_i - \theta_j), \tag{22}$$

where $\chi(\cdot)$ is a pairwise attractive potential defined as:

$$\chi(z) = \begin{cases} 0, & \text{if } |z| \leq \frac{5\pi}{9}; \\ c_1\left(z^2 - \frac{25\pi^2}{81}\right)^2, & \text{otherwise,} \end{cases} \tag{23}$$

where c_1 is a parameter to control the penalty level induced. It can be shown that $\chi(\cdot)$ is continuously differentiable and thus the gradient $\boldsymbol{\Phi}$ can be represented as follows:

$$\boldsymbol{\Phi}(i) = \sum_{j=1,j\neq i}^{C} \phi(\theta_i - \theta_j), \tag{24}$$

$$\phi(z) = \begin{cases} 0, & \text{if } |z| \leq \frac{5\pi}{9}; \\ 4c_1 z\left(z^2 - \frac{25\pi^2}{81}\right), & \text{otherwise.} \end{cases}$$

3.2.2 Effective Laplacian

As illustrated in [117], we deduce that a sufficient condition to ensure $\widetilde{\mathbf{L}}$ is PSD is $\widetilde{l}_{ij} < 0$ where $i \neq j$. Furthermore, to simplify the controller design, we assume that the cyber communication graph is undirected which (coupled with the fact that the Kron-reduced physical graph is undirected) implies that the integrated cyber-physical graph is undirected thus constraining $\widetilde{\mathbf{L}}$ to be symmetric. Therefore, in our framework, we design the ijth element of the effective Laplacian matrix $\widetilde{\mathbf{L}}$ as follows:

$$\widetilde{l}_{ij} = \begin{cases} c_2, & \text{if } i = j; \\ \frac{c_2}{C-1}, & \text{otherwise.} \end{cases} \tag{25}$$

3.2.3 Linear Navigation Feedback

To reduce complexity, we assign the following linear navigation feedback term: $F(\omega_l, \omega_r) = c_3\omega_l$, where c_3 is a cyber control parameter.

Based on the above analysis, we have the following result:

$$\widetilde{\mathbf{u}} = -\boldsymbol{\Phi} - \widetilde{\mathbf{L}}\omega_l - c_3\omega_l, \tag{26}$$

Using Eqs. (20) and (26), we obtain the following result:

$$\dot{\mathbf{u}} = -\boldsymbol{\Phi} + \left(\widehat{\mathbf{R}} + \widehat{\mathbf{S}}\Psi\right)\omega_l - \widetilde{\mathbf{L}}\omega_l - c_3\omega_l - \mathbf{B}\dot{\omega}_l. \tag{27}$$

By integrating both sides of Eq. (27), we can formulate $\mathbf{u}$, which represents the power transmission $\mathbf{P}_u$ between the fast-reacting power source and the synchronized generators, as:

$$\mathbf{u} = -\Gamma + \int_{t_0}^{t} \left(\widehat{\mathbf{R}} + \widehat{\mathbf{S}}\Psi\right) \omega_l d\tau - \widetilde{\mathbf{L}}\theta_l - c_3\theta_l - \mathbf{B}\omega_l, \tag{28}$$

where θ_0 is the constant term in $\int_{t_0}^{t} \omega d\tau$ and $\Gamma = \int_{t_0}^{t} \Phi d\tau$ whose element is represented as follows:

$$\Gamma(i) = \sum_{j=1, j\neq i}^{C} \left[\int_{t_0}^{t} \phi\left(\theta_i - \theta_j\right)\right] d\tau. \tag{29}$$

Let the C-dimension column vector η denote $\int_{t_0}^{t} \left(\widehat{\mathbf{R}} + \widehat{\mathbf{S}}\Psi\right) \omega_l d\tau$. Since $\widehat{\mathbf{R}}$, $\widehat{\mathbf{S}}$, and θ are time-varying, and the information of them is available, using Eqs. (7) and (9) we obtain:

$$\begin{aligned} \eta(i) &= \sum_{j=1, j\neq i}^{C} \int_{t_0}^{t} \left[P_{ij}\cos\left(\theta_{ij} + \varphi_{ij}\right) + \sum_{k\in\mathscr{I}_j} P_{ik}\cos\left(\theta_{ij} - \varepsilon_k^0 + \varphi_{ik}\right)\right] \omega_{ij} d\tau, \\ &= \sum_{j=1, j\neq i}^{C} P_{ij}\sin\left(\theta_{ij} + \varphi_{ij}\right) + \sum_{j=1, j\neq i}^{C} \sum_{k\in\mathscr{I}_j} P_{ik}\sin\left(\theta_{ij} - \varepsilon_k^0 + \varphi_{ik}\right) - \eta_i^0, \end{aligned} \tag{30}$$

where $\mathscr{I}_j$ denotes the index set of the secondary agents belonging to the jth cluster, $\omega_{ij} = \omega_i - \omega_j$, $\theta_{ij} = \theta_i - \theta_j$, and

$$\eta_i^0 = \left[\sum_{j=1, j\neq i}^{C} P_{ij}\sin\left(\theta_{ij} + \varphi_{ij}\right) + \sum_{j=1, j\neq i}^{C} \sum_{k\in\mathscr{I}_j} P_{ik}\sin\left(\theta_{ij} - \varepsilon_k^0 + \varphi_{ik}\right)\right]_{t=t_0}.$$

3.3 Hierarchical Control Protocol Stability Analysis

We define the following Lyapunov function H:

$$H = \frac{1}{2}\omega_l^T \mathbf{M}\omega_l + \mathbf{V} \tag{31}$$

for which $H(\mathbf{0},\mathbf{0}) = 0$ and $H(\theta,\omega_l) > 0$ for $\forall(\theta,\omega_l) \neq (\mathbf{0},\mathbf{0})$. Calculating the derivative of H along the dynamics derived in Eqs. (19) and (26) we obtain:

$$\dot{H} = \omega_l^T \mathbf{M}\dot{\omega}_l + \omega_l^T \nabla V = -\omega_l^T \left(\widetilde{\mathbf{L}} + c_3\mathbf{I}\right) \omega_l - \omega_l^T \widehat{\mathbf{S}}\Delta. \tag{32}$$

Based on our proposed framework, $\widetilde{\mathbf{L}}$ is the effective Laplacian matrix which is PSD and $c_3 > 0$. Therefore, $\left(\widetilde{\mathbf{L}} + c_3\mathbf{I}\right)$ is a Positive Definite (PD) matrix. Using the property of PD matrices, we deduce $\omega_l^T \left(\widetilde{\mathbf{L}} + c_3\mathbf{I}\right) \omega_l > 0$.

Since information on ω_l and $\widehat{\mathbf{S}}$ is available and $\widetilde{\mathbf{L}}$ and c_3 are designable, using Lyapunov redesign, we obtain:

$$\dot{H} \leq -\lambda_m \|\omega_l\|^2 + \rho \|\omega_l\| \|\widehat{\mathbf{S}}\| = -\|\omega_l\| \left(\lambda_m \|\omega_l\| - \rho \|\widehat{\mathbf{S}}\|\right) \tag{33}$$

where λ_m is the smallest eigenvalue of $\left(\widetilde{\mathbf{L}} + c_3 \mathbf{I}\right)$.

Since $\widetilde{\mathbf{L}}$ is a Laplacian with minimum eigenvalue 0, $\lambda_m = c_3$. Therefore, $\dot{H} < 0$ is guaranteed when:

$$\|\omega_l\| \geq \frac{\rho \|\widehat{\mathbf{S}}\|}{c_3}. \tag{34}$$

The high physical coherency between intra-cluster agents ensures that ρ is sufficiently small. In practice, the tolerance interval of the normalized relative frequency is $[-0.02, 0.02]$, and thus $\rho < 0.02$ Therefore, we can design c_3 to satisfy:

$$c_3 \geq \frac{\rho \|\widehat{\mathbf{S}}\|}{0.02}. \tag{35}$$

Based on Eqs. (34) and (35), we deduce that $\dot{H} < 0$ if $\|\omega_i\| > 0.02$, where $i = 1, 2, \ldots, C$. Thus, the frequencies of all the lead agents are bounded within the required tolerance interval $[-0.02, 0.02]$. Thus, our proposed distributed control guarantees transient stability given the existence of an accurate and efficient coherent cluster identification algorithm.

4 Timely Dynamic Agent Coherency Identification

In order to efficiently implement the hierarchical control framework illustrated above, it is necessary to rapidly identify the agent coherency with high accuracy. In this section, we propose a timely dynamic agent coherency identification scheme which requires very short observation window. Our scheme transforms the data of agents' state from the observation space to an information space whereby the agents' frequencies and phases characterize the movement and dynamics of boids within multiple flocks with different features. Boid i carries three-dimensional information describing the ith agent's status at time $t = k$ as:

$$\begin{cases} \mathscr{I}_i^1(k) = \theta_i(k) \\ \mathscr{I}_i^2(k) = \omega_i(k) \\ \mathscr{I}_i^3(k) = \delta_i(k) \end{cases}, \tag{36}$$

where $\mathscr{I}_i(k) = [\mathscr{I}_i^1(k) \ \mathscr{I}_i^2(k) \ \mathscr{I}_i^3(k)]^T$, $\theta_i(k)$ and $\omega_i(k)$ are the phase angle and the normalized frequency, respectively, of the ith generator at the time step $t = k$ that are obtained directly from PMU information, and $\delta_i(k)$ is the acceleration of the ith generator at the time step $t = k$ estimated from the current and historical values of $\omega_i(k)$.

The state of Boid i is described as follows:

$$\mathscr{S}_i(k) = [\mathbf{p}_i(k), \mathbf{v}_i(k)]^T, \tag{37}$$

where $\mathbf{p}_i(k), \mathbf{v}_i(k) \in \mathbb{R}^{2\times 1}$ denote the boid's position and the velocity, respectively. Two boids are considered to be neighbors at time step $t = k$ if the distance between them is less than the predetermined threshold d_c. Therefore, we define the set of neighbors for the ith boid as follows: $\mathscr{N}_i(k) = \{\forall j \,|\, \|\mathbf{p}_i(k) - \mathbf{p}_j(k)\| < d_c\}$.

We compute the *informational* (feature) *similarity* between neighboring Boids i and j as follows. For $j \in \mathscr{N}_i(k)$,

$$\zeta_{ij}(k) = \left| \sum_{n=1}^{3} \alpha_n \times \left(\mathscr{S}_i^n(k) - \mathscr{S}_j^n(k)\right) \right|, \tag{38}$$

where $\{\alpha_n\}$ is a scalar weight determining the impact of specific information on boid interaction. Given a threshold value $\zeta_{th}(k)$, if $\zeta_{ij}(k) \le \zeta_{th}(k)$ they are assumed to be in the same flock and hence are called *flockmates*. Otherwise, they are assumed to be in different flocks.

As illustrated in detail in [115], we model the dynamics of Boid i based on their feature similarity and flocking rules as:

$$\begin{cases} \mathbf{v}_i(k+1) = \mathbf{v}_i(k) + \Delta t \sum_{l=1}^{3} w_l \mathbf{g}_{i,l}(k), \\ \mathbf{p}_i(k+1) = \mathbf{p}_i(k) + \Delta t \mathbf{v}_i(k), \end{cases} \tag{39}$$

where $\mathbf{g}_{i,1}, \mathbf{g}_{i,2}, \mathbf{g}_{i,3}$ represent the accelerations calculated based on the flocking rules *flock centering*, *velocity matching*, and *obstacle avoidance*, respectively, w_l denotes the weight representing the impact of the component $\mathbf{g}_{i,l}$, and Δt is the algorithm time step for coherence identification.

Based on dynamic model in Eq. (39), we can plot the boids' trajectories in the information space and achieve the multiple flocks constituted by the boids, which corresponds to the clusters constituted by the agents having high physical coherency in the observation space.

5 Witness-Based Verification and Estimation Protocol

The PMUs of the lead agents in our two-tier framework provide critical measurements for maintaining smart grid stability. Therefore, detection of possible lead PMU data corruption and subsequent real-time estimation are necessary for smart grid stability maintenance. In order to address this problem, we propose a cyber-physical verification and estimation protocol developed under the following threat model, as illustrated in Fig. 4.

Threat Model: Let H_k be the number of agents in the kth cluster of our proposed two-tier hierarchical framework. An attack can corrupt up to $\left\lfloor \frac{1}{2} H_k \right\rfloor$ PMU measurements where $\lfloor\cdot\rfloor$ denotes the floor function. Corruption constitutes biasing PMU

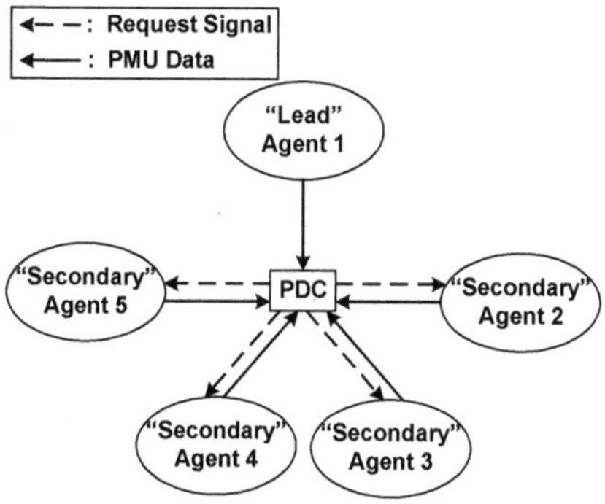

Fig. 4 Communication between PDC and agents locally within each cluaster

readings or equivalently replacing true values with fabricated quantities over a verification *period.*

As described in Eq. (8), the states of the secondary agents can be considered noisy estimates of the states of their lead. Based on this fact, our verification protocol treats the secondary agents as "witnesses" with their PMU data representing redundant information to measure the trustworthiness of the PMU readings of the lead agents.

In the intra-cluster LAN, the PDC must therefore probe the PMU data from secondary agents (at a lower data rate than for lead PMUs called the *verification* rate). Using the received data, the PDC measures the trustworthiness of a lead agent's PMU using the verification scheme described in Table 1. Since our proposed flocking-based control protocol is robust to the biases on the measurement of the lead agents' frequency [96], we address detection and mitigation of the compromised reading on the lead agents' phase angle.

At the end of each verification procedure, if the PDC concludes that the lead agent's PMU is valid, it stores the ℓ most recent bias samples $\{\xi_i | i \in \mathscr{I}\}$ for possible future estimation use. Otherwise, it estimates the true value using the proposed cyber-physical estimation scheme of Table 2.

The PDC then uses the estimated value for calculation of P_u and increases the verification probe rate to that of the sampling rate of the lead agent PMUs until it concludes the reading of the lead agent's PMU is valid for two consecutive verification periods or an operator deems the lead PMU reading authentic. Convergence of the algorithm of Eq. (11) is guaranteed analytically [116], but witness-based protocol performance is studied empirically.

Therefore, our proposed cyber-physical verification and estimation schemes both aim to leverage the hierarchy of the physical interaction amongst agents to achieve low computational complexity, which facilitates scalability and real-time implementation. Our verification scheme adopts a dynamically adjustable verification rate to optimally reduce bandwidth usage. When the PDC reports an attack on the lead agent's PMU, our estimation scheme employs a short Hamming window to estimate the true value of the attacked PMU's readings, which includes the historical information to improve the estimation accuracy and also assigns a higher priority

Table 1 Proposed Cyber-Physical Verification Scheme

Let the lead agent PMU reading be θ^c. Let the secondary agents be represented with indices from the set $i \in \mathscr{I}$ and their readings be denoted θ_i. Let $\Delta\theta_i$ be the phase angle difference between θ_i and θ^c at static state (i.e., pre-fault).
We assign $H_k = |\mathscr{I}| + 1$.
1. Initialize $Count = 0$ and set the threshold τ_p.
2. For each $i \in \mathscr{I}$
   $\xi_i = \theta_i - \Delta\theta_i - \theta^c$,
   If $\xi_i \leq \tau_p$
      $Count = Count + 1$,
   End
 End
3. If $Count < \left\lfloor \frac{1}{2} H_i \right\rfloor + 1$
   The PDC reports the lead agent's PMU as being attacked,
 Else
   The PDC reports the lead agent's PMU as valid,
 End

Table 2 Proposed Cyber-Physical Estimation Scheme

Let the secondary agents be represented with indices from the set $i \in \mathscr{I}$.
Let $\xi_i \in \mathbb{R}^\ell$ be a vector containing the ℓ most recent sample values of ξ_i in chronological order. Let $a(n)$ be an ℓ-point Hamming window.
1. For each $i \in \mathscr{I}$
   Secondary agent estimates lead agent phase angle using Eq. (8).
   Secondary agent reports the estimation result $\widehat{\theta}_i^c$ to the PDC.
 End
2. The PDC evaluates estimation accuracy for $i \in \mathscr{I}$ by computing:

$$\widehat{\sigma}_i = \sqrt{\frac{\sum_{n=1}^{\ell} a(n-1)\xi_i(n)^2}{\sum_{n=1}^{\ell} a(n-1)}}. \tag{40}$$

3. The PDC forms $\widehat{\theta}_l$ consisting of elements $\widehat{\theta}_i^c, i \in \mathscr{I}$ ordered to reflect monotonically increasing values in $\widehat{\sigma}_i$.
4. The PDC estimates θ^c from a median-like value from the elements of $\widehat{\theta}_l$ to avoid extreme biases:

$$\widehat{\theta}^c = \begin{cases} \widehat{\theta}_l\left(\frac{1}{2}H_k\right), & \text{if } H_k \text{ is even;} \\ \frac{1}{2}\left[\widehat{\theta}_l\left(\frac{1}{2}(H_k-1)\right) + \widehat{\theta}_l\left(\frac{1}{2}(H_k-1)+1\right)\right], & \text{otherwise} \end{cases}$$

to the current data. Moreover, our estimation achieves high robustness to potential attacks on the secondary agents' PMUs by choosing the median-like value rather than a weighted average for the final estimation result.

6 Simulations and Performance Assessment

We demonstrate the performance of our flocking-based two-tier hierarchical control framework with dynamics in Eqs. (11) and (12) collectively also described by Eq. (5) in achieving smart grid stability for two case studies on the New England 39 Bus system as shown in Fig. 5 and detailed in [13] consisting of $C = 10$ generators. MATLAB/Simulink is employed for simulations. In each case, we illustrate the efficiency of our proposed two-tier hierarchical control framework in selectively leveraging physical couplings to apply cyber data and control selectively. In all (non-hierarchical and hierarchical) cases the cyber control parameters of Eq. (28) are set to $c_1 = 5$, $c_2 = \frac{1}{10}$ and $c_3 = 3$, and the PMU sampling rate is 50 Hz. The power transmission limit for the fast-acting grid is set to $\mu = P_{u,i}/P_{r,i} \leq 1$ where $P_{r,i}$ is the rated power.

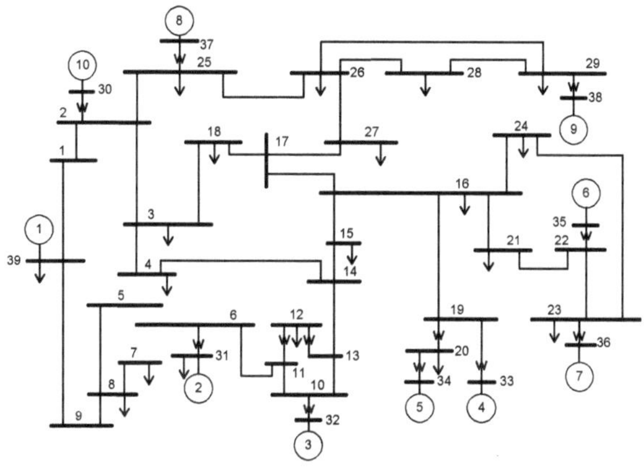

Fig. 5 New England 39-bus power system

We compare our results to situations when no control is computed nor applied (corresponding to minimum information use and control) and when non-hierarchical control is applied (corresponding to maximum information use and control). An efficient hierarchical framework would have comparable stabilizing performance to the latter case without the associated overhead. In each case, we also evaluate the performance of our hierarchical framework when experiencing cyber communication delay and practical constraints of fast-acting EES.

6.1 Ideal Environment

6.1.1 Case I

The system disturbance consists of a 3-phase short circuit in the middle of Line $14-15$ of Fig. 5 which occurs at time $t = 0$ s. The Line $14-15$ is removed at $t = 0.1$ s. Fig. 6 shows the normalized rotor frequencies and phase angles over a period of 10 s when no control is applied corresponding to Eq. (6) for $\alpha_i = 0$ for all i. Instability is clearly evident in all plots.

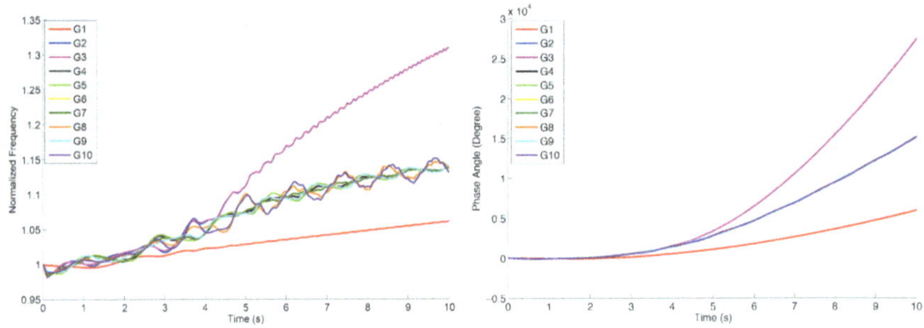

(a) (b)

Fig. 6 (a) Normalized rotor frequencies and (b) phase angles without cyber control

Fig. 7 (note: scale differs from Fig. 6) demonstrates performance for *non-hierarchical* cyber-physical control activated at time $t = 0.15$ s, in which the PMU of each agent is activated and cyber control works at each agent. This *non-hierarchical* framework can be mathematically described by using Eq. (6) which $\alpha_i = 0$ for all Agent i. The EES power absorbtion/injection to each generator bus, determined by the control signal $\mathbf{u}$, is shown in Fig. 8. Even though the clipping of the control signal occurs due to the capacity limit previously discussed, smart grid stability is still achieved.

Our two-tier hierarchical cyber-physical control framework is implemented in the following three steps. 1) the proposed timely dynamic agent coherency identification scheme is implemented immediately after Line $14-15$ is removed at time $t = 0.1$ s. The corresponding boid trajectories introduced by the flocking analogy used in our agent coherency identification scheme is presented in Fig. 9(a) for a very brief observation period of $t = 0.05$ s; as described in Eq. (36), each boid carries the information describing the associated agent's status. The neighboring boids interact with each other based on the informational similarity which is defined in Eq. (38) and their dynamics are modeled in Eq. (39). From Fig. 9(a), we observe that the agent coherency involving the following groups:$\{Agent_1\}$, $\{Agent_2, Agent_3\}$, and $\{Agent_4, \ldots, Agent_{10}\}$. 2) Based on the achieved result on agent coherency, we determine that our two-tier hierarchical framework consists of the clusters $\{Agent_1\}$, $\{Agent_2, Agent_3\}$, and $\{Agent_4, \ldots, Agent_{10}\}$, and the lead agents for these three

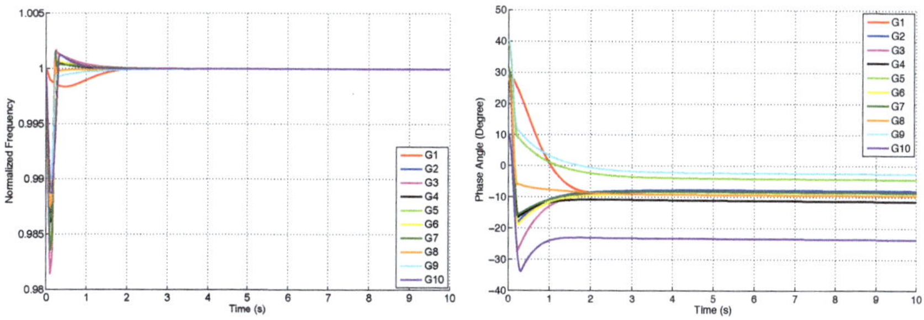

(a) (b)

Fig. 7 (a) Normalized rotor frequencies and (b) Phase angles with the non-hierarchical control

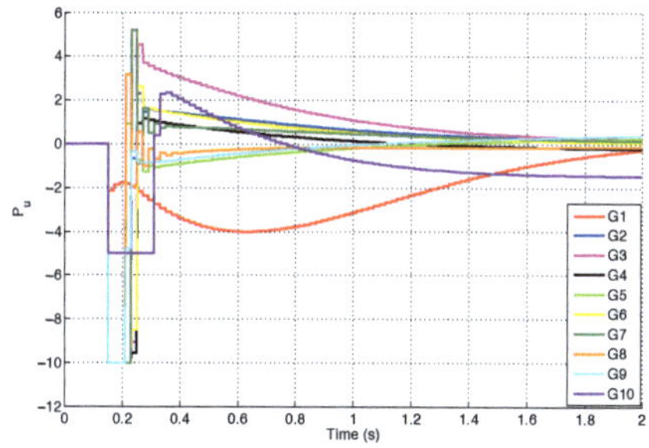

Fig. 8 Power transfer P_u by fast acting energy storage at generator buses in the presence of non-hierarchical control for Case Study I

clusters are *Agent* 1, *Agent* 3 and *Agent* 4, which have larger inertia compared with other agents belonging to the same cluster. 3) After determining the hierarchical framework, at time $t = 0.15$ s, our proposed two-tier hierarchical cyber-physical control framework is activated which controls the fast-acting EES associated with each lead agent to absorb/inject power to the generator buses of the associated agents. Based on our proposed control framework, the power absorption/injection is calculated based on Eq. (28) and is plotted in Fig. 9(b). As shown in Fig. 9(b), the EESs of the Lead Agents 1, 3, 4 are activated to absorb power from the system at time $t = 0.15$ s and then adjust their power output at each time step $\Delta t = 20$ ms to track the command given by the associated local controllers. After time $t = 0.25$ s the power output of each EES sinusoidally decays to zero. In contrast to the EES power outputs for nonhierarchy, the sinusoidal oscillations are higher in frequency. This is because the hierarchical case represents an "under-actuated" version of the nonhierarchical such that the control applied to select generators must stabilize all of them. This requires that the associated control signals to be more "reactionary" and faster-moving.

Figure 10 presents the generator frequencies and phase angles by using our proposed two-tier hierarchical cyber-physical control framework. In contrast to Fig. 6 in which there is no control, smart grid stabilization is evident. In contrast to the non-

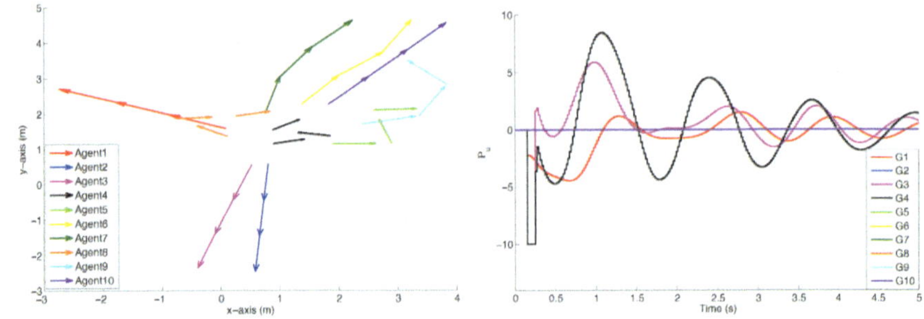

(a) (b)

Fig. 9 (a) The trajectories of the boids for Case Study I and (b) power transfer P_u by fast acting energy storage at generator buses in the presence of hierarchical control for Case Study I

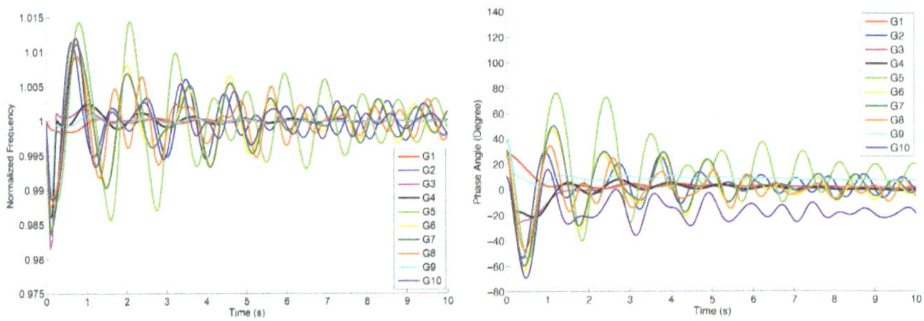

(a) (b)

Fig. 10 (a) Normalized rotor frequencies and (b) phase angles with hierarchical control

hierarchical case shown in Fig. 7, there is more high frequency oscillatory behavior due to the nature of the activated EES power outputs. From Fig. 10, we deduce that although the information acquisition and control is selectively applied to lead agents only, the high physical coherency between the secondary agents and their associated agents ensures maintaining the smart grid stability of all the agents.

6.1.2 Case II

The system disturbance consists of a 3-phase short circuit occurs at time $t = 0$ s in the middle of Line $17-27$ of Fig. 5. The Line $17-27$ is removed at $t = 0.1$ s. Figure 11 shows the normalized rotor frequencies and phase angles over a period of 10 s when no control is applied. Based on Fig. 11, we assess smart grid stability of the system by calculating the power angle-based stability margin ξ [78], and achieve $\xi_1 = 57.1$ which implies that the system smart grid security is low and very sensitive to perturbation. Parameter $\xi = \frac{360-\delta_{max}}{360+\delta_{max}} \times 100$ where δ_{max} is the maximum angle separation of any two generators at the same time in the post-fault response, and $-100 < \xi < 100$.

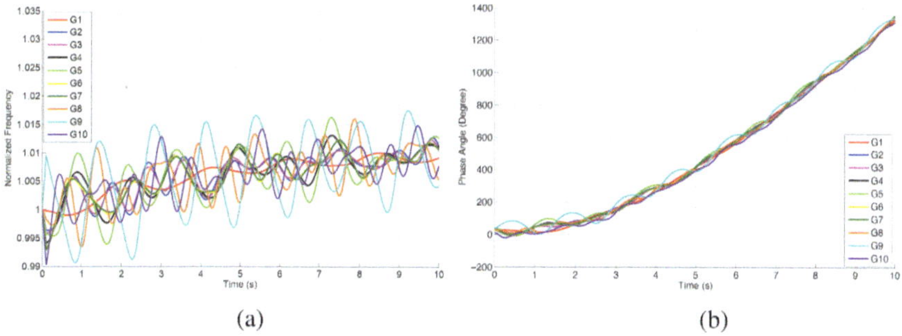

Fig. 11 (a) Normalized rotor frequencies and (b) phase angles without cyber control

To relax the angle-based stability margin to improve the system's smart grid security after fault, we implement our hierarchical control framework, in which the outputs of fast-acting energy storage are controlled to compensate for demand power's fluctuations caused by the 3-phase short circuit fault. Our timely dynamic agent coherency identification scheme is implemented immediately after Line $17-27$ is removed at time $t = 0.1$ s. The corresponding boid trajectories introduced by the flocking analogy used in our agent coherency identification scheme is presented in Fig. 12 for a very brief observation period of $t = 0.05$ s. From Fig. 12, we determine that our two-tier hierarchical framework consists of the clusters $\{Agent_1, Agent_{10}\}$, $\{Agent_2, \ldots, Agent_8\}$, and $\{Agent_9\}$, and the lead agents for these three clusters are *Agent* 1, *Agent* 4 and *Agent* 9. After determining the hierarchical framework, our proposed two-tier hierarchical cyber-physical control framework is implemented during time $t = 0.15$ s to 3 s, which is critical maintenance duration.

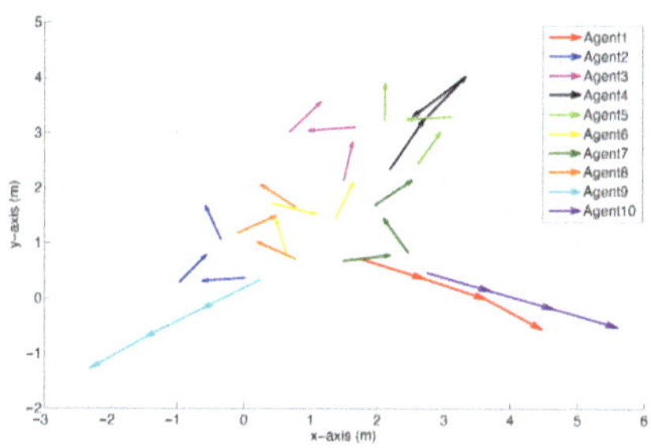

Fig. 12 The trajectories of the boids for Case Study II

Figure 13 evaluate the performance of normalized rotor frequencies and phase angles by using our proposed hierarchical control framework and Figure. 14 shows the power transfer from the fast-acting energy storage to each generator bus. We achieve the angle-based stability margin $\xi_2 = 70.7$, which validates our framework is efficient in improving the smart grid security of the power system after severe fault.

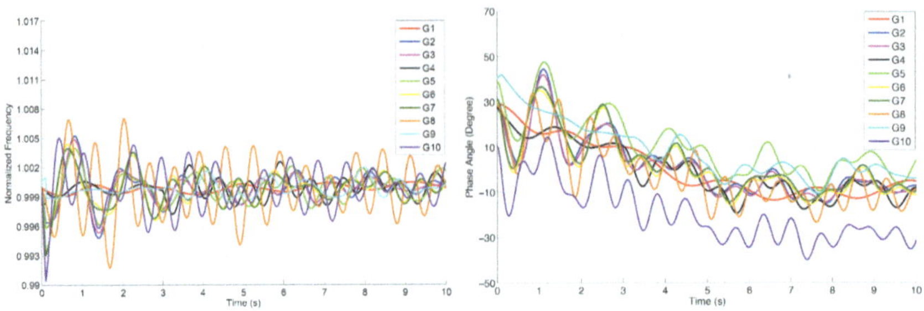

(a) (b)

Fig. 13 (a) Normalized rotor frequencies and (b) phase angles with hierarchical framework

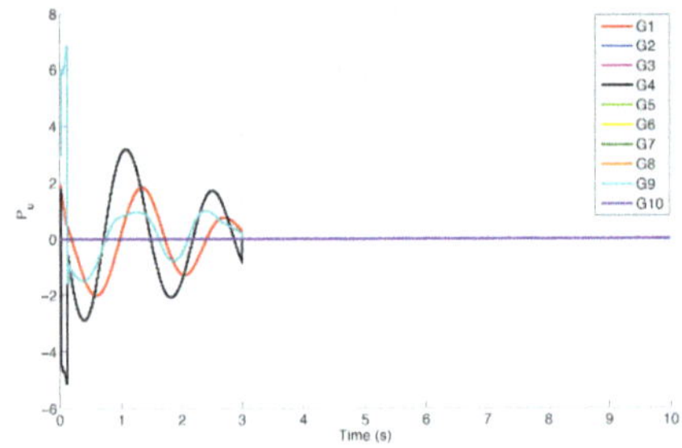

Fig. 14 Power transfer P_u

6.2 Environment with Practical Constraints of Energy Storage

We evaluate the performance of our hierarchical framework by considering the practical constraints of fast-acting energy storage on power output P_u. We assume the energy storage associated with each Agent i has two constraints: 1) the power output $|P_{u,i}| \leq \rho_1$ p.u., and 2) the rate of the power change $|\Delta P_{u,i}| \leq \rho_2$ p.u./Δt, where $\Delta t = 20$ ms denotes the time step for calculating the control signal for $P_{u,i}$. In the simulation, we consider the same two cases in previous section. Figure 15 evaluates, given different values of ρ_1, the minimum value of ρ_2 required for maintaining smart grid stability by using hierarchical and non-hierarchical frameworks in Case I. Figure 15 also evaluates the minimum value of ρ_2 required for improving ξ equivalent to ensuring $\xi > 57.1$ versus different values of ρ_1 by using hierarchical and non-hierarchical frameworks in Case II. From Fig. 15, it is clear that in Case I, compared to the non-hierarchical framework, the hierarchical framework requires higher but comparable physical requirement for energy storage when $\rho_1 \leq 8$. Figure 15 also indicates that in Case II, the hierarchical and non-hierarchical framework desire the same physical requirement for energy storage.

In order to analyze the performance of our proposed control framework under the two constraints in more detail, Fig. 16 evaluates the power angle-based margin ξ achieved by implementing our control framework when $\rho_1 \in [1,5]$ and $\rho_2 \in [0.1, 0.5]$. From Fig. 16(a), it is clear that in Case I the proposed hierarchi-

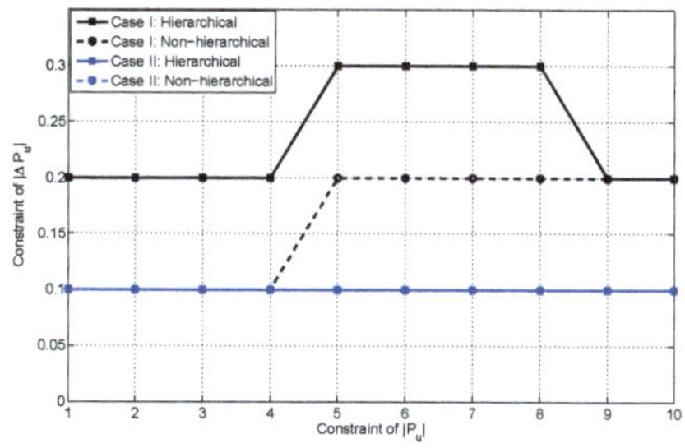

Fig. 15 Performance evaluation of Cases I and II by considering physical constraints of fast-acting energy storage

cal framework is able to maintain smart grid stability when $\rho_1 \leq 4.5$ and $\rho_2 \geq 0.2$, or $\rho_1 = 5$ and $\rho_2 \geq 0.2$. From Fig. 16(b), it is clear that in Case II the proposed hierarchical framework is able to improve the stability margin when $\rho_1 \leq 5$ and $\rho_2 \geq 0.1$. Based on the above observation, we can get that the constraints of the power output and the rate of the power output jointly impact on the performance of the proposed control framework. Furthermore, in both cases, better stability margin can be achieved by implementing the non-hierarchical control framework, but the performance of the hierarchical control framework is comparable with that of the non-hierarchical framework. Therefore, the conclusions obtained from Fig. 16 are consistent with the conclusion got from Fig. 15. We believe it is reasonable that under the practical physical constraints the non-hierarchical control framework achieves slightly better results than the hierarchical framework. This is because that in the non-hierarchical framework, more fast-acting ESSs are activated which mitigates the impact of the constraints associated with each ESS.

Figure. 17 shows the transfer power P_u between the ESS and the power system by implementing our proposed hierarchical control framework under the constraints $\rho_1 = 2$ and $\rho_2 = 0.3$ in Case I, and Fig. 18 shows the generators' normalized rotor

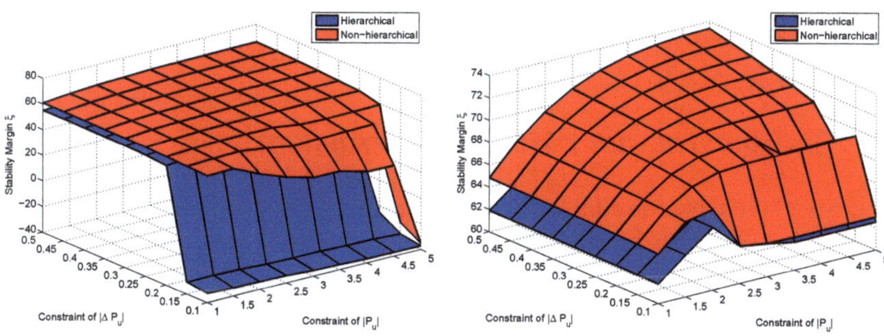

(a) (b)

Fig. 16 The stability margin achieved under the practical constraints in (a) Case Study I and (b) Case Study II frequency and phase angle in this case study. From the simulation results, it is clear that our proposed framework is able to efficiently to maintain smart grid stability under the practical physical constraints of the fast-acting ESSs.

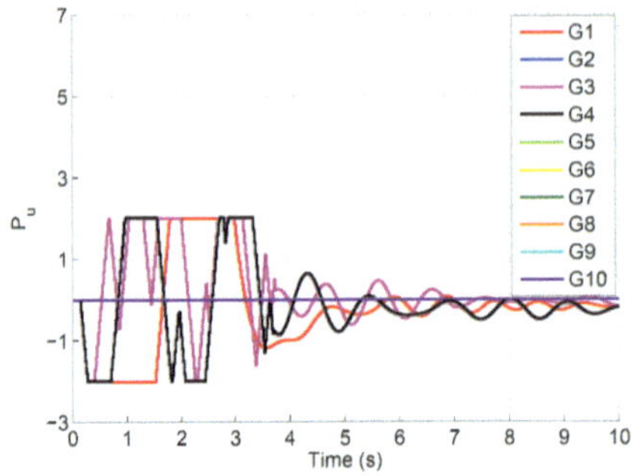

Fig. 17 Power transfer P_u by fast acting energy storage at generator buses

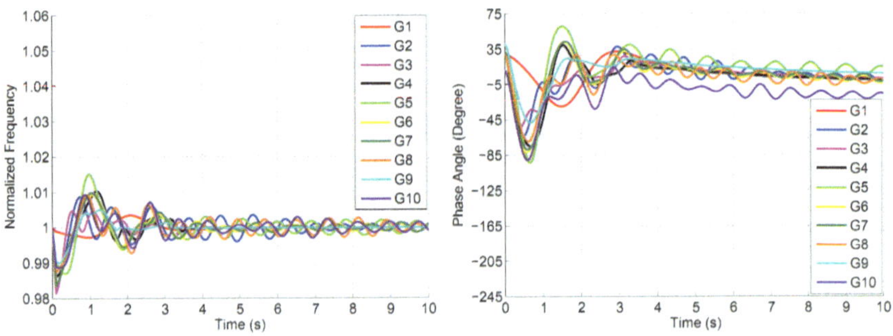

(a) (b)

Fig. 18 (a) Normalized rotor frequencies and (b) phase angles versus time

6.3 Environment with PMU Data Corruption

In our simulation, we consider the practical constraints with the energy storage $\rho_1 = 2$ and $\rho_2 = 0.3$. Furthermore, the PMU sampling rate is assigned as 50 Hz, the verification probe rate is initially set to 5 Hz (no-attack condition) and then raised to 50 Hz after lead generator attack detection, and $\ell = 50$. The threshold $\tau_p = 35^\circ$. Figures 19 and 20 show the normalized frequencies, rotor phase angles, and P_u in the presence of information corruption on Agent 4, 6 and 7 when no witness-based cyber protection is applied. The compromised PMUs of Agent 4, 6 and 7 collude and report the same biased readings (bias = -257.8°) starting at $t = 0.5$ s for duration 2.5 s, 3 s, and 2 s, respectively. From Figs. 19 and 20, it is clear that the corrupted readings mislead the PDC of the third cluster, result in a miscomputation of $P_{u,3}$ and subsequent instability results.

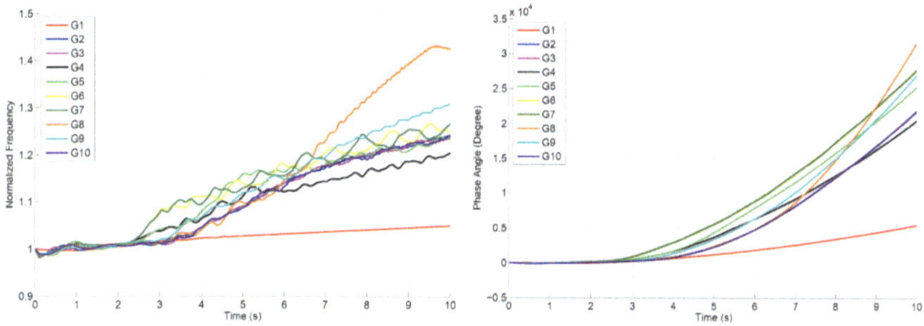

(a) (b)

Fig. 19 (a) The normalized frequencies and (b) the rotor phase angles versus time without proposed cyber-physical security protocol in presence of random attack.

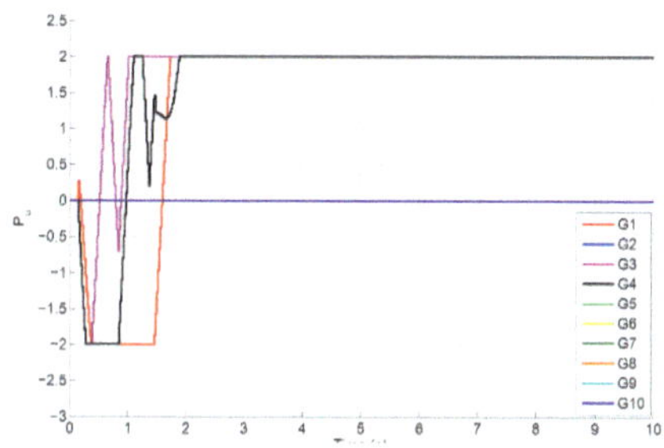

Fig. 20 P_u versus time without proposed cyber-physical security protocol in presence of random attack

Figures 21 and 22 show the normalized rotor frequencies, phase angles and P_u when our cyber-physical control and witness-based protection protocol is applied. We observe the stabilizing performance of our proposed protocol in verifying the validity of the readings of the lead agents' PMUs and estimating their true values. Smart grid stability is still maintained in the presence of the random attack.

These simulation results illustrate that our proposed cyber-physical verification and estimation schemes can efficiently identify and correct the corrupted lead agents' PMUs' readings to aid in successful maintenance of the smart grid stability. The simulation results also help demonstrate robustness against attacks on the secondary agents' PMUs as long as our threat model of Section 5 is satisfied.

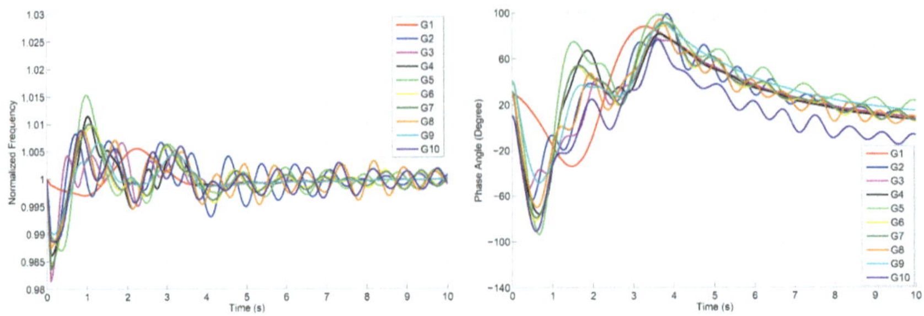

(a) (b)

Fig. 21 (a) The normalized frequencies and (b) the rotor phase angles versus time with proposed cyber-physical security protocol in presence of random attack

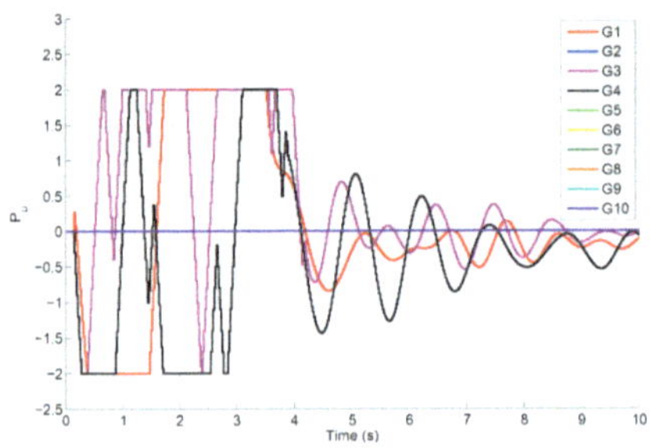

Fig. 22 P_u versus time with proposed cyber-physical security protocol in presence of random attack

7 Conclusions

The last few years have witnessed the radical transformation in structure and functionality of electrical energy systems. Such systems were traditionally executed in the physical world and are now also cyber-enabled. This cyber-enabled energy system, called smart grid, can be envisioned as the marriage of information technology with the electricity network. While its increased dependence on cyber infrastructure aims to enable greater reliability, efficiency and capacity of power delivery, this reliance also creates a host of unfamiliar vulnerabilities. Due to the highly integrated and connected nature of smart grids, it is important to account for their salient cyber-physical coupling when making critical design decisions and identifying solutions to promote security.

In this chapter, we present a biologically-inspired cyber-physical multi-agent distributed control framework for maintaining smart grid stability under various forms of physical and cyber attacks. Through this multi-agent control framework, we demonstrate real-time cyber-physical integrated strategies using "wisely"-placed Phasor Measurement Units (PMUs) and energy storages. Our research has evolved in three stages. We first propose a cyber-physical multi-agent dynamical systems paradigm to model the cyber-physical interactions in smart grids, in which each agent is modeled as having dynamics that synergistically describe physical and information couplings with neighboring agents. Inspired by the analogy between the flocking rules and the smart grid stability requirements, we develop a flocking-based scheme to formulate the cyber-physical integrated action for each agent. In the second stage, we extend the multi-agent dynamical systems paradigm to a two-tier hierarchical framework which reduces information acquisition by leveraging physical couplings between the agents and applying cyber controls selectively on critical agents. In the context of the hierarchical framework, we develop a novel witness-based cyber-physical protocol whereby physical coherence is leveraged to probe and identify phasor measurement unit data corruption and estimate the true information values for attack mitigation.

References

1. NERC CIP standards, `http://www.nerc.com`
2. Reliability considerations from the integration of smart grid. North American Electric Reliability Corporation (2010)
3. Roadmap to achieve energy delivery system cyber security. Energy Sector Control Systems Working Group (ESCSWG) (2011)
4. Intelligrid program: 2012 annual review. Electric Power Research Institute (EPRI) (2013)
5. Smart grids and renewables: A guide for effective deployment. International Renewable Energy Agency (IRENA) (2013)
6. How much electricity does an american home use? (2014), `http://www.eia.gov/tools/faqs/faq.cfm?id=97&t=3`
7. Adeodu, O., Chmielewski, D.: Design of massive energy storage systems within electric transmission networks. In: 2013 AIChE Annual Meeting, San Francisco, CA (2013)
8. Almond, S.J., Baird, S., Flynn, B.F., Hawkins, D.J., Mackrell, A.J.: Integrated protection and control communications outwith the substation: Cyber security challenges. In: Proc. IET 9th International Conference on Developments in Power System Protection, pp. 698–701 (2008)
9. Amin, S., Cárdenas, A.A., Sastry, S.S.: Safe and secure networked control systems under denial-of-service attacks. In: Majumdar, R., Tabuada, P. (eds.) HSCC 2009. LNCS, vol. 5469, pp. 31–45. Springer, Heidelberg (2009)
10. Amin, S.M.: Energy infrastructure defense systems. Proceedings of the IEEE 93(5), 861–875 (2005)
11. Amina, M., Stringer, J.: The electric power grid: Today and tomorrow. MRS Bulletin 33, 399–407 (2008)
12. Ananad, M., Cronin, E., Sherr, M., Blaze, M., Ives, Z., Lee, I.: Security challenges in next generation cyber physical systems. In: Proc. Beyond SCADA: Cyber Physical Systems Meeting (HCSS-NEC4CPS), Pittsburgh, Pennsylvania (2006)
13. Athay, T., Podmore, R., Virmani, S.: A practical method for the direct analysis of transient stability. IEEE Transactions on Power Apparatus and Systems PAS-98, 573–587 (1979)
14. Bakken, D.E., Hauser, C.H., Gjermundrod, H., Bose, A.: Toward more flexible and robust data delivery for monitoring and control of the electric power grid. Technical Report EECS-GS-009, Washington State University, Pullman, Washington (2007)
15. Bergen, A.R., Vittal, V.: Power Systems Analysis. Prentice Hall (1999)

16. Bobba, R., Khurana, H., AlTurki, M., Ashraf, F.: PBES: A policy based encryption system with application to date sharing in the power grid. In: Proc. ACM Sympoisum of Information, Computer and Communications Security, ASIACCS 2009, pp. 262–275 (2009)
17. Bobba, R., Rogers, K.M., Wang, Q., Khurana, H., Nahrstedt, K., Overbye, T.J.: Detecting false data injection attacks on DC state estimation. In: Proc. First Workshop on Secure Control Systems, Stockholm, Sweden (2010)
18. Byres, E., Chauvin, B., Hoffman, J., Kube, N.: The special needs of SCADA/PCN firewalls: Architectures and test results. In: Proc. 10th IEEE Conference on Emerging Technologies and Factor Automation, vol. 2, pp. 877–884 (2005)
19. C1 Working Group Members of Power System Relaying Committee: Cyber security issues for protective relays. In: Proc. IEEE Power Engineering Society General Meeting, pp. 1–8 (2007)
20. Cárdenas, A.A., Amin, S., Sastry, S.: Research challenges for the security of control systems. In: Proc. 3rd USENIX Conference on Hot Topics in Security, p. Article 6 (2008)
21. Cárdenas, A.A., Amin, S., Sastry, S.: Secure control: Towards survivable cyber-physical systems. In: Proc. 28th International Conference on Distributed Computing Systems Workshops, pp. 495–500 (2008)
22. Cárdenas, A.A., Amin, S., Sastry, S.: Secure control: Towards survivable cyber-physical systems. In: Proc. First International Workshop on Cyber-Physical Systems (2008)
23. Cárdenas, A.A., Roosta, T., Taban, G., Sastry, S.: Cyber security basic defenses and attack trends. In: Franceschetti, G., Grossi, M. (eds.) Homeland Security Technology Challenges, ch. 4, pp. 73–101. Artech House (2008)
24. Cleveland, F.M.: Cyber security issues for advanced meter infrastructure (AMI). In: Proc. IEEE Power and Energy Society General Meeting – Conversion and Delivery of Electrical Energy in the 21st Century, pp. 1–5 (2008)
25. Constable, G., Somerville, B.: A Century of Innovation: Twenty Engineering Achievements That Transformed Our Lives. Joseph Henry Press, Washington, DC (2003)
26. Conte de Leon, D., Alves-Foss, J., Krings, A., Oman, P.: Modeling complex control systems to identify remotely accessible devices vulnerable to cyber attack. In: Proc. First Workshop on Scientific Aspects of Cyber Terrorism, Washington, D.C. (2002)
27. Dán, G., Sandberg, H.: Stealth attacks and protection schemes for state estimators in power systems. In: Proc. First IEEE International Conference on Smart Grid Communications (SmartGridComm), Gaithersburg, MD, pp. 214–219 (2010)
28. Dán, G., Sandberg, H., Ekstedt, M., Björkman, G.: Challenges in power system information security. IEEE Security & Privacy 10(4), 62–70 (2012)
29. Darby, J., Phelan, J., Sholander, P., Smith, B., Walter, A., Wyss, G.: Evidence-based techniques for evaluating cyber protection systems for critical infrastructures. In: Proc. IEEE Military Communications Conference, pp. 1–10 (2006)
30. Davis, C.M., Tate, J.E., Okhravi, H., Grier, C., Overbye, T.J., Nicol, D.: SCADA cyber security testbed development. In: Proc. 38th North American Power Symposium, pp. 483–488 (2006)
31. Dawson, R., Boyd, C., Dawson, E., Manuel Gonzàlez Nieto, J.: SKMA – A key management architecture for SCADA systems. In: Proc. Fourth Australasian Workshops on Grid Computing and E-Research, vol. 54, pp. 183–192 (2006)
32. Depoy, J., Phelan, J., Sholander, P., Smith, B., Varnado, G.B., Wyss, G.: Risk assessment for physical and cyber attacks on critical infrastructures. In: Proc. IEEE Military Communications Conference, vol. 3, pp. 1961–1969 (2005)

33. Dondossola, G., Garrone, F., Szanto, J.: Supporting cyber risk assessment of power control systems with experimental data. In: Proc. IEEE Power Systems Conference and Exposition, pp. 1–3 (2009)
34. Dörfler, F., Bullo, F.: Synchronization and transient stability in power networks and non-uniform kuramoto oscillators. In: Proc. American Control Conference, pp. 930–937 (2010)
35. Draney, B., Cambell, S., Walter, H.: NERSC cyber security challenges that require doe development and support. Technical Report LBNL–62284, Ernest Orlando Lawrence Berkeley National Laboratory, Berkeley, California (2007)
36. Dudenhoeffer, D.D., Permann, M.R., Woolsey, S., Timpany, R., Miller, C., McDermott, A., Manic, M.: Interdependency modeling and emergency response. In: Proc. 2007 Summer Computer Simulation Conference, pp. 1230–1237 (2007)
37. Eberle, W., Holder, L.: Insider threat detection using graph-based approaches. In: Proc. Cybersecurity Applications and Technology Conference for Homeland Security, pp. 237–241 (2009)
38. Edwards, D., Srivastava, S.K., Cartes, D.A., Simmons, S., Wilde, N.: Implementation and validation of a mult-level security model architecture. In: Proc. International Conference on Intelligent Systems Applications to Power Systems, pp. 1–4 (2007)
39. Ekstedt, M., Sommestad, T.: Enterprise architecture models for cyber security analysis. In: Proc. IEEE Power Systems Conference and Exposition, pp. 1–6 (2009)
40. Falliere, N., Murchu, L., Chien, E.: W32.stuxnet dossier, version 1.3. Symantec (2010)
41. Farris, J.F., Nicol, D.M.: Evaulation of secure peer-to-peer overlay routing for survivable SCADA systems. In: Proc. 36th Conference on Winter Simulation, pp. 300–308 (2004)
42. Fernandez, E.B., Wu, J., Larrondo-Petrie, M.M., Shao, Y.: On building secure SCADA systems using security patterns. In: Proc. 5th Annual Workshop on Cyber Security and Information Intelligence Research: Cyber Security and Information Intelligence Challenges and Strategies (2009)
43. Fleury, T., Khurana, H., Welch, V.: Towards a taxonomy of attacks against energy control systems. In: Second Annual IFIP Working Group 11.10 International Conference on Critical Infrastructure Protection (2008)
44. Flick, T., Morehouse, J.: Securing the Smart Grid: Next Generation Power Grid Security. Syngress (2011)
45. Gellings, C.: The Smart Grid: Enabling Energy Efficiency and Demand Response. Fairmont Press (2009)
46. Giani, A., Karsai, G., Roosta, T., Shah, A., Sinopoli, B., Wiley, J.: A testbed for secure and robust SCADA systems. SIGBED Review 5(2), Article No. 4 (2008)
47. Gilchrist, G.: Secure authentication for DNP3. In: Proc. IEEE Power and Energy Society General Meeting – Conversion and Delivery of Electrical Energy in the 21st Century, pp. 1–3 (2008)
48. Gonen, T.: Electric Power Distribution System Engineering. Mcgraw-Hill College (1985)
49. Grid, N.: Operating the electricity transmission networks in 2020 (2011)
50. GridWise Alliance: GridWise(TM) accelerates efforts to develop a smart grid in the U.S. In: GridWeek, Washington DC, MD (2007)
51. Grochocki, D., Huh, J., Berthier, R., Bobba, R., Sanders, W., Cardenas, A., Jetcheva, J.: AMI threats, intrusion detection requirements and deployment recommendations. In: Proc. Third IEEE International Conference on Smart Grid Communications (SmartGridComm), Tainan, pp. 395–400 (2012)

52. The Cyber Security Coordination Task Group: Smart Grid Cyber Security Strategy and Requirements. National Institute of Standards and Technology
53. Hadeli, H., Schierholz, R., Braendle, M., Tuduce, C.: Generating configuration for missing traffic detector and security measures in industrial control systems based on the system description files. In: Proc. IEEE Conference on Technologies for Homeland Security, pp. 503–510 (2009)
54. HadjSaid, N., Tranchita, C., Rozel, B., Viziteu, M., Caire, R.: Modeling cyber and physical interdependencies – application in ICT and power grids. In: Proc. IEEE Power Systems Conference and Exposition, pp. 1–6 (2009)
55. Hasan, R., Bobba, R., Khurana, H.: Analyzing NASPInet data flows. In: Proc. IEEE Power Systems Conference and Exposition, pp. 1–6 (2009)
56. Holcomb, J.: Auditing cyber security configuration for control system applications. In: Proc. IEEE Conference on Technologies for Homeland Security, pp. 7–13 (2009)
57. Holstein, D.K., Diaz, J.: Cyber security management for utility operations. In: Proc. 39th Annual Hawaii International Conference on Systems Sciences, vol. 10, p. 241c (2006)
58. Hughes, T.: Networks of Power: Electrification in Western Society, 1880-1930. JHU Press (1993)
59. Hull, J., Khurana, H., Markham, T., Staggs, K.: Staying in control: Cyber security and the modern electric grid. IEEE Power & Energy Magazine 10(1), 41–48 (2012)
60. Jones, P.: The role of new technologies: A power engineering equipment supply base perspective. In: Grid Policy Workshop, Paris, France (2010)
61. Kang, D.J., Kim, H.M.: A method for determination of key period using QoS function. In: Proc. Future Generation Communication and Networking, vol. 2, pp. 532–535 (2007)
62. Kang, D.J., Kim, H.M.: A proposal for key policy of symmetric encryption application to cyber security of KEPCO SCADA network. In: Proc. Future Generation Communication and Networking, vol. 2, pp. 609–613 (2007)
63. Khaitan, S., McCalley, S.: Cyber physical system approach for design of power grids: A survey. In: Proc. IEEE Power & Energy Society General Meeting, Vancouver, BC, pp. 1–5 (2013)
64. Khaitan, S., McCalley, S.: Design techniques and applications of cyber physical systems: A survey. IEEE Systems Journal (2014)
65. Khalil, H.: Nonlinear Systems. Prentice-Hall (2002)
66. Khurana, H., Hadley, M., Lu, N., Frincke, D.: Smart-grid security issues. IEEE Security Privacy 8(1), 81–85 (2009)
67. Khurana, H., Khan, M.M.H., Welch, V.: Leveraging computational grid technologies for building a secure and manageable power grid. In: Proc. Hawaii International Conference on System Sciences, pp. 115–124 (2007)
68. Khurana, H., Koleva, R., Basney, J.: Performance of cryptographic protocols for high-performance high-bandwidth and high-latency grid systems. In: Proc. Third IEEE International Conference on e-Science and Grid Computing, pp. 431–439 (2007)
69. Kim, H.M., Kang, D.J., Kim, T.H.: Flexible key distribution for SCADA network using multi-agent system. In: Proc. ECSIS Syposium on Bio-inspired, Learning, and Intelligent Systems for Security, pp. 29–34 (2007)
70. Klein, S.A.: An open source IEC-61850 toolkit for utility automation and wind power applications. In: Proc. IEEE/PES Transmission and Distribution Conference and Exposition, pp. 1–4 (2008)
71. Klein, S.A.: A secure IEC-61850 toolkit for utility automation. In: Proc. Cybersecurity Applications and Technology Conference for Homeland Security, pp. 245–250 (2009)

72. Kosut, O., Jia, L., Thomas, R.J., Tong, L.: Limiting false data attacks on power system state estimation. In: Proc. 44th Annual Conference on Information Sciences and Systems (CISS), Princeton, NJ, pp. 1–6 (2010)
73. Kosut, O., Jia, L., Thomas, R.J., Tong, L.: Malicious data attacks on smart grid state estimation: Attack strategies and countermeasures. In: Proc. First IEEE International Conference on Smart Grid Communications (SmartGridComm), Gaithersburg, MD, pp. 220–225 (2010)
74. Kundur, D.: Cyber-physical security of the smart grid. Lecture conducted from University of Toronto, Toronto, Canada (2013)
75. Kundur, D., Feng, X., Liu, S., Zourntos, T., Butler-Purry, K.: Towards a framework for cyber attack impact analysis of the electric smart grid. In: Proc. IEEE International Conference on Smart Grid Communications (SmartGridComm), Gaithersburg, Maryland, pp. 244–249 (2010)
76. Kundur, D., Feng, X., Mashayekh, S., Liu, S., Zourntos, T., Butler-Purry, K.: Towards modeling the impact of cyber attacks on a smart grid. International Journal of Security and Networks 6(1), 2–13 (2011)
77. Kundur, P.: Power System Stability and Control. McGraw-Hill Professional (1994)
78. Kundur, P.: Power System Stability and Control. McGraw-Hill (1994)
79. Kundur, P., Paserba, J., Ajjarapu, V., Andersson, G., Bose, A., Canizares, C., Hatziargyriou, N., Hill, D., Stankovic, A., Taylor, C., Cutsem, T., Vittal, V.: Definition and classification of power system stability: Ieee/cigre joint task force on stability terms and definitions. IEEE Transactions on Power Systems 19, 1387–1401 (2004)
80. Lin, H., Sambamoorthy, S., Shukla, S., Thorp, J., Mili, L.: Power system and communicaiton network co-simulation for smart grid applications. In: Proc. IEEE PES Conference on Innovative Smart Grid Technologies (ISGT), Anaheim, California, pp. 1–6 (2011)
81. Liu, C.C., Ten, C.W., Govindarasu, M.: Cybersecurity of SCADA systems: Vulnerability assessment and mitigation. In: Proc. IEEE Power Systems Conference and Exposition, pp. 1–3 (2009)
82. Liu, S., Liu, X., El-Saddik, A.: Denial-of-service (DoS) attacks on load frequency control in smart grids. In: Proc. IEEE PES Innovative Smart Grid Technologies (ISGT), Washington DC, MD, pp. 1–6 (2013)
83. Liu, Y., Ning, P., Reiter, M.: Generalized false data injection attacks against state estimation in electric power grids. ACM Transactions on Information and System Security (TISSEC) 14(1) (2011)
84. Liu, Y., Ning, P., Reiter, M.K.: False data injection attacks against state estimation in electric power grids. In: Proc. 16th ACM Conference on Computer and Communications Security, Chicago, IL, pp. 21–32 (2009)
85. Liu, Y., Ning, P., Reiter, M.K.: False data injection attacks against state estimation in electric power grids. ACM Transactions on Information and System Security (2011) (to appear)
86. Mander, T., Nabhani, F., Wang, L., Cheung, R.: Integrated network security protocol layer for open-access power distribution systems. In: Proc. IEEE Power Engineering Society General Meeting, pp. 1–8 (2007)
87. McDaniel, P., McLaughlin, S.: Security and privacy challenges in the smart grid. IEEE Security Privacy 7(3), 75–77 (2009)
88. McMillin, B.: Complexities of information security in cyber-physical power systems. In: Proc. IEEE Power Systems Conference and Exposition, pp. 1–2 (2009)

89. McMillin, B., Gill, C., Crow, M.L., Liu, F., Niehaus, D., Potthast, A., Tauritz, D.: Cyber-physical systems distributed control: The advanced electric power grid. In: Proc. National Workshop on Beyond SCADA: Networked Embedded Control for Critical Physical Systems, HCSS:NEC4CPS (2006)
90. McQueen, M.A., Boyer, W.F.: Deception used for cyber defense of control systems. In: Proc. 2nd Conference on Human System Interactions, pp. 624–631 (2009)
91. McQueen, M.A., Boyer, W.F., Flynn, M.A., Beitel, G.A.: Quantitative cyber risk reduction estimation methodology for small SCADA control system. In: Proc. 39th Annual Hawaii International Conference on Systems Sciences, vol. 9, pp. 226–236 (2006)
92. Meyer, C.D.: Matrix Analysis and Applied Linear Algebra. SIAM (2001)
93. Mohsenian-Rad, A., Leon-Garcia, A.: Distributed internet-based load altering attacks against smart power grids. IEEE Transactions on Smart Grid 2(4), 667–674 (2011)
94. Moslehi, K., Kumar, R.: A reliability perspective of the smart grid. IEEE Transactions on Smart Grid 1(1), 57–64 (2010)
95. Olfati-Saber, R.: Flocking for multi-agent dynamic systems: Algorithms and theory. IEEE Transactions on Automatic Control 51(3), 401–420 (2006)
96. Olfati-Saber, R., Fax, J., Murray, R.: Consensus and cooperation in networked multi-agent systems. Proceedings of the IEEE 95(1), 215–233 (2007)
97. Patel, S.C., Bhatt, G.D., Graham, J.H.: Improving the cyber security of SCADA communication networks. Communications of the ACM 52(7), 139–142 (2009)
98. Piètre-Cambacédès, L., Sitbon, P.: Cryptographic key management for SCADA systems – issues and perspectives. In: Proc. International Conference on Information Security and Assurance, pp. 156–161 (2008)
99. Reynolds, C.: Flocks, herds, and schools: a distributed behavioral model. Computer Graphics 21(4), 25–34 (1987)
100. Risley, A., Carson, K.: Low- or no-cost cybersecurity solutions for defending the electric power system against electronic intrusions. Schweitzer Engineering Laboratories, Inc. (2006)
101. Rozel, B., Viziteu, M., Caire, R., Hadjsaid, N., Rognon, J.P.: Towards a common model for studying critical infrastructure interdependencies. In: Proc. IEEE Power and Energy Society General Meeting – Conversion and Delivery of Electrical Energy in the 21st Century, Pittsburgh, Pennsylvania, pp. 1–6 (2008)
102. Sauer, P., Pai, M.: Power System Dynamics and Stability. Prentice Hall (1997)
103. Sioshansi, F.: Smart Grid: Integrating Renewable, Distributed & Efficient Energy. Academic Press (2011)
104. Sologar, A., Moll, J.: Developing a comprehensive substation cyber security and data management solution. In: Proc. IEEE/PES Transmission and Distribution Conference and Exposition, pp. 1–7 (2008)
105. Sou, K., Sandberg, H.: Detection and identification of data attacks in power system. In: American Control Conference (ACC), Montreal, QC, pp. 3651–3656 (2012)
106. Stamp, J., McIntyre, A., Ricardson, B.: Reliability impacts from cyber attack on electric power systems. In: Proc. IEEE Power Systems Conference and Exposition, pp. 1–8 (2009)
107. Takano, M.: Sustainable cyber security for utility facilities control system based on defense-in-depth concept. In: Proc. SICE Annual Conference, pp. 2910–2913 (2007)
108. Tan, H.: Security analysis of a cyber-physical system. Master's thesis, University of Missouri-Rolla (2007)
109. Tang, H., McMillin, B.: Security property violation in CPS through timing. In: Proc. 28th International Conference on Distributed Computing Systems Workshops, pp. 519–524 (2008)

110. Ten, C.W., Liu, C.C., Govindarasu, M.: Vulnerability assessment of cybersecurity for SCADA systems using attack trees. In: Proc. IEEE Power Engineering Society General Meeting, pp. 1–8 (2007)
111. Ton, D.: DOE's perspectives on smart grid technology, challenges, & research opportunities. In: UCLA Engineering SmartGrid Seminar, Los Angeles, CA (2009)
112. Tuzzo, S.: A PlugN'Play platform independent solution that eliminates unauthorized access without the use of passwords or encryption keys. In: Proc. IEEE Conference on Technologies for Homeland Security, pp. 79–85 (2008)
113. Vijayan, J.: Stuxnet renews power grid security concerns. Computerworld (2010)
114. Wang, Y., Chu, B.T.: sSCADA: Securing SCADA infrastructure communications (2004), http://eprint.iacr.org/2004/265.pdf
115. Wei, J., Kundur, D.: A multi-flock approach to rapid dynamic generator coherency identification. In: Proc. IEEE Power & Energy Society General Meeting, Vancouver, Canada, pp. 1–5 (2013)
116. Wei, J., Kundur, D., Zourntos, T.: On the use of cyber-physical hierarchy for smart grid security and efficient control. In: Proc. IEEE Canadian Conference on Electrical and Computer Engineering (CCECE), Montreal, Canada (2012)
117. Wei, J., Kundur, D., Zourntos, T., Butler-Purry, K.: A flocking-based dynamical systems paradigm for smart power system analysis. In: Proc. IEEE Power & Energy Society General Meeting, San Diego, California (2012)
118. West, A.: Securing DNP3 and Modbus with AGA12-2J. In: Proc. IEEE Power and Energy Society General Meeting – Conversion and Delivery of Electrical Energy in the 21st Century, pp. 1–4 (2008)
119. Xiangjun, Z.: Context information-based cyber security defense of protection system. IEEE Transactions on Power Delivery 22(3), 1477–1481 (2007)
120. Xiao, K., Chen, N., Ren, S., Shen, L., Sun, X., Kwiat, K., Macalik, M.: A workflow-based non-intrusive approach for enhancing the survivability of critical infrastructures in cyber environment. In: Proc. Third International Workshop on Software Engineering for Secure Systems (2007)
121. Xie, L., Mo, Y., Sinopoli, B.: False data injection attacks in electricity markets. In: Proc. IEEE International Conference on Smart Grid Communications, Tainan, Taiwan, pp. 226–231 (2010)
122. Yamada, T., Maruyama, T.: Study on a security framework for a plant level network. In: Proc. 2006 SICE-ICASE International Joint Conference, Bexco, Busan Korea, pp. 1063–1066 (2006)

Cyber-Physical Security Testbed for Substations in a Power Grid

Junho Hong, Ying Chen, Chen-Ching Liu, and Manimaran Govindarasu

Abstract. The physical system of the power grids relies on the cyber system for monitoring, control, and operation. As a result, the reliable operation of power grids is highly dependent on the associated cyber infrastructures. The integrated cyber and physical system of power grids creates a large and complex infrastructure. Due to the high penetration of Information and Communications Technology (ICT), Supervisory Control And Data Acquisition (SCADA) systems are highly interconnected with one another, resulting in higher vulnerability with respect to cyber intrusions. Recent reports indicate that cyber-attacks are increasingly likely for the critical infrastructures, e.g., control centers, nuclear power plants, and substations. These attacks may cause significant damages on the power grid. Cyber security research for the power grid is a high priority subject for the emerging smart grid environment.

Substations in the power grid are critical as they are installed with power system components such as transformers, busbars, circuit breakers, and Intelligent Electronic Devices (IEDs). Measurements from substations are used as input to Energy Management System (EMS) software applications, including state estimation and optimal power flow. These cyber and physical devices can be physically or electrically connected. For example, a protection and control unit of a transformer is connected to the user-interface via the substation local area network.

Junho Hong
ABB US Corporate Research Center, Raleigh, NC, USA

Junho Hong · Ying Chen · Chen-Ching Liu
Washington State University, Pullman, WA, USA

Chen-Ching Liu
University College Dublin, Dublin, Ireland

Ying Chen
Tsinghua University, Beijing, China

Manimaran Govindarasu
Iowa State University, Ames, IA, USA

S.K. Khaitan, J.D. McCalley, and C.-C. Liu (eds.), *Cyber Physical Systems Approach to Smart Electric Power Grid*, Power Systems, DOI: 10.1007/978-3-662-45928-7_10

Remote access to substation networks is a common way for maintenance of substation facilities. However, there are many potential cyber security issues including remote access connection. Simultaneous cyber intrusions to important substations may trigger multiple, cascaded sequences of events, leading to a blackout. As a result, it is crucial to enhance the cyber security of substations and analyze cyber and physical security as one integrated structure in order to enhance the resilience of power grids. The mitigation strategy is vital to cyber-physical security of substations in order to stop the attack, disconnect the intruder, and restore the power system to a normal state. Mitigation methods can be taken on the cyber (ICT) side and physical (power system) side. The key to cyber mitigation is to find anomaly activities or malicious behaviors, and disconnect or stop the intrusion.

A cyber-physical testbed is critical for the study of cyber-physical security of power systems. For reason of security by power companies, real measurements (e.g., voltages, currents and binary status) and ICT data (e.g., communication protocols, system logs, and security logs) are not available. A testbed is a good alternative to acquire realistic cyber (i.e., ICT data) and physical (i.e., power system measurements) system data for research and demonstration purposes. The cyber-physical testbed provides a realistic environment to study the interactions between a complex power system and the ICT system. It is important to study the cause-effect relationships of cyber intrusions, vulnerability and resilience of power systems, as well as the performance and reliability of applications in a realistic environment provided by a testbed.

1 Introduction to Cyber and Physical System in a Power Grid

Power grids are complex cyber and physical systems [1]. The physical system of power grids includes power plants, substations, and transmission and distribution systems [2]. Electric power is produced by generators, while substations convert Alternating Current (AC) voltage from a voltage level to another for delivery from power plants to the load. Transmission systems deliver electric power to distribution substations through transmission networks. Distribution systems deliver electric energy to customers. The physical system of power grids relies on the cyber system for monitoring, control, and operation. The cyber system of power grids is formed by the Information and Communications Technology at the substations and the SCADA system at the control center [3]. The SCADA system supports the EMS that includes a number of power system software applications. Measurements (e.g., current and voltage) from Current Transformers (CTs) and Voltage Transformers (VTs) at the substations are delivered to control centers through ICT networks, e.g., TCP/IP based wide area networks. Control commands, such as opening of a Circuit Breaker (CB), can be sent from the SCADA system at a control center to the Remote Terminal Units (RTUs) or gateways in the substations. The integrated cyber and physical systems of power grids create a large and complex infrastructure.

SCADA systems have evolved from independent systems to networked-systems [4]. The first SCADA systems were isolated from other systems. As the requirements for data points are increased, more ICT networks are implemented in the substations and SCADA systems. Due to the high penetration of ICT systems, modern SCADA systems are highly interconnected with one another and, as a result, become more vulnerable than before with respect to cyber intrusions. Unsecured web servers in the user-interfaces, default passwords of IEDs and mis-configured firewalls are among the potential cyber vulnerabilities [5].

Substations in the power grid are critical since it has power system components such as transformers, bus bars, circuit breakers and IEDs, and the measurements from substations are used for input to EMS applications, e.g., state estimation and optimal power flow. The traditional power grid is designed based on the N-1 security criterion[1] [6]. However, a well coordinated cyber attack may compromise multiple substations. Therefore, simultaneous cyber intrusions to important substations may trigger multiple, cascaded sequences of events, leading to a power grid blackout. As a result, it is crucial to enhance the cyber security of substations and analyze cyber and physical security as one integrated structure in order to enhance resilience of power grids.

Cyber security concerns and potential threats to the power infrastructures have been reported by governments and other organizations, e.g., General Accounting Office (GAO), National Institute of Standards and Technology (NIST) or NISTIR (NIST Internal Reports) and Department of Energy (DOE) [7, 8, 9]. North American Electric Reliability Corporation (NERC) Critical Infrastructure Protection (CIP) published reliability standards 002-009 that covers the security of critical cyber assets, physical and cyber security, electronic security perimeters as well as personnel training and security management [10].

There are different cyber-physical systems in a smart grid, e.g., substation automation system, distribution automation system, advanced metering infrastructure, and electricity market. Moreover, many other critical infra structures, e.g., transport, health, water, gas, are critically dependent on ICT systems. Due to the vulnerability of these critical systems, a successful cyber intrusion may cause serious damages to the cyber system or physical components.

This chapter is concerned with cyber security of substations. A real-time cyber-physical security testbed is proposed for simulation of potential cyber intrusions and validation of cyber and physical mitigation methods. In the remaining of this chapter, Section III provides the concepts and design of the cyber-physical system. Section IV describes the IEC 61850 standard and multicast messages in a substation automation system. In Section V, vulnerabilities and hypothesized intrusion and mitigation scenarios of substation automation systems are explained. Section

[1] The power system has to be designed to absorb a loss of one or more system elements that occurs first, e.g., loss of a single generator or a transmission line. However, the expression "-1" may refer the failure of multiple elements since there could be physically or electrically linked elements as one, e.g., multiple feeders that are connected to one transformer.

VI provides architectures and components of the proposed cyber-physical security testbed and mitigation strategies. Simulation results involving cyber attacks and intrusion detection in the testbed environment are reported.

2 Cyber-Physical System Testbeds

By gradual deployment of advanced information and power technologies, the power system is in transition to a smart grid. This important transition creates the need for Research and Development (R&D) on the enabling technologies, such as integrated communication, advanced metering infrastructure (AMI), demand response, distribution automation and integration of large scale renewable devices. However, recent reports indicate that cyber attacks are increasingly likely for the critical infrastructures (e.g., control centers, nuclear power plants, and substations). Therefore, cyber security research for the power grid is an important priority for the emerging smart grid [11]. The work of [52, 53] show the survey of cyber-physical security research of power grid for researchers and system operators.

Advanced communication technologies and integration of a large number of Phasor Measurement Units (PMUs) enable the reliable and dependable Wide-Area Monitoring, Protection and Control (WAMPAC) system of the power grids. This system has the potential to analyze the power system condition, predict problems that may arise, and prevent worsening system conditions. The Automatic Generation Control (AGC) and Automatic Voltage Regulator (AVR) are representative applications that use the WAMPAC system. For instance, AVR uses measurements from substation RTUs periodically (e.g., 10 seconds or 2 minutes). It also sends control commands from the control center to the substation RTUs in order to increase or decrease the reactive power output of the generators, turn on/off switches of the capacitor banks, and change the positions of On-Load Tap Changers (OLTCs). The information is managed by the application in EMS. However, the performance of WAMPAC depends on the ICT networks. It also generates a lot of measurements and control messages to the ICT network (Specially, PMUs accumulate large amounts of data into the wide-area communication network between substations and control centers). All measurements and controls are transferred, stored and managed by the same ICT system in order to reduce the operation and implementation costs. As a result, there are potential threats and cyber security concerns for the EMS applications that are supported by ICT systems. These problems can be analyzed by the cyber-physical security testbed.

The cyber attack and defense studies cannot be conducted on the real systems due to the potential risk of service disruption. Furthermore, communication data in a substation and a control center may not be available for the cyber security research. As a result a realistic testbed is the best alternative for study of cyber security for power systems. The conventional power system simulation software or hardware tools do not incorporate the ICT systems (e.g., communication protocols and SCADA). Similarly, communication simulation tools do not support power system simulations (e.g., power system stability analysis and optimal power flow).

In order to analyze the cause- effect scenarios of the cyber and physical system (e.g., what is the consequence and impact to a power grid upon a successful cyber attack to a substation), integration of the cyber and physical system is needed. Therefore, a cyber-physical testbed is required for study of cyber-physical security of power systems. This testbed can be used for complex power system analysis with ICT systems and also provide a methodology to study the interaction between a power system and the ICT system that cannot be performed by a traditional power system testbed [11].

Several testbeds for cyber-physical security of power systems have been developed by a number of institutions. Idaho National Laboratory (INL) developed a National SCADA Testbed (NSTB) that can be used to identify and mitigate existing vulnerabilities [12, 13, 14]. The Virtual Control System Environment (VCSE) is developed by Sandia National Laboratory (SNL) that can be used to model and simulate cyber-physical system security [15], [16]. Iowa State University established the PowerCyber testbed using Real Time Digital Simulators (RTDS), and ISEAGE WAN emulation [1]. The Virtual Power System Testbed (VPST) was developed by the University of Illinois with the PowerWorld power system simulator and a Real-Time Immersive Network Simulation Environment (RINSE) [17]. The work of [18] proposes anomaly-based intrusion detection on the SCADA Control Systems (TASSCS) at the University of Arizona. The CRUTIAL testbeds are proposed to analyze the ICT resilience of power control systems in Europe [19], [20]. The testbed at the University College Dublin (UCD) has the capability to simulate cyber attacks and its impact on the power grids. This testbed is based on the commercial EMS and DIgSILENT power system simulator [21]. Royal Melbourne Institute of Technology (RMIT) developed the SCADASim testbed for testing of different attack and security solutions on actual devices and applications using a simulated environment [22].

In Sections 3.1, the common architecture and research applications of cyber-physical power system testbeds will be presented. Then the common design of a Cyber-Physical Substation Testbed (CPST) will be introduced in Section 3.2.

2.1 Common Framework and Applications of the Cyber-Physical Testbeds

As described in the last Section, there are various types of cyber-physical testbeds. Although the components and configurations of the testbeds are different, they do share some similarity. The figures 1 ~ 4 will explain the common modeling and simulation platforms of the cyber-physical testbeds and their applications.

The integrated system is composed of cyber (ICT systems) and physical (power system elements) systems as shown in Fig. 1. The cyber-physical system can be divided into three parts, i.e., control center with EMS and SCADA system, substations with critical devices (e.g., transformers, buses, generators, feeders, capacitors) and ICT systems that link them.

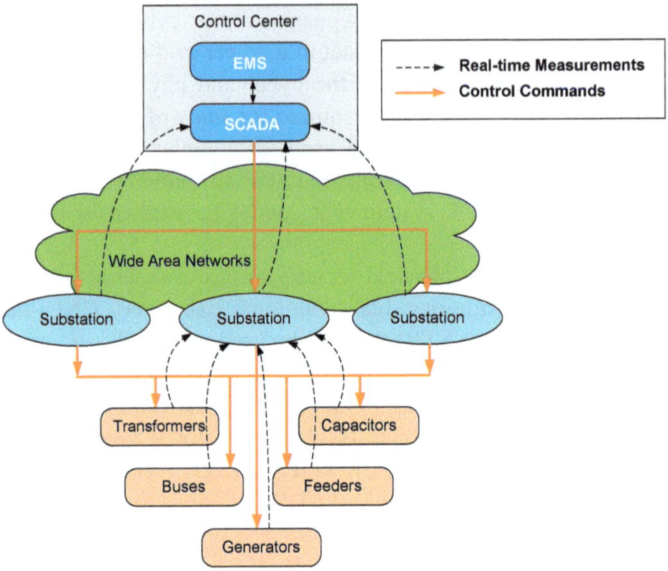

Fig. 1 General structure of the integrated system

The power flow snapshots can be obtained in two ways, i.e., acquired raw data and results from the state estimation software. Generally, the power flow snapshot from the raw data provides enough information to dispatchers to enable basic operations, e.g., monitoring and emergency warning. However, if dispatchers need to determine the controls for reconfiguring the power system, they need to use a credible power flow snapshot from the state estimation. The optimal or contingency dispatch strategies are generated by EMS in order to optimize the operation of the power system or prevent outages that are caused by power system faults and disturbances. These actions are performed by predefined algorithms or experienced dispatchers. Control commands are sent to the gateways or RTUs in the substations to operate the appropriate devices.

To enhance the robustness of a power system, double layered control systems are proposed as shown in Fig. 2. Remote controls are used to enhance the efficiency and stability of the power system whereas local controls are designed to maintain the integrity of the devices and reliability of the power system. Some local controls, e.g., excitation and governor controls for the generators, are responsible for enhancing the steady and dynamic stability of the power system.

Fig. 3 is an illustration of a typical structure of the cyber-physical testbed. Each module represents a simulation software package, hardware device or routine program. The functional modules include four features, i.e., the physical system, ICT, cyber system and system management module.

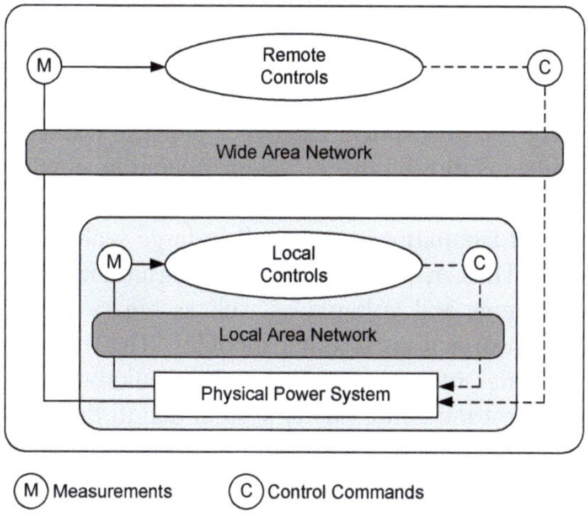

Fig. 2 Double layered controls within cyber-physical system

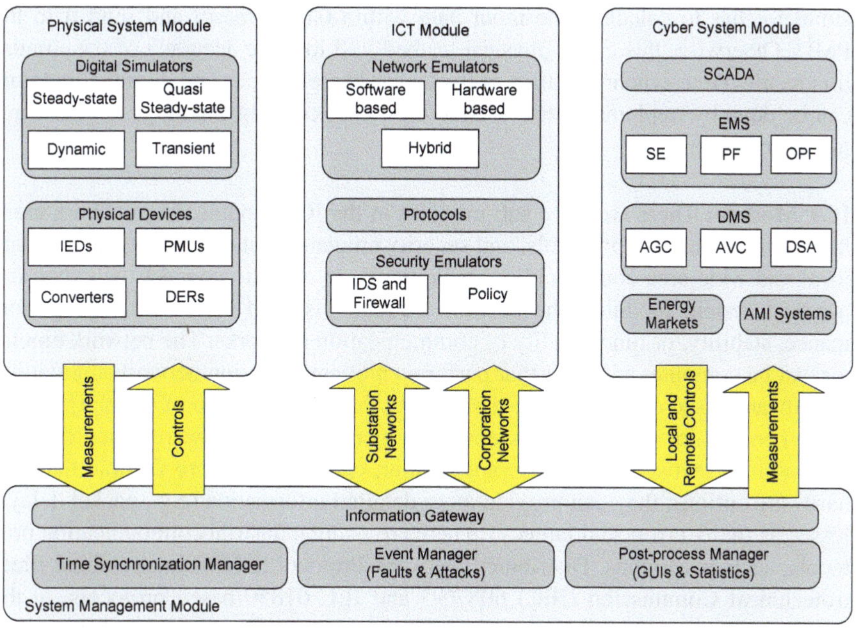

Fig. 3 Typical structure of the cyber-physical testbed

Each module's functions and interfaces are described in the following:

Physical System Module: This module is to simulate the complex power system behaviors accurately and efficiently. The cyber-physical testbed needs to have the capability to simulate various power system operating conditions, e.g., steady state, quasi-steady state (i.e., time series analysis based successive power flow), dynamic and transient simulations with small or large-scale power system models as shown in Fig. 3. Thus, it provides various measurements such as active and reactive powers, currents and voltages as well as binary status of switches and lines before/after control actions to other modules. The transient simulation produces detailed dynamics of power systems within microseconds. However, it is not feasible to represent the entire power system due to the modeling complexity and heavy computational burden. It is important to determine the tradeoffs between different levels of modeling and simulation methods.

In general, the hardware-in-loop refers to hybrid simulations that combine hardware devices and digital simulators. Hardware devices (e.g., IEDs, PMU and converters) need input data from the digital simulators to process its own algorithms. Digital simulators have to finish all calculations within a data sampling period. For instance, if PMU required 30 data samples every second, the digital simulator has to calculate the input data within 0.033 [mesc] and send it to the PMU. Otherwise, this cyber-physical testbed will lose the accuracy of the simulation result. Time synchronization between hardware devices and digital simulators can be done by implementation of time tags, e.g., creating time tags for measurements and controls.

ICT Module: There are three sub-modules in the ICT module, i.e., network emulator, communication protocols, and security emulator that can be used to simulate local and wide area controls in power systems. In order to connect the cyber and physical system modules, the network emulator is used to evaluate the performance, stability, or functionality of communication networks. The network emulator can be computer software that performs a communication network simulation, a hardware based device or hybrid simulator that includes both of them. There are some benefits for using network emulators compared to hardware based devices. For instance, network emulators are smaller, cheaper and more flexible than the hardware options; they can provide more detailed information (e.g., packet delays, losses, network errors and latency). There are many industrial communication protocols, such as Modbus, Distributed Network Protocol (DNP), International Electrotechnical Commission (IEC) 60870-5 and IEC 61850 based protocols, in the power grids. They may not be suited for evaluation of cyber related intrusions since they do not adopt up to date security measures. False data injection attacks may lead to wrong control or protection actions. Large scale denial of service attacks may cause disruptions to the communication networks. Potential cyber threats, attack models and its mitigations have to be studied using a proper cyber-physical testbed. Security emulators can be used to detect network anomalies and malicious behaviors with predefined rules or algorithms. The Intrusion Detection System (IDS) and firewall are representative security devices in the ICT systems. The Intrusion Prevention System (IPS) can disconnect or stop the intruders whereas a passive IDS will trigger an alarm to the operators. The ICT module has two interfaces, substation and corporate networks. The time synchronized measurements are sent from the physical system module to the cyber system module whereas time synchronized controls come from the cyber system module to the physical system module through the ICT module, as depicted in Fig. 3.

Cyber System Module: This module includes software, applications and systems that are developed for a study of the management, operation, and control of power systems in the cyber-physical testbed, e.g., EMS, Distribution Management System (DMS), AMI and energy markets. A solution to enable the cyber system module is to deploy commercial products from industry vendors. In fact, most of existing cyber-physical power system testbeds follow this approach since the commercial products provide both simplified ICT networks integration and a general level of system-in-the-loop tests. On the other hand, the cost of this solution is a barrier. In addition, it is hard to implement customized or developed algorithms into the commercial products since vendors may not allow users to access or modify their source code. Open source code based emulators can be customized and incorporated into existing platforms. They are suitable for cyber-physical security research where it is desirable to evaluate security algorithms or applications against cyber intrusions or threats. Examples include the impact of cyber attacks on the secured state estimation, security enhanced automatic generation controls and automatic voltage regulations, and analysis of resilience of power systems against cyber attacks. The measurements and controls have to be synchronized by the time synchronization manager as shown in Fig. 3.

System Management Module: This module is in charge of the overall operation of cyber-physical power system testbeds. The responsibilities include maintaining time synchronization between modules, managing all events and attacks that are generated from external sources, and analyzing/visualizing the test steps and results on the Graphic User Interfaces (GUIs). An important part in building the cyber-physical testbed is to balance between the physical power grid module and communication network and cyber system module. The physical power grid module produces continuous dynamic data (i.e., measurements), whereas the communication network and cyber system module generates discrete dynamic data (i.e., controls). The information gateway can create, add, delete and query all exchanged data in the system by the key-value mapping scheme. As all data and information that are generated from four modules (e.g., the physical system module, ICT module, cyber system module, and system management module) are exchanged at the information gateway, time synchronization becomes the only feature that determines the consistency of these interactions. Therefore, significant effort has been made by researchers to design and model the proper time synchronization mechanism and function modules in order to build a cyber-physical testbed. There are three types of synchronization mechanisms that can serve to provide the backbone time line for data exchange such as Global Positioning System (GPS) signal for real-time simulation, virtual-time stamp for off-line simulation or hybrid simulation that includes both of them. If the power system simulation tool (physical system module) calculates the power flow and sends measurements to the state estimation (cyber system module), this system can be synchronized by a combination of real-time and virtual-time mechanisms after ignoring the power flow calculation time at the power system simulation tool. Recently, the Functional Mock-up Interface (FMI) protocol standard has been proposed by the FMI research group. This independent standard is a tool to support both model exchange and co-simulation of dynamic models using a combination of xml-files and compiled C code. Moreover, it can reduce the system configuration times and efforts by offering the synchronized Function Mock-up Unit (FMU) that contains descriptions of interface data and functionalities [23].

The cyber-physical testbeds generate a large amount of data so it is important to manage and analyze all information and results by the post-process manager, as shown in Fig. 3. GUI can manage the global inputs and outputs of the testbed, configuration of physical, and ICT and cyber system modules. It can also be used for the external events due to cyber attacks, abnormal weather conditions, intensive fluctuation of loads and physical device failures. If there is an uncertain event, the statistical application will calculate the probability distributions to be used for stochastic analysis. Data mining techniques are used to perform the multi-round tests with different parameters and scenarios.

A cyber-physical testbed consists of various types of components such as software, hardware, ICT networks, emulators, and communication protocols. As shown in Fig. 4, the research of [1] identifies various applications of cyber-physical security testbed for power grids. The cyber-physical testbed is beneficial as a realistic platform for modeling and simulation of the power system and ICT applications, as well as the attack and defense strategies.

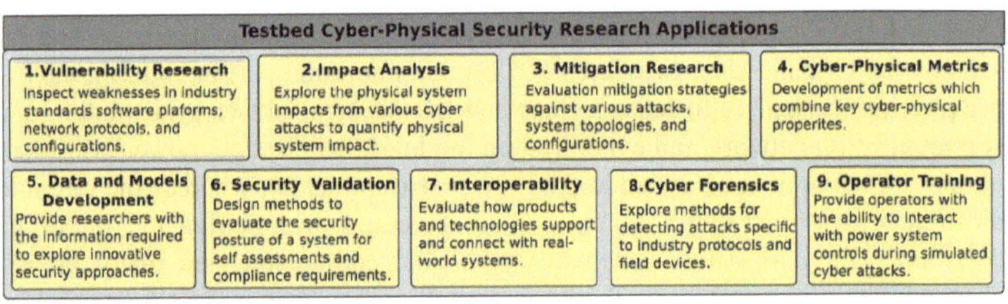

Fig. 4 Testbed applications [1]

2.2 *Design of the Off-line Cyber-Physical Testbed*

This section provides details of cyber-security testbed with examples of the actual implementation, e.g., off-line testbed. The off-line testbed has been used for the cyber security of AVR applications between substations and control centers. In the off-time testbed, all components are synchronized with a virtual-time stamp. The proposed off-line testbed requires more time and effort compared to the real-time testbed since this testbed does not use commercial products. It uses only freely available software. The ability of this testbed is limited since it can only perform the off-line based simulations and tests. However, it can produce reliable results and the implementation cost is low. The figures 5 ~ 9 will explain the framework and architecture of the off-time testbed for cyber security of the substations.

As shown in the Fig. 5, the testbed consists of open-source software packages and interfaces for physical, ICT, and cyber modules. Since this testbed adopted off-line testbed with the virtual time stamp, it can simulate large-scale power grids and ICT systems for the cyber-physical security research.

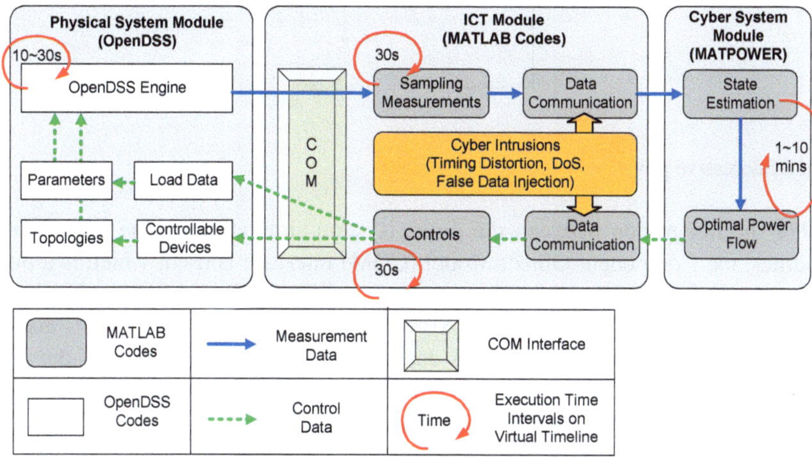

Fig. 5 Application flow of the off-line testbed

Physical System Module: Wide area control applications in SCADA systems can significantly affect the dynamics of a power system. Therefore, this testbed uses the OpenDSS[2] simulation tool to model the physical power systems. The OpenDSS (The Open Distribution System Simulator) is a toolbox for power system simulation, especially for the simulation of distribution networks with

[2] The OpenDSS is a comprehensive electrical power system simulation tool primarily for electric utility power distribution systems. It supports nearly all frequency domain (sinusoidal steady-state) analyses commonly performed on electric utility power distribution systems [24].

DER integrations. The functions of the OpenDSS include power flow, successive power flow with controls and load variations, common multi-phase circuit analysis, distributed generator analysis, harmonic analysis, and solar energy analysis. In this testbed, the OpenDSS is used as a simulation engine to generate quasi-steady dynamics of the power system. As shown in Fig. 6, the time series of system states are obtained by successive power flow calculations with inputs of the controls and load variations, where T is the current time, T_{end} is the end of simulation time, and Δt is the time difference between the previous and the current power flow.

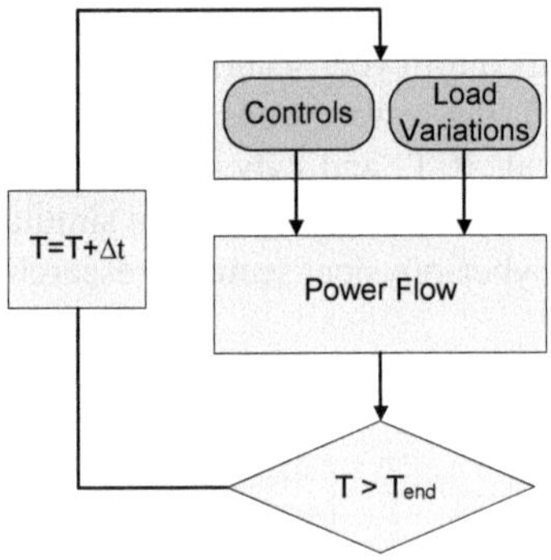

Fig. 6 Successive power flows

For the integration between the power system simulation engine and external modules, the Component Object Model (COM) interface is used, which is a built-in feature of OpenDSS and can be executed by MATLAB and Python languages. Through the COM interface, MATLAB can access the power flow results and adjust the parameters of the OpenDSS data (e.g., load data and status of devices). Fig.7 shows an example of the MATALB source code that controls the time-based simulation in OpenDSS. Therefore, MATLAB is acting as the wide area ICT module that connects the cyber system and physical system modules.

```
while(present_step <= num_pts)
    DSSCircuit.Solution.Solve;

    t = sample_dss_values(present_step);
    t = do_custom_control_loop();
    if (t == 0)
        disp('Error running the pv control loop')
        k = 0;
        return
    end

   %increment the present step
    present_step = present_step + 1;
end
result = dss_Command('CloseDI');
```

Fig. 7 Pseudo codes for the time-based simulation in Matlab

ICT and Cyber System Module: MATLAB is used for the ICT module. For the cyber system module, Matpower[3] has been used to simulate the EMS applications, including optimal power flow and state estimation. There are six steps for the ICT and cyber system module, e.g., sampling measurements, data communications, state estimation, optimal power flow, controls and cyber intrusions, as shown in Fig. 5. First, every 30 seconds, the function of measurement sampling has the responsibility to obtain all required measurements from the OpenDSS engine, such as the three phase voltages of all buses, power injections on each side of branches and transformers, and outputs of the generators. Measurements are tagged with time stamps, and they will be stored at the measurement data pool (i.e., database). Then, the data communication function will transfer time synchronized measurements to the state estimation function in the cyber system module. During this process, the network latency will be tagged to each time stamp according to the predefined configurations of communication channels. Also, the state estimation function imports the latest measurements from the data pool to produce the power flow snapshot. For every predefined time interval, if there is any limitation violation, the optimal power flow function will be executed using the data from state estimation. The optimal power flow results provide the necessary control commands, which are tagged with time stamps to controllable devices. Finally, the data communication function will transfer all controls from the optimal power flow function to the control function. Again, all control commands will be handled by the data communication function, which will create time tags and network latency information and store them in the command data pool. The control function will check the commands from the optimal power flow function periodically. Once it receives a new command, the control function changes the status of the corresponding devices, which are modeled in the OpenDSS engine via the COM interfaces.

Cyber Intrusion Module: In order to study cyber-security of the power system with wide area control applications, the cyber intrusion module is modeled and implemented in the testbed. Although there are numerous possibilities of attacks on the cyber system, the impact of these cyber attacks is related to three important features such as function integrity, service availability, and information confidentiality. The consequences caused by the availability and integrity attacks on the ICT module are explored with proper models and simulations. Note that if the testbed has a simulation function of the power system market, it can also simulate information confidentiality attacks on electricity market prices.

---

[3] MATPOWER is a package of MATLAB M-files for solving power flow and optimal power flow problems. It is intended as a simulation tool for researchers and educators. MATPOWER is designed to give the best performance possible while keeping the code simple to understand and modify [25].

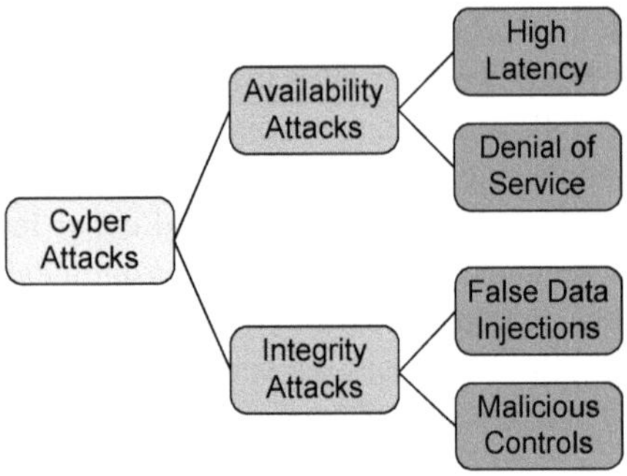

Fig. 8 Classification of cyber attacks

As shown in Fig. 8, wide area networks of a power system can be jammed by a large amount of data requests or injections as an attempt to reduce the availability of the cyber system. Moreover, if the ICT network links and communication protocols are compromised , the integrity of the wide area control systems can be damaged by injecting false data. Moreover, attackers may generate fabricated control messages to critical devices such as generators, transformers and breakers to create large scale outages. To model the cyber attacks, three dimensions are given, e.g., time, space, and style as illustrated in Fig. 9.

The execution time intervals for each function are shown in Fig. 5. They represent the operation period of each module. For example, the OpenDSS engine performs power flow calculations to generate measurements of the power system every 10 ~ 30 seconds. Since the proposed testbed is working in the off-line mode, the virtual timeline can be manipulated by adding, removing or adjust events. The operation periods of each module are important parameters for the study of wide area control cyber-security since it can influence the performance of EMS applications. Therefore, the off-line testbed is suitable for the parameter sensitivity analysis of the cyber-physical system security of power systems.

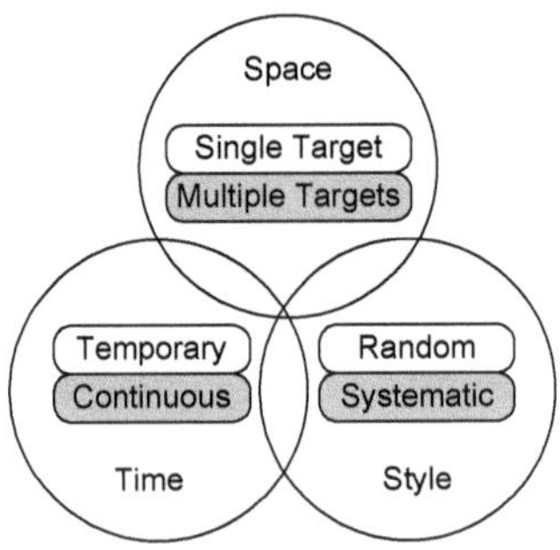

Fig. 9 Implementation strategies of cyber attacks

3 Substation Automation System

The concept and design of substation automation system was proposed by the IEC Technical Committee (TC) 57, Working Group (WG) 10. IEC TC 57 published IEC 61850 which is a standard for the design of substation automation system. The main purposes of IEC 61850 standard can be divided into four parts, (1) Lower configuration and installation cost, (2) Multi-vendor interoperability, (3) Long term stability, and (4) Minimal impact to the existing system.

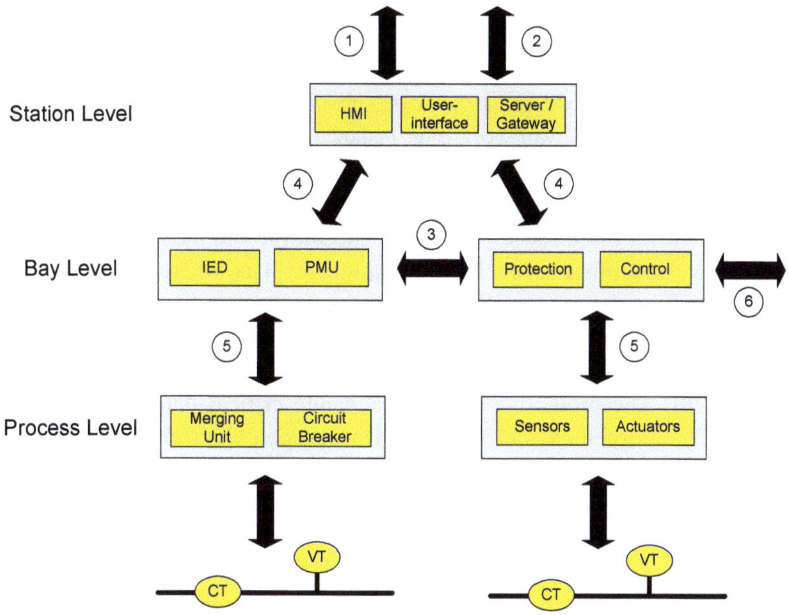

Fig. 10 Communication topology of the substation automation system (cyber system)

The installation and engineering cost of IEC 61850 based devices are drastically reduced since all hardwired connections from CTs and VTs to relays are changed to Ethernet based communications using Sampled Measured Value (SMV) messages which contain sampled data of currents and voltages. The Generic Object Oriented Substation Event (GOOSE) enables IEC 61850 based devices to quickly exchange critical data (e.g., a trip signal to a circuit breaker), i.e., less than 4 [msec], over the Ethernet based communication. This also significantly reduces the cost of wire installation. The Substation Configuration Language (SCL) contains device configuration information. Therefore, IEC 61850 based devices do not need any manual configurations; they import the configured SCL file through the ICT network. Standardized communication protocols and logical nodes enhance multi-vendor interoperability. Therefore, substation operators can use IEDs and user-interfaces from different vendors in a substation. The concept of IEC 61850 is extended to Distributed Energy Resources (DERs) and distribution automation. Hence, IEC 61850 enables devices from different manufacturers to exchange information in the substation level as well as system level [26]. The ICT technologies have been fast evolving over the last decade and the trend is continuing. However, the evolving cycle of power substation functions and software applications are slow compared to that of ICTs. The long term stability allows upgrading of ICT at a substation without re-engineering of the entire substation system. Since multi-vendor interoperability significantly reduced the gaps of device configuration between different vendors, substation engineers can add or remove existing devices at a lower cost. For instance, substation engineers can set up new devices and applications in a substation by sending SCL files via the ICT network [27].

Fig. 10 shows the three levels of the substation automation system, i.e., the station, bay, and process levels. The station level is where the user-interface, Human Machine Interface (HMI), substation server and gateway are located. The server and gateway exchange data coming from/to substation, e.g., remote access points (interface 1), control centers (interface 2) using DNP 3.0 or IEC 60870-5 [28]. The protective devices exchange critical data, e.g., interlocking (interface 3), between bays using GOOSE messages. Control and protection data are exchanged between the station and bay level using Manufacturing Message Specification (MMS) message (interface 4). Measurements such as currents and voltages are sent to the station level from the process level to bay level whereas control data are sent from the bay level to process level (interface 5) using SMV and GOOSE, respectively. Interface 6 shows the remote control and protection features between substations [29].

A substation includes various types of critical physical equipment, e.g., transformers, circuit breakers (52), bus bars, disconnect switches, and feeders, as shown in Fig. 11. The substation in Fig. 11 has two main transformers, and single busbars. When a fault occurs at a transformer or a busbar, the faulted area can be isolated by switching actions. The substation equipment will be protected by different types of protective relays. For instance, the transformer and busbar are protected by differential relays while the feeder is protected by overcurrent relays.

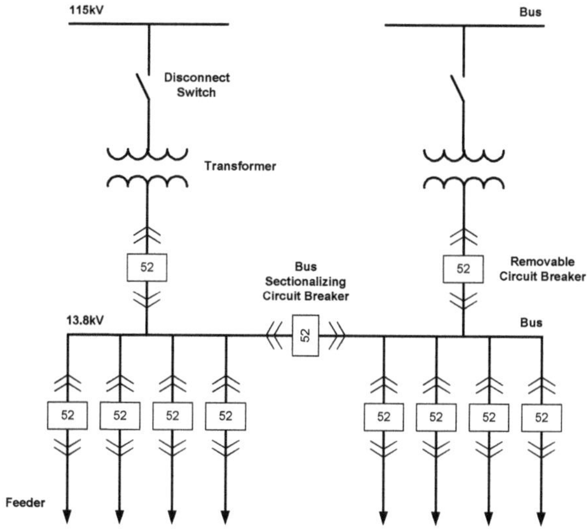

Fig. 11 The one line diagram of a substation (physical system) [30]

3.1 IEC 61850 Standard

The IEC 61850 is divided into 10 sections and 7 sub-sections as shown in Table I. Part 1 is an overview of the IEC 61850 standard series, basic interface and reference model of a substation automation system. Part 2 provides an explanation of the abbreviations and terms that are used in IEC 61850 series. Part 3 describes the general requirements of the ICT networks and guidelines for environmental conditions and recommendations. Part 4 is concerned with the system and project management with respect to the engineering process, life cycle of the overall system and supporting tools for engineering and testing. The scope of part 5 covers the communication requirements of the functions that are performed in the substation automation system. It also explains the Logical Nodes (LNs) for each function, e.g., PTOC is an AC time overcurrent relay that is able to trip the circuit breaker when the input current exceeds the predetermined threshold. The IED related configuration languages are shown in part 6, e.g., SCL, IED Capability Description (ICD), System Exchange Description (SED), Instantiated IED Description (IID), System Specification Description (SSD) and Configured IED Description (CID) that are based on the Extensible Markup Language (XML). Part 7 deals with the basic communication structure for substation and feeder equipment. Part 7-1 explains the principles of the modeling method, communication and information models that are used in IEC 61850-7-x. The definition and structure of Abstract Communication Service Interface (ACSI) communication in substations are introduced in part 7-2. Part 7-3 provides details of the layered substation communication architecture. The ICT models of functions and devices that are related to substation automation are described in part 7-4. Specially, this part of the standard includes details of logical node names and data names for communication between substation devices, e.g., IEDs and user-interfaces. Part 8-1 describes a method for data exchange between ACSI and MMS communication. Finally, part 9-1 and part 9-2 explain the structure and mapping of the SMV. Part 10 covers the subject of conformance testing for IEC 61850 systems.

Table 1 Sections of IEC 61850 standards

Section	Title
IEC 61850-1	Introduction and overview
IEC 61850-2	Glossary
IEC 61850-3	General requirements
IEC 61850-4	System and project management
IEC 61850-5	Communication requirements for functions and device models
IEC 61850-6	Configuration language for communication in electrical substations related to IEDs
IEC 61850-7	Basic communication structure for substation and feeder equipment
├ IEC 61850-7-1	├ Principles and models
├ IEC 61850-7-2	├ Abstract communication service interface (ACSI)
├ IEC 61850-7-3	├ Common Data Classes
└ IEC 61850-7-4	└ Compatible logical node classes and data classes
IEC 61850-8	Specific communication service mapping (SCSM)
└ IEC 61850-8-1	└ Mappings to MMS (ISO/IEC9506-1 and ISO/IEC 9506-2)
IEC 61850-9	Specific communication service mapping (SCSM)
├ IEC 61850-9-1	├ Sampled values over serial unidirectional multidrop point to point link
└ IEC 61850-9-2	└ Sampled values over ISO/IEC 8802-3
IEC 61850-10	Conformance testing

3.2 *Multicast Messages in a Substation Automation System*

The communication protocols in IEC 61850 can be classified into seven types. Due to the requirement of type 1, 1A and 4 messages, e.g., GOOSE and SV, they use three communication stacks, i.e., physical, data link and application layer as shown in Fig. 12. GOOSE supports critical data exchange such as interlocking between IEDs, trip messages from IED to circuit breakers or the status of circuit breakers to IED. The basic concept of information exchange is that a publisher writes values in a GOOSE packet and subscriber receives and reads the values from the GOOSE packet. GOOSE uses Media Access Control (MAC) address for the multicast[4] scheme. Due to the real-time requirement, GOOSE applies a re-transmission[5] scheme in order to achieve the appropriate level of communication speed and reliability. As shown in Fig. 10, the merging unit receives voltage and current values from CT and VT through the hard wire. Then the merging unit sends measured current and voltage values to protection IEDs using SMV messages. A merging unit can send SMV messages to multiple IEDs since SMV supports the multicast scheme. There are three types of resolution (bits) amplitude for SMV messages such as bits (P1 class), 16 bits (P2 class) and 32 bits (P3 class) [31].

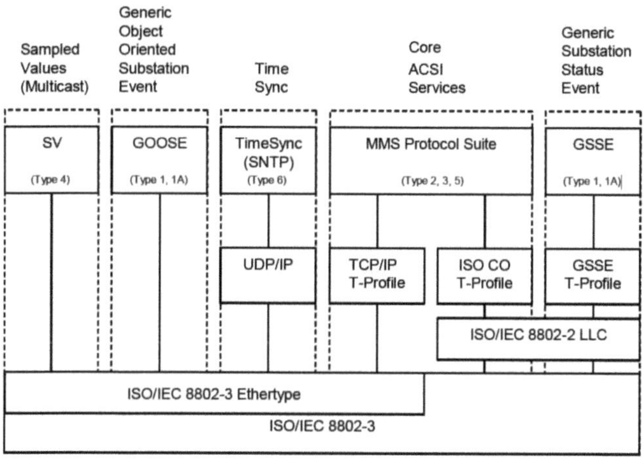

Fig. 12 Communication protocols in IEC 61850 [32]

- Type 1: Fast messages
- Type 1A: Trip
- Type 2: Medium speed messages
- Type 3: Low speed messages
- Type 4: Raw data messages
- Type 5: File transfer functions
- Type 6: Time synchronization messages

---

[4] Multicast is the delivery of data or information in a single host to multiple receivers simultaneously.

[5] The receiver does not send any response to the sender.

4 Vulnerability, Intrusion and Mitigation Scenarios of the Substations

The cyber security of substations has been recognized as a critical issue since it consists of various types of critical physical and cyber devices as explained in Section 3. They can be physically or electrically connected, e.g., a protection and control unit of a transformer is connected to user-interface via the substation local area network. The remote access to substation networks, e.g., IED or user-interface, is a common way for maintenance of the substation facilities. However, there are many potential cyber security issues, such as: (1) Well-trained intruder(s) compromise the remote access points for cyber attacks, (2) Standardized communication protocols allow intruders to analyze the substation communications, (3) Unencryptable multicast messages (e.g., GOOSE and SMV) due to the requirements, (4) Mis-configured firewalls, and (5) IEDs and user-interfaces with default passwords. .

4.1 Substation Vulnerabilities

4.1.1 Unsecured Industrial Protocols

Communication protocol is an important element for the operation of a power grid. The protocol must not be modified, fabricated or monitored except by system operators. Despite their importance, cyber security features are not included in most industrial protocols since cyber security was not a major concern when industrial communication protocols were published, e.g., DNP 3.0, IEC 61850, IEC 60870-5 and Inter-Control Centre Communication Protocol (ICCP). Therefore, IEC TC 57 WG 15 established the IEC 62351 standard. The primary objective is to develop standards for security of the communication protocols defined by IEC TC 57. The GOOSE and SMV messages contain critical information and use the multicast scheme as explained in Section 3. The multicast scheme has potential cyber vulnerabilities, e.g., group access control and group center trust. Most encryption schemes or other cyber security features that delay the transmission time are not applicable for these protocols since the performance requirement of GOOSE and SMV messages is within 4 [msec]. Therefore, IEC 62351 standard recommends an authentication scheme with a digital signature using Hash-based Message Authentication Code (HMAC) for GOOSE and SMV. However, the performance test to apply the authentication scheme to GOOSE and SMV is yet to be performed. The existing intrusion and anomaly detection systems do not normally support IEC 61850 based protocols since they are more focused on general cyber intrusions such as Distributed Denial of Service attack (DDoS). In order to mitigate the communication based cyber attacks to substation automation networks, the work of [33] proposed an Intrusion Detection System (IDS) for IEC 61850 based substation automation system. An intrusion detection system for serial communication based MODBUS and DNP3 in the substations is proposed in [34]. Reference [35] proposes a temporal anomaly detection method and [36] reports an integrated anomaly detection method for detecting malicious activities of IEC 61850 based multicast protocols (e.g., GOOSE and SMV) in the substation ICT network.

4.1.2 Remote Access Points

Power system components are located in wide-spread and remote sites. Remote access to substation networks using Virtual Private Network (VPN), dial-up or wireless is a common way to monitor and maintain the substation. The main problem of the remote access point is that remote access points may not be installed with adequate security features, e.g., poorly configured firewall, weak ID and password policy, bad key management for cryptography, and use of un-secured external memory (e.g., USB flash drive). Therefore, substation security managers have to consider the following actions in order to enhance the cyber security: (a) Check firewall policies and logs periodically to identify security breaches, (b) Change ID and password frequently and enhance the password policy (e.g., including numerical digits and special characters), (c) Enhance security of the key server against attacker(s), and (d) Provide security practice education for operators.

4.1.3 Default Password and Built-in Web Server

A typical substation may have a number of IEDs and it is difficult to manage the different passwords for each IED. Therefore, substation operators may use the default or same password for all IEDs. In addition, some IEDs and user interfaces have a built-in web server and hence it may be vulnerable to cyber intrusions, e.g., remote configuration change and control with default passwords. Substation security managers have to check the security and system logs of IEDs and user-interface to detect unauthorized access.

4.2 Hypothesized Intrusion Scenarios to Substations

Security threats to the substation automation system can be divided into two parts based on the physical and cyber assets. The physical assets are the hardware components, e.g., GPS (A4), IED (A5) and circuit breaker (A8), whereas cyber assets include physical and cyber resources, e.g., firewall (A2), communication network (A3) and software applications in the user-interface (A6), as illustrated in Fig. 13. Mitigation actions against security threats have to consider both physical and cyber intrusions. More details about the mitigation will be discussed in Section 5.

Security threats to substations can be inadvertent events as well as deliberate attacks. Inadvertent events include animal intrusions, equipment failures and natural disasters [37]. Animal intrusion is a major concern for substation operators [38]. A significant amount of research has been undertaken over the last decade concerning monitoring of the health condition for substation components.

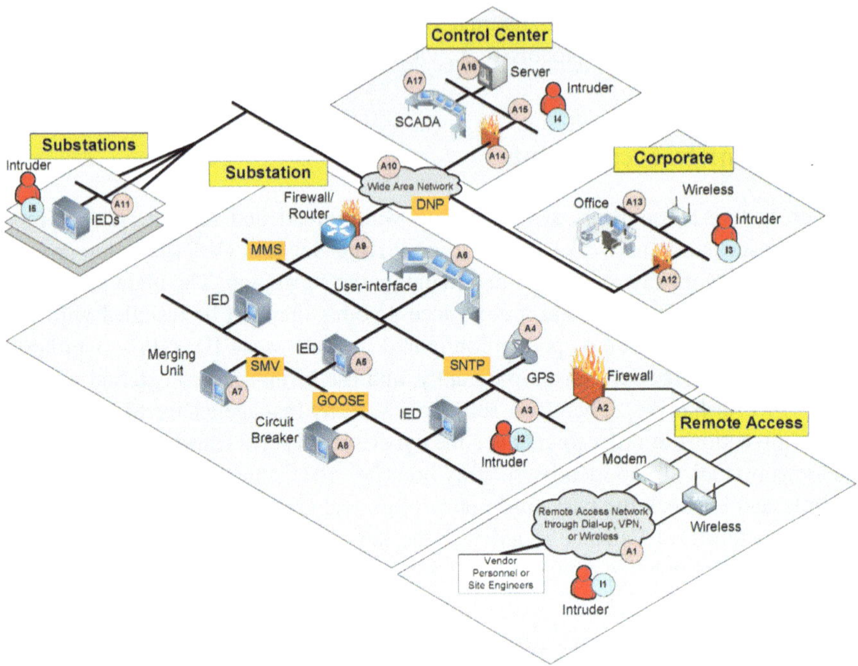

Fig. 13 Overview of substation ICT network diagram and security threats

Natural disasters such as flood, volcanic eruption, earthquake and tsunami, are rare but, in a severe scenario, can lead to cascading events and catastrophic outages. The work of [39] proposes weather-related power outages and enhancement of the system resiliency. Deliberate threats can be caused by disgruntled employees, cyber attackers, and malwares. Disgruntled employees can be threats for the substation security as they are familiar with the substation systems. The threats of cyber attacks are higher than before since substations need remote access connections for maintenance. *Stuxnet* is a relevant example of cyber threats (malwares) that are aimed at control systems of critical power infrastructure [40].

4.2.1 Single Substation Attack

As shown in Fig. 13, potential cyber security threats and locations of intruders in a substation automation network include:

A1: Compromise remote access points (e.g., dial-up, VPN and wireless)
A2/A9/A12/A14: Compromise firewall
A3: Gain access to substation network
A4: Interrupt GPS time synchronization
A5: Gain access to bay level devices or change protective device settings
A6: Gain access to user-interface
A7: Compromise process level devices (e.g., merging unit)

A8: Change the status of circuit breaker (e.g., close to open or vices versa)
A10: Gain access to wide area network (e.g., DNP 3.0)
A11: Gain access to neighbor substation network
A13: Gain access to corporate network
A15: Gain access to control center network
A16: Compromise the server in a control center
A17: Compromise the user-interface in a control center

I1: Intruder from outside of substation network via remote access points
I2: Intruder from inside of substation network
I3: Intruder from outside of substation network via corporate network
I4: Intruder from outside of substation network via control center network
I5: Intruder from outside of substation network via neighbor substation network

As depicted in Fig. 13, possible intrusions to the substation local area network can originate from outside or inside a substation network.

The following combinations represent the possible intrusion paths from outside to a local area network at a substation. Intrusions can originate from remote access points (A1) or neighbor substation network (A11) or corporate network (A13) or control center network (A15) all the way to the substation local area network (A3), e.g.,

from A1-A2-A3;
from A11-A10-A9-A3;
from A13-A12-A10-A9-A3;
from A15-A14-A10-A9-A3

Cyber attacks from inside the substation can originate from the substation network (A3) or user-interface (A6) then gain access to other facilities in the substation. An inside attack can be performed by social engineering [41]. One of the realistic examples of this attack is that intruder(s) send an email to substation operators that appears to come from a credible source. However, this email contains a fabricated website link or malware software so once operators open this email, their desktops or laptops will be infected. After that, this malware will infect the external flash drive that plugged into compromised devices. Finally, operator(s) may use the infected flash drive at the substation network to copy documentation. Then this malware will find a path to external communication, and send all information to intruder(s) or change the setting of the protection devices (e.g., IEDs).

It is crucial to protect the substation automation ICT network against cyber attacks as a successful cyber intrusion can cause significant damages on the power grid. Once an intruder can access the substation communication network, (s)he can access other facilities in the substation. For instance, the result of cyber attack, A4, may disrupt time synchronization of all communication protocols in the substation ICT network, and operators may lose the availability of substation communications. Upon successfully cracking an user name and a password and gaining an access to the user-interface (A6), the intruder may control or modify the settings of the IEDs (A5). Then they can operate circuit breakers through the connection of IEDs. Another possibility is to gain access to the ICT network of a neighbor substation, e.g., from A9-A10-A11, then multiple cyber attacks can be carried out. More details about simultaneous cyber attacks to the multiple substations will be discussed in Section 5.2.2.

4.2.2 Simultaneous Attacks to Multiple Substations

Each substation has a different level of importance in a power grid. Since generally, a high voltage substation carries more power. The level of cyber security is also different at each substation. For instance, substation A uses firewall, IDS and cryptography features for cyber security mitigation whereas substation B only uses firewalls. In this example, the security level of substation A is higher than substation B whereas the cost of security implementation at substation B is lower. By analyzing the security level of each substation and importance in a power grid, an intruder may find the optimal combination (considering cost-benefit model) of target substation(s) that can trigger a sequence of cascading events, leading to a system blackout. Therefore, the impact of simultaneous cyber attacks to multiple substations can be much higher than that of a single substation attack.

Table 2 Cost for cyber intrusion

Substation	Cyber security level	Physical importance	Cost for attack
1	Low	Medium	5
2	Medium	Medium	4
3	High	High	10
Successful attack combinations		(1, 2), (3), (1, 3), (2, 3), (1, 2, 3)	

For instance, there are 3 substations in the power system shown in Table II. If an attacker knows the cyber security level, physical importance, costs of an attack, and attack combinations that lead to a power system collapse, they may find the optimal attack combination. In this example, the lowest cost combination that can cause a collapse of the power system is (1, 2). Therefore, the attacker is likely to choose this combination to achieve the goal.

4.2.3 Attack Tree

In the field of computer science and information technology, attack trees have been used to analyze potential threats and attack paths against cyber attacks [42, 43, 44]. However, the concept of attack trees is broadened and applied to other systems, e.g., cyber security of power systems [45, 46]. Although there are numerous concepts and definitions of attack trees, the most commonly occurring concepts are nodes (root or leaf), edges, connectors and attributes [47]. Fig. 14 shows a simplified attack tree for the substation automation system. Root node (T1) is the ultimate goal (i.e., open circuit breakers) with combinations of leaf nodes (T3) that do not have any predecessor. Leaf nodes (T3) contain sub goals or steps to archive the final goal (T1). Edges (T2) are connectors for all nodes. There are two types of connectors (T4) in Fig. 14, "AND" and "OR." AND connector shows different steps (nodes) toward achieving the same goal. For instance, an intruder has to complete two steps, *Social Engineering* and *Compromise Operator Laptop*, in order to achieve *Obtain ID and Password*. Attributes represent features or properties relevant for numerical analysis of security models, e.g., attack probability and cost of an attack. Fig. 14 shows an example model of cost of an attack. If the first priority is to minimize the attack cost, the combination of (9)-(10)-(5)-(2) is the best way to achieve the final goal. However, if the priority of attack is to minimize attack steps, (4)-(1) is the best way to open circuit breakers.

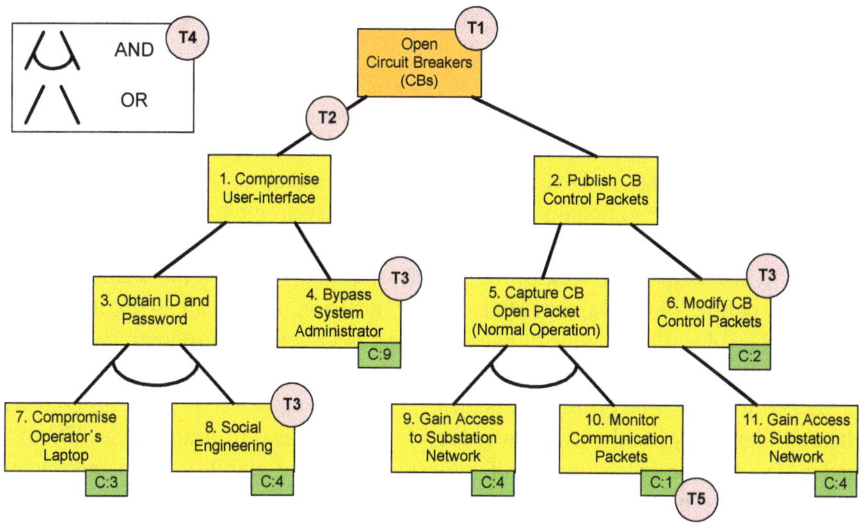

Fig. 14 Attack tree diagram for substation automation systems

4.3 *Mitigation Strategies That Include Cyber and Physical Aspects*

The mitigation strategy is vital to cyber-physical security of substations in order to stop the attack, disconnect the intruder, and restore the power systems to a normal state. Mitigation methods can be divided into two sides, e.g., cyber (ICT) and physical (power system) side. On the cyber side, real-time network monitoring, intrusion detection system, encryption, authentication and enhanced firewall are common practices in industry. The key to cyber mitigation is to find anomaly activities or malicious behaviors, and disconnect or stop the intrusion. A remedial action for mitigation can be performed on the physical side. For example, the Optimal Power Flow (OPF) algorithm, with an objective function that minimizes load shedding, can be used to calculate the mitigation actions. Substation operators need to determine an optimal cost-benefit solution. For instance, they can implement new security measures at important substations (e.g., high voltage substations).

4.3.1 Framework

Confidentiality, Integrity and Availability (CIA) are essential concepts in information security [48]. Confidentiality is to prevent access to data or information by unauthorized individuals. Communication data in a substation network must be protected since any successful eavesdropping attack can capture critical packets that contain important information to be used for an attack. Integrity refers to the ability to maintain authenticity, accuracy and provenance of recorded and reported information over its life cycle, i.e., data cannot be modified by an unauthorized person [49]. Upon successfully modification of fabrication of control messages (e.g., circuit breaker control or transformer tap change) in a substation network, an attacker may trigger an outage of the power system. Availability refers to the timely delivery of functional capability. Communications from/to a substation must be available all the time. A Denial-of-service (DoS) attack to a substation network can disrupt the communication for controls and measurements from/to the control center. Therefore, these security objectives (i.e., CIA) must be met.

Fig. 15 shows a framework of mitigation strategies based on the status of intrusion, i.e., before, on-going, and after a cyber intrusion. Before an attack is encountered at the substations, security managers and operators need to analyze potential vulnerabilities using the system and security logs, penetration test, etc. Encryption is needed for non-time critical messages, e.g., MMS and DNP in order to enhance the confidentiality. Authentication is used for time critical messages, e.g., GOOSE and SMV, for the enhancement of integrity. Transient stability and contingency analysis will be performed to check whether the power system can maintain stability when it undergoes hypothesized cyber attacks. During the intrusions, it is important to find the intrusion point and type of attack. Then the intruder(s) can be disconnected from the substation network through an IDS and a firewall. The impact analysis will be performed to find the most critical attack that can cause the worst case damage to the power grid. Once intruders are blocked or disconnected from the substation network, the security manager has to analyze the security breach using security and system logs.

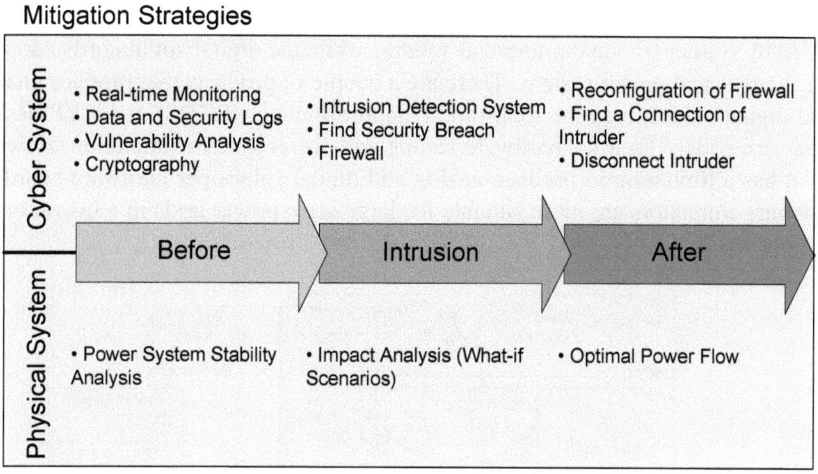

Fig. 15 The framework of mitigation strategies

5 Real-time Testbed for the Cyber Security of the Substations

5.1 Objectives and Requirements

As mentioned in previous Sections, a cyber-physical power system testbed is helpful for the study of the cause-effect relationships of cyber intrusions, resilience of power systems, as well as the performance and reliability of applications in a realistic environment. In a real-time testbed, all components that include software, hardware, communications and emulators are synchronized with GPS or time protocol. Real-time dynamics of communication and information processing are required when cyber intrusions, detections and mitigations are studied. The following Section explains the framework and architecture of a real-time testbed for cyber security of substations.

5.2 Architecture and Components

The work of [50] proposes a Real-time monitoring, Anomaly detection, Impact analysis, and Mitigation strategies (RAIM) framework. Real-time monitoring allows tracking of activities on the cyber-power system. The objective of Anomaly detection is to identify the events on cyber systems that indicate potential cyber intrusions. The tasks of impact analysis are to evaluate the intrusion behaviors and consequences on the power system operating condition. Impact analysis can be achieved by computer simulations in a way similar to the contingency evaluation for online security assessment. The mitigation module serves to illustrate the preventive, remedial or restorative actions to mitigate potential damages caused by cyber intrusions. Analytical techniques can be evaluated on a software based cyber-power system testbed with the control center and substation models.

As shown in Fig. 16, the testbed consists of real-time substation ICTs and the SCADA system from a commercial vendor, while the digital simulator is adopted for the physical power system. There are a couple of products that produce analog and digital values, e.g., the Real-time Digital Simulator (RTDS) and HYPERsim. They are widely used for hardware testing and power system simulation. However, it has a limitation to produce analog and digital values per hardware board so software simulators are more suitable for large scale power grids in a hybrid or an off-line mode.

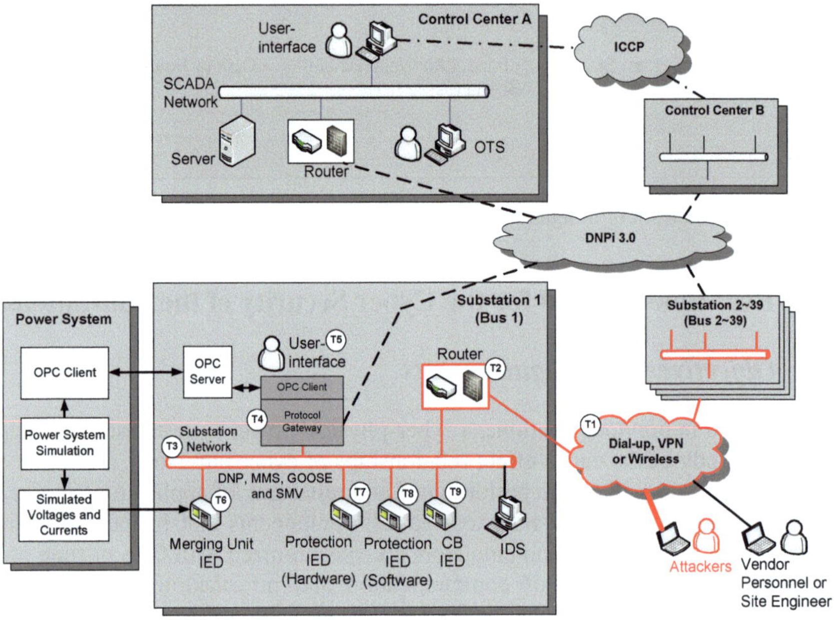

Fig. 16 Real-time testbed for cyber-physical substations

Physical System Module: This testbed comprises two control centers and thirty nine substations, as shown in Fig. 16. The DIgSILENT power factory is a suite of software simulation tools for power systems incorporating applications for system dynamics, transient analysis, optimal power flow, and state estimation. It is used as a real-time simulator for power systems in this testbed. Institute of Electrical and Electronics Engineers (IEEE) 39-bus system has been modeled and implemented in the power factory for research on cyber security of substations and transmission systems. In fact, the power factory is not a real-time simulator. However, it can be used to generate real-time simulation results that include electromechanical transient dynamics of a small system (e.g., IEEE 39-bus) by advanced multi-core based microprocessors and fast Solid-State Drivers (SSDs). Four types of IEDs are installed at the substation networks such as merging unit IED, hardware type protection IED, software type protection IED and circuit breaker IED. Both hardware and software types of IEDs have the capability to deliver control commands (GOOSE messages) of a circuit breaker whereas the circuit breaker IED is designed to subscribe to GOOSE messages, and publish the status (open/close) to software type protection IED. The merging unit IED can send calculated currents and voltages values to software type protection IED. The IDS is designed to detect anomalies and malicious behaviors in a substation automation system [51].

ICT Module: ICCP is used for communication between control centers. Each control center belongs to a different power company. Therefore, control center A only monitors the measurements of control center A but control center A does not have the jurisdiction to control devices supervised by control center B. DNPi (DNP over TCP/IP) protocol enables communications between control centers and substations. All measurements are sent from substations to control centers whereas control commands come from the control centers to substations via DNPi protocol. Object Linking and Embedding for Process Control (OPC) enables the communications between the physical and cyber systems. As illustrated in Fig. 16, all measured status data and analogue values from the power system simulator (i.e., powerfactory) are mapped with the OPC client and linked to an OPC server. The gateway is also mapped with an OPC client and connected to server. Therefore, the control center user interface is able to supervise and control the power system. IEC 61850 based protocols (e.g., SMV, MMS and GOOSE) have been implemented for the substation's communication network using SISCO MMS EASE Lite which can be used to simulate real-time substation automation communication. As described in Section 3, MMS messages are used for the communication between the user-interface and IEDs; SMV messages that include currents and voltages are sent from MU IED to the software type protection IED; GOOSE messages are used for communications between IEDs and circuit breakers, i.e., when an IED sends a tripping signal to CB, and CB sends a status to an IED. The role of the gateway is to convert different communication protocols in a substation network, e.g., convert DNPi to MMS and vice versa. The integrated IDS and firewalls are deployed for cyber security measures. In order to evaluate cyber intrusions from remote access points to substation networks, the IDS and routers need to be connected to the same ICT network. There are three types of remote access points (e.g., dial-up, VPN or wireless) in the testbed environment.

Cyber System Module: SCADA systems can collect, store and visualize the measurements and events on multiple screens. A commercial SCADA system is installed to maximize the accuracy of data and enhance interoperability between devices. The SCADA system consists of network devices, computer servers, databases, user interfaces and the Operator Training Simulator (OTS). The OTS enables operators to simulate and analyze how realistic cyber intrusions can cause damages to a power system and how to defend against the intrusions. Therefore, it can help operators to be prepared for emergency situations. The EMS supports power system applications such as state estimation and optimal power flow.

5.3 *Case Study*

A real-time cyber-physical testbed enables users to study realistic scenarios of cyber attacks and defense strategies. In this Section, scenarios that include possible intrusion paths to the substation systems and cyber attacks that compromise the substation will be discussed.

Table 3 Cyber intrusions and mitigations

No.	Intrusions	Results	IT mitigations
1	GOOSE replay attack	Open CB	Network based IDS
2	GOOSE data modification	Open CB	Network based IDS
3	DoS attack using GOOSE	Lost availability of protection IEDs and CB	Network based IDS
4	Generate fabricated GOOSE packets	Open CB	Network based IDS
5	SMV replay attack	Open CB	Network based IDS
6	SMV data modification	Open CB	Network based IDS
7	DoS attack using SMV	Lost availability of protection IEDs and MU	Network based IDS
8	Generate fabricated SMV packets	Open CB	Network based IDS
9	Modify control values at gateway	Open CB / change transformer tap position	Host based IDS
10	Modify measurement values at gateway	Send wrong data to control center	Host based IDS
11	Man-in-the-middle attack	Lead wrong operation action	Network / host based IDS
12	Compromise user-interface	Change password / open CB / change transformer tap position	Host based IDS
13	Compromise protection IED	Change protection setting / open CB	Host based IDS

As depicted in Fig. 16, a possible intrusion path is from the remote access point (T1) to the substation systems (T2-9). This intrusion path is protected and monitored by security enhanced firewalls with confidential security rules. However, once an intruder successfully compromises the site engineer computer, (s)he can install back door software and is able to acquire the user name and password of the VPN (T1) connection. Using a legitimate ID and password, the firewall (T2) cannot detect the intruder. Once they access the substation network (T3), all network devices can be detected by ping and port scanner software tools that are publicly available. Therefore, intruders can find the substation's user interface (T5), IEDs (T6~9) and protocol gateway (T4). Unfortunately, these critical devices are sometimes mis-configured or improperly protected against cyber intrusions since they may be perceived as part of an isolated communication network. It has been reported that many computer servers and systems use default user IDs and passwords, and operators may not know the configurations of their firewalls. Through these security breaches, the intruder could compromise the substation system. After compromising the substation system, attacks can be launched based on their scenarios as follows.

(1) The intrusion scenario, GOOSE replay attack, is to capture the normal operation of GOOSE packets that contain a CB trip signal, and then retransfer them to the substation network without any modification using free available software, as shown in Fig. 17. The software has the capability to capture and retransfer packets from/to chosen network. This attack can open the circuit breaker (T9).

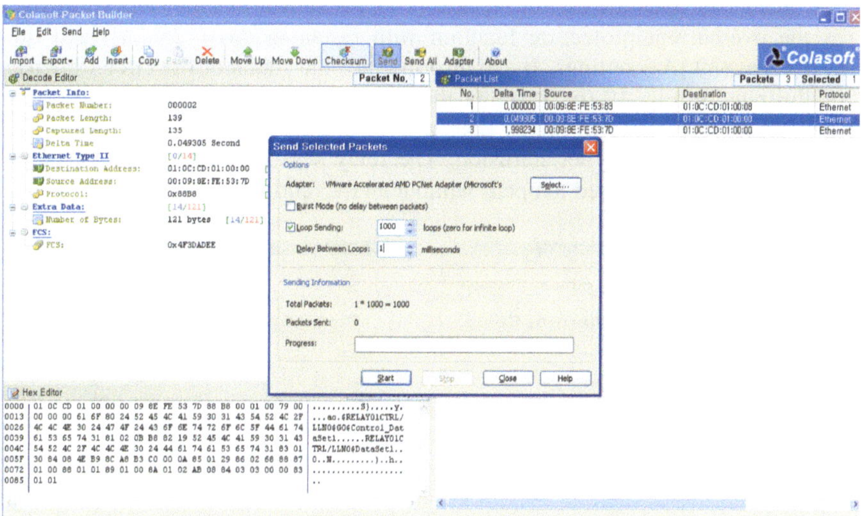

Fig. 17 Retransfer captured GOOSE packet to the substation network

(2) Fig. 18-(a) shows an HMI of the circuit breaker before the attack. In this status, the circuit breaker is closed, and status of relay is normal. After the GOOSE data modification attack, the circuit breaker is opened with the associated relay alarm status as shown in Fig. 18-(b). This alarm indicates that no overcurrent is sensed by the relay but the circuit breaker is tripped.

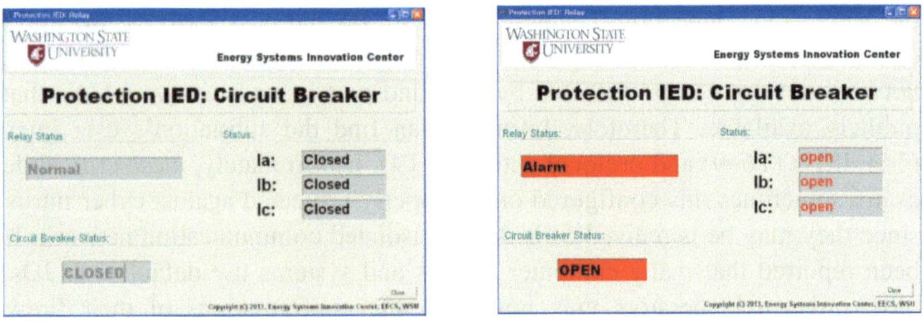

(a) Before attack (b) After attack

Fig. 18 Consequence of a GOOSE related cyber attack to the circuit breaker

(3) The intrusion scenario, DoS attack using GOOSE, is to generate a huge amount of GOOSE packets into the substation network. This attack can disrupt the availability of substation ICT network so operators will lose controls and measurements.

(4) The attack scenario, generate fabricated GOOSE packets, is to capture, modify, and transfer fabricated GOOSE packets to the substation ICT network. This attack can open the circuit breaker (T9). As shown in Fig. 19-(a), this relay has the overcurrent protection function with instantaneous (125 [A]) and time overcurrent (30 [A]) settings. It can also monitor the status of the circuit breaker. As illustrated in Fig. 19-(b), it shows the consequence of a GOOSE modification attack since the overcurrent relay does not sense any fault current but the circuit breaker is opened by cyber intrusion. The relay sensed a change of the circuit breaker status (from closed to open) without an overcurrent condition.

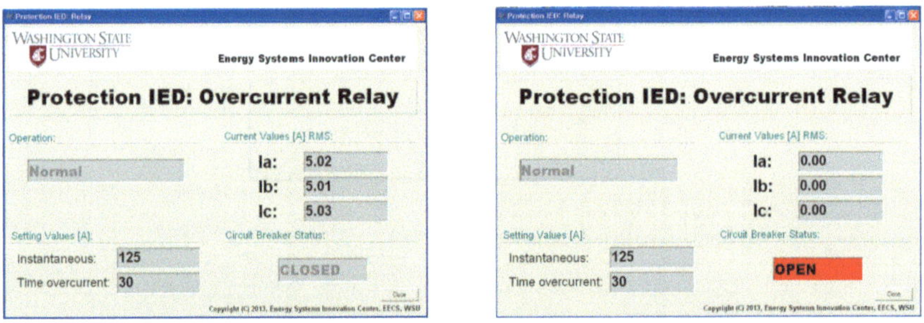

(a) Before attack (b) After attack

Fig. 19 Consequence of GOOSE related cyber attack to overcurrent relay

(5) The intrusion scenario, SMV replay attack, is to capture normal SMV packets when a fault occurred, and then retransfer them to the substation network without any modification. This attack will execute the overcurrent protection (since captured SMV packets contain overcurrent data) and relay (T8) will trip the circuit breaker (T9).

(6) The intrusion, SMV data modification, is to capture normal SMV packets from substation ICT network, modify the measurement data (e.g., low current value to high current value), and then retransfer them to the substation ICT network.

This attack will execute the protection functions at relay (T8), and open the circuit breaker (T9).

(7) The scenario, DoS attack using SMV, is to generate a large amount of SMV packets into the substation network. This attack can disrupt the availability of a substation network so operators will lose controls and measurements of network connected devices.

(8) The scenario, generate fabricated SMV packets, is to capture, modify, and transfer fabricated SMV packets to the substation ICT network. This attack will execute the protection functions at relay (T8) and open the circuit breaker (T9).

(9) The intrusion scenario, modify control values at gateway, involves compromising the substation gateway (T4). An attacker can monitor, modify and generate all measured analog and status values using the compromised gateway. A false signal is generated and a trigger open (T7 and T8) command is sent to substation circuit breaker (T9).

(10) The scenario, modify measurement values at gateway, is to generate a forged CB status at the gateway. As a result, control center operators will be presented with fabricated data for the CB status or current and voltage values. However, the actual status has not changed.

(11) The attack scenario, man-in-the-middle attack, is to generate fabricated analog values to the control center using a man-in-the-middle attack. Once an intruder successfully compromises the substation Local Area Network (LAN) (T3) or OPC client, (s)he is able to monitor and capture all measured data from field devices. Attackers send fabricated data to the control center as illustrated in Fig. 20. Once data passes through the SCADA system, system operators will observe an operational emergency. As a result, operators may take emergency controls such as reducing voltage set points at generators, while the power system is actually in a normal operation condition. In the worst case, these (logical) actions based on fabricated data can drive the system into a sequence of cascading events, leading to a power outage.

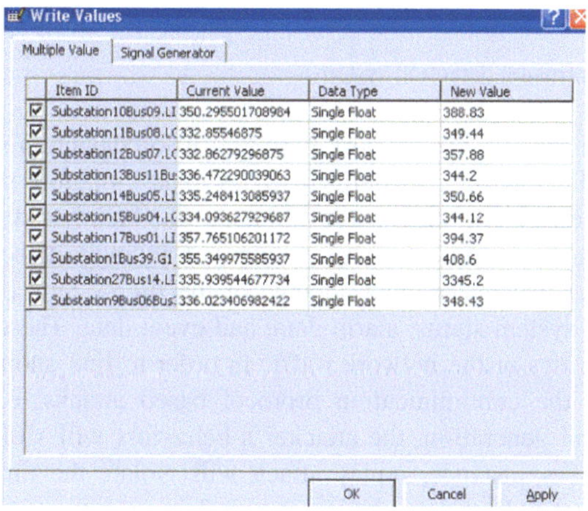

Fig. 20 Generating fabricated analog values to the control center

(12) The attack scenario, compromise user-interface, is to find and compromise the substation user-interface. If an attacker has sufficient knowledge about the substation automation system, they may find all network connected devices using ping and port scanner. Once intruders compromise (i.e., find user name and password) the substation user-interface, they may change the password of user-interface, execute opening command to circuit breaker, or change the transformer tap position.

(13) After compromising the substation protection IED, intruders can access the IED with authorized user name and password, and then change the protection settings to execute the system protection functions.

Mitigation actions are needed for the substation IT as well as the power grid. For IT mitigation, a host-based and network-based IDS have been proposed [51]. The host-based IDS uses an anomaly detection algorithm based on the logs of temporal events whereas the network-based IDS monitors malicious behaviors that violates the predefined rules as illustrated in Fig. 21 and 22.

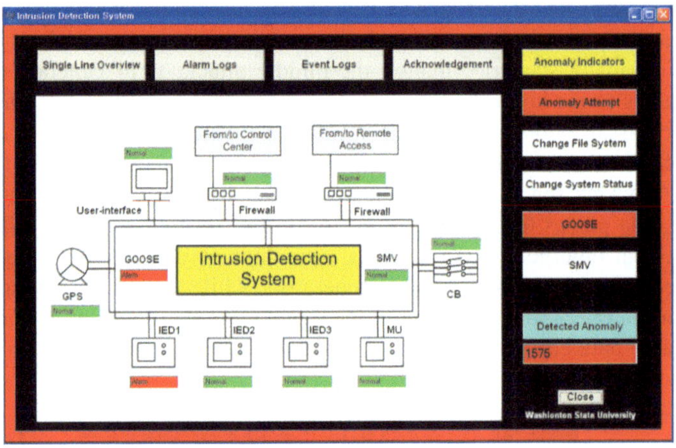

Fig. 21 HMI of intrusion detection system

As shown in Fig. 22, the network-based intrusion detection system has 6 types of anomaly indicators, i.e., predefined logics, data violation, security constrains, detected intrusions, alarm data, and event data. The host-based intrusion detection system has 8 anomaly indicators, i.e., temporal anomaly detection, unauthorized control actions, intrusion attempt, change of the file system, change of IED setting, change of system status, alarm data, and event data. These modules monitor all system activities or the network traffic in order to find anomalies or abnormal behaviors. For the communication protocol based attacks, e.g., replay, packet modification and generation, the attacker's behaviors will violate the predefined security rules. For instance, replay attack will violate the time synchronization since the attacker will use previously captured control messages that contains an incorrect time stamp. When the attacker tries to access the substation gateway which is on the user interface, logs of intrusion attempts (user interface) will be generated. Any intrusion that attempts to change of the target system's status (e.g., circuit breaker status and change settings of IED) will generate system logs. For the gateway intrusion scenario, the attacker will create logs of changes of the file system (gateway) and intrusion attempts (user interface). In the user-interface intrusion scenario, the attacker triggers logs of intrusion attempts (user interface) and changes of file system logs. The host-based IDS is able to detect the intrusions by analyzing the log files. The proposed collaboration scheme between IDS and the firewall is able to disconnect intruders from the substation network.

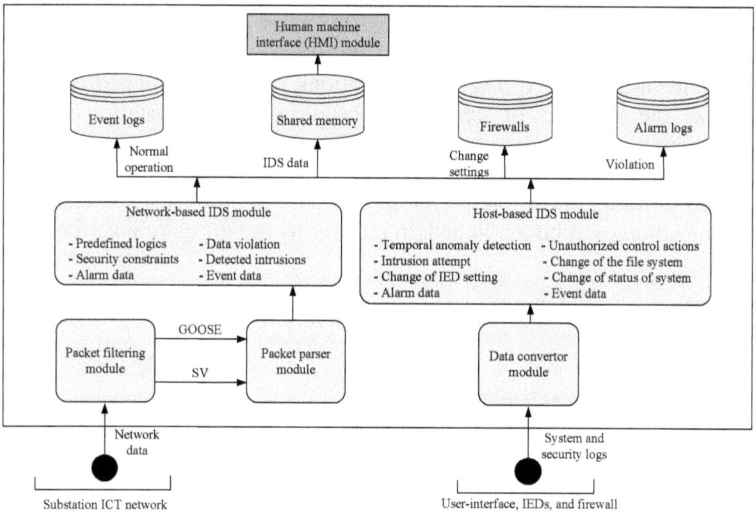

Fig. 22 Host- and network-based intrusion detection system

The IDS has been validated under different types of attack packet intervals, e.g., 1, 10, 20 and 30 [msec], in order to check the performance of IDS. The false negative ratio (FNR) is defined as the number of misclassified abnormal packets divided by the total number of abnormal packets. Table IV shows the mean value of FNR of each test case: 1 ms: 0.95%, 10 ms: 0.62%, 20 ms: 0.29% and 30 ms: 0.11%, respectively. The FNR performance of the proposed intrusion detection system depends on the interval between packets. This is due to the fact that IDS may lose packets when the interval between packets is too small [36]. The false positive ratio (FPR) is defined as the number of misclassified normal packets divided by the total number of normal packets. As shown in Table IV, the mean value of FPR of each test case are 1 ms: 0.79%, 10 ms: 0.56%, 20 ms: 0.18% and 30 ms: 0.027%, respectively.

Emergency control actions are taken to mitigate the effects of cyber intrusions as an attempt to restore the system back to a normal condition. Fig. 23-(a) and 23-(b) show the consequence of multiple cyber attacks to substation 27 and 28 at the

Table 4 False ratio of the substation intrusion detection system

Attack packet interval	1 [msec]	10 [msec]	20 [msec]	30 [msec]
False negative ratio	0.95 %	0.62 %	0.29 %	0.11 %
False positive ratio	0.79 %	0.56 %	0.18 %	0.027 %

same time (attack time is at 5 second) whereas Fig. 23-(c) shows the results of different time (attack times are at 5 and 10 second, respectively) based cyber intrusions. Both attacks consider the worst case scenario such as opening all circuit breakers at a substation as shown in Fig. 24. After compromising substations 27 and 28, intruders open all circuit breakers at target substations. As the consequence of this attack, the voltages of substations 27 and 28 dropped to 0 pu. In the mean time, the voltages of neighbor substations, e.g., substations 24 and 26, dropped dramatically. However, Optimal Power Flow (OPF), with an objective function that minimizes load shedding and constraints that include generator maximum and minimum allowable P and Q, executed as a physical mitigation. At 5.5 second, the voltages at buses 24 and 26 recover by the mitigation actions.

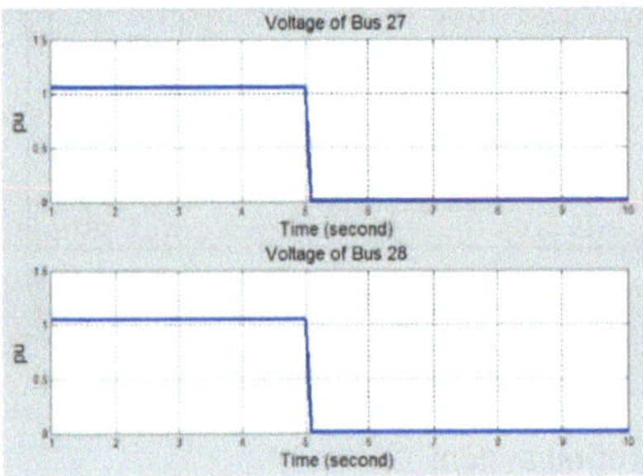

Same time attack

(a) Bus voltages for target substations

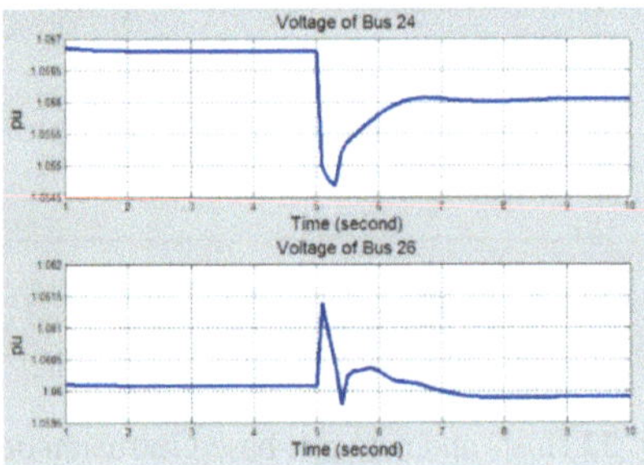

Same time attack

(b) Bus voltages for neighbor substations

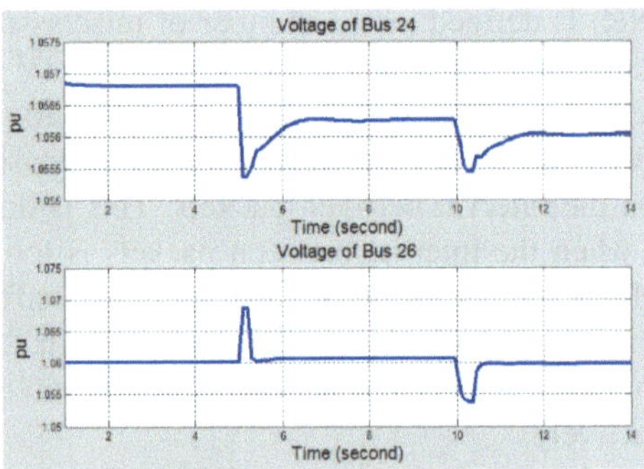

different time attacks

(c) Bus voltages for neighbor substations

Fig. 23 Consequence of cyber attacks on IEEE 39 bus system

Fig. 23-(c) and Fig. 24 illustrate how voltages of neighbor substations vary after multiple cyber intrusions (substation 27 at 5 second and substation 28 at 10 second, respectively). The first cyber intrusion is executed at 5 second which leads to a voltage drop, and then physical mitigation (i.e., OPF) leads to the sharp voltage rise. Another 5 seconds later, the second attack on substation 28 is executed, and then the amount of load shedding is determined by OPF and power systems are back to a normal status.

The time domain dynamic calculation has been used for the case study. For power grid mitigation, an optimal power flow based generation control with an objective function that minimizes the network losses is used for power system recovery after cyber attacks. Coordination between mitigations in the cyber and physical systems enhances the security of a substation.

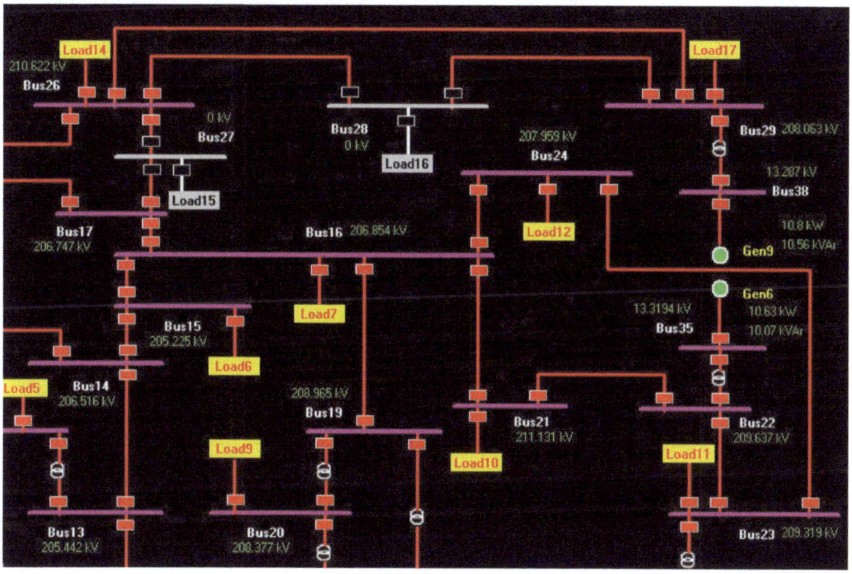

Fig. 24 Consequence of physical system after simultaneous cyber attacks to multiple substations on the IEEE 39-bus system (buses 27 and 28)

Acknowledgement. This research is sponsored by U.S. National Science Foundation, "Collaborative Research: Resiliency Against Coordinated Cyber Attacks on Power Grid," EECS-1202229. This work is partially sponsored by National Natural Science Foundation of China, Grant No. 51277104, and Tsinghua University.

References

[1] Hahn, A., Ashok, A., Sridhar, S., Govindarasu, M.: Cyber-Physical Security Testbeds: Architecture, Application, and Evaluation for Smart Grid. IEEE Trans. on Smart Grid 4(2), 847–855 (2013)
[2] Glover, J.-D., Sarma, M.-S., Overbye, T.-J.: Power system analysis and design. Thomson (2011)
[3] Li, F., Qiao, W., Sun, H., Wan, H., Wang, J., Xia, Y., Xu, Z., Zhang, P.: Smart Transmission Grid: Vision and Framework. IEEE Trans. Smart Grid 1(2), 168–177 (2010)
[4] Igure, V.-M., Laughter, S.-A., Williams, R.-D.: Security Issues in SCADA Networks. Computers & Security 25(7), 498–506 (2006)
[5] Liu, C.-C., Stefanov, A., Hong, J., Panciatici, P.: Intruders in the Grid. IEEE Power Energy Magazine 10(1), 58–66 (2012)
[6] Milano, F., Canizares, C.-A., Invernizzi, M.: Voltage Stability Constrained OPF Market Models Considering Contingency Criteria. Electric Power Systems Research 74(1), 27–36 (2005)
[7] Govindarasu, M., Hann, A., Sauer, P.: Cyber-Physical Systems Security for Smart Grid. Future Grid Initiative White Paper, PSERC (February 2012), http://www.pserc.wisc.edu/documents/publications/papers/fgwhitepapers/Govindarasu_Future_Grid_White_Paper_CPS_May_2012.pdf
[8] GAO-11-117, Electricity Grid Modernization: Progress Being Made on Cyber Security Guidelines, but Key Challenges Remain to be Addressed. Government Accountability Office (GAO) (January 2011), http://www.gao.gov/new.items/d11117.pdf
[9] Guidelines for Smart Grid Cyber Security, National Institute for Standards and Technology (August 2010), http://csrc.nist.gov/publications/nistir/ir7628/nistir-7628_vol2.pdf
[10] North American Electric Reliability Corporation (NERC) Critical Infrastructure Protection (CIP) Reliability Standards 002-009, http://www.nerc.com/pa/Stand/Pages/CIPStandards.aspx
[11] Govindarasu, M., Liu, C.-C.: Cyber Physical Security Testbed for the Smart Grid: Fidelity, Scalability, Remote Access, and Federation. Position Paper to National CPS Energy Workshop (2013)
[12] National SCADA test bed: Fact sheet, Idaho National Laboratory, INL (2007)
[13] Common Cyber Security Vulnerabilities Observed in Control System Assessments by the INL NSTB Program, Idaho National Laboratory (INL) (November 2008)
[14] Rohde, M.-R.-P.: Cyberassessment Methods for SCADA Security. Instrumentation, Systems and Automation Society (ISA), Tech. Rep. (2005)
[15] McDonald, M.-J., Conrad, G.-N., Service, T.-C., Cassidy, R.H.: Cyber Effects Analysis Using VCSE. Promoting Control System Reliability, Sandia National Laboratories, SAND, 2008-5954 (September 2008)
[16] McDonald, M.-J.: Modeling and Simulation for Cyber-Physical System Security Research. Development and Applications, Sandia National Laboratories, SAND2010-0568 (February 2010)

[17] Bergman, D.C., Jin, D., Nicol, D.M., Yardley, T.: The Virtual Power System Testbed and Inter-Testbed Integration. In: Proc. 2nd Workshop Cyber Security Exp. Test (August 2009)
[18] Mallouhi, M., Al-Nashif, Y., Cox, D., Chadaga, T., Hariri, S.: A Testbed for Analyzing Security of SCADA Control Systems (TASSCS). In: Proceedings of IEEE PES Innov. SmartGrid Technol. (ISGT) (January 2011)
[19] Dondossola, G., Garrone, G., Szanto, J., Deconinck, G., Loix, T., Beitollahi, H.: ICT Resilience of Power Control Systems: Experimental Results from the CRUTIAL Testbeds. In: Proceedings of IEEE/IFIP Int. Conf. Dependable Syst. Netw. (DSN), pp. 554–559 (July 2009)
[20] Dondossola, G., Deconinck, G., Garrone, F., Beitollahi, H.: Testbeds for Assessing Critical Scenarios in Power Control Systems. In: Setola, R., Geretshuber, S. (eds.) CRITIS 2008. LNCS, vol. 5508, pp. 223–234. Springer, Heidelberg (2009)
[21] Hong, J., Wu, S.-S., Stefano, A., Fshosha, A., Liu, C.-C., Gladyshev, P., Govindarasu, M.: An Intrusion and Defense Testbed in a Cyber-power System Environment. In: IEEE Power and Energy Society General Meeting (July 2011)
[22] Queiroz, C., Mahmood, A., Tari, Z.: SCADASim A Framework for Building SCADA Simulations. IEEE Trans. Smart Grid 2(4), 589–597 (2011)
[23] Blochwitz, T., Otter, M., Akesson, J., Arnold, M., Clauß, C., Elmqvist, H., Friedrich, M., Junghanns, A., Mauss, J., Neumerkel, D., Olsson, H., Viel, A.: Functional Mockup Interface 2.0: The Standard for Tool independent Exchange of Simulation Models. In: Proceedings of 9th International Modelica Conference, Munich (2012), https://www.fmi-standard.org/start
[24] Simulation Tool - OpenDSS, Smart Grid Resource Center, Electric Power Research Institute (EPRI), http://www.smartgrid.epri.com/SimulationTool.aspx
[25] MATPOWER, A MATLAB Power System Simulation Package, Power Systems Engineering Research Center (PSERC), http://www.pserc.cornell.edu//matpower/
[26] Vyatkin, V., Zhabelova, G., Higgins, N., Schwarz, K., Nair, N.C.: Towards Intelligent Smart Grid Devices with IEC 61850 Interoperability and IEC 61499 Open Control Architecture. In: IEEE PES Transmission and Distribution Conference (April 2010)
[27] Mackiewicz, R.E.: Overview of IEC 61850 and Benefits. In: IEEE PES Transmission and Distribution Conference, pp. 376–383 (May 2006)
[28] Clarke, G., Reynders, D., Wright, E.: Practical Modern SCADA Protocols, IDC technologies (2004)
[29] Communication Networks and Systems for Power Utility Automation, IEC 61850-90-1 Standard: Use of IEC 61850 for the Communication between Substations, 1st edn. (March 2010)
[30] Electrical Single Line Diagram - Part Two, Electrical Knowhow, http://www.electrical-knowhow.com/2012/12/electrical-single-line-diagram-part-two.html
[31] Communication Networks and Systems in Substations, IEC 61850-5 Standard: Communication Requirements for Functions and Device Models, 1st edn. (July 2003)
[32] Specific Communication Service Mapping (SCSM), IEC 61850 8-1 Standard: Mapping to MMS (ISO/IEC9506-1 and ISO/IEC 9506-2), 1st edn. (May 2004)

[33] Premaratne, U.-K., Samarabandu, J., Sidhu, T.-S., Beresh, R., Tan, J.-C.: An Intrusion Detection System for IEC 61850 Automated Substations. IEEE Trans. Power Del. 25(4), 2376–2383 (2010)

[34] Morris, T., Pavurapu, K.: A Retrofit Network Transaction Data Logger and Intrusion Detection System for Transmission and Distribution Substations. In: IEEE International Conference on Power and Energy (PECon), pp. 958–963 (November 2010)

[35] Ten, C.-W., Hong, J., Liu, C.-C.: Anomaly Detection for Cybersecurity of the Substations. IEEE Trans. Smart Grid 2(4), 865–873 (2011)

[36] Hong, J., Liu, C.-C., Govindarasu, M.: Detection of Cyber Intrusions Using Network-Based Multicast Messages for Substation Automation. In: Proceedings of IEEE Innovative Smart Grid Technologies (ISGT) Conference (2014)

[37] Power Systems Management and Associated Information Exchange - Data and Communications Security, IEC TS 62351-1 Standard: Part 1: Communication Network and System Security - Introduction to Security Issues, 1st edn. (May 2007)

[38] Pender, T.: When Power Goes Out, a Squirrel is Likely to Blame, The Record (October 2013), http://www.therecord.com/news-story/4164925-when-power-goes-out-a-squirrel-is-likely-to-blame/

[39] Campbell, R.-J.: Weather-Related Power Outages and Electric System Resiliency, Congress Research Service 7-5700, http://www.fas.org/sgp/crs/misc/R42696.pdf

[40] Kushner, D.: The Real Story of Stuxnet. IEEE Spectrum 50(3), 48–53 (2013)

[41] Orgill, G.-L., Romney, G.-W., Bailey, M.-G., Orgill, P.-M.: The Urgency for Effective User Privacy-Education to Counter Social Engineering Attacks on Secure Computer Systems. In: Proceedings of the 5th Conference on Information Technology Education (CITC5), pp. 177–181 (2004)

[42] Schneier, B.: Attack Trees: Modeling Security Threats. Dr. Dobb's Journal (December 1999)

[43] Dawkins, J., Hale, J.: A Systematic Approach to Multi-stage Network At-tack Analysis. In: Second IEEE International Information Assurance Workshop, pp. 48–56 (April 2004)

[44] Moore, A.-P., Ellison, R.-J., Linger, R.-C.: Attack Modeling for Information Security and Survivability. Survivable Systems, Technical Note CMU/SEI-2001-TN-001 (March 2001)

[45] Ten, C.-W., Liu, C.-C., Govindarasu, M.: Vulnerability Assessment of Cybersecurity for SCADA Systems Using Attack Trees. In: IEEE Power and Energy Society General Meeting (June 2007)

[46] North American Electric Reliability Corporation, Cyber Attack Task Force, Final Report (May 2012), http://www.nerc.com/docs/cip/catf/12-CATF_Final_Report_BOT_clean_Mar_26_2012-Board%20Accepted%200521.pdf

[47] Kordy, B., Pietre-Cambacedes, L., Schweitzer, P.: DAG-Based Attack and Defense Modeling: Don't Miss the Forest for the Attack Trees. arXiv preprint arXiv:1303.7397 (2013)

[48] Ericsson, G.N.: Management of Information Security for an Electric Power Utility-On Security Domains and Use of ISO/IEC17799 Standard. IEEE Transactions on Power Delivery 20(2), 683–690 (2005)

[49] Bayuk, J.-L., Healey, J., Rohmeyer, P., Sachs, M.-H., Schmidt, J., Weiss, J.: Cyber Security Policy Guidebook. Wiley (2012)

[50] Hong, J., Stefano, A., Liu, C.-C., Govindarasu, M.: Cyber-Physical Security in a Substation. In: IEEE Power and Energy Society General Meeting (July 2012)

[51] Hong, J., Liu, C.-C., Govindarasu, M.: Integrated Anomaly Detection for Cyber Security of the Substations. IEEE Trans. Smart Grid 5(4), 1643–1653 (2014)

[52] Khaitan, S.K., McCalley, J.D.: Cyber physical system approach for design of power grids: A survey. In: IEEE Power and Energy Society General Meeting (July 2013)

[53] Khaitan, S.K., McCalley, J.D.: Design Techniques and Applications of Cyber-physical Systems: A Survey. IEEE Systems Journal (2014)

Cyber-Attacks in the Automatic Generation Control

Maria Vrakopoulou, Peyman Mohajerin Esfahani, Kostas Margellos, John Lygeros, and Göran Andersson

Abstract. Power systems are traditionally monitored and controlled by an IT infrastructure, referred to as Supervisory Control and Data Acquisition (SCADA) system. The cyber-physical interaction of power systems (physical) and SCADA systems (cyber) rises security issues, since the links between those systems are vulnerable to cyber-attacks that can potentially lead to catastrophic economical and societal effects. In this chapter we focus on a specific cyber-physical link, the Automatic Generation Control (AGC), which is an automatic frequency control loop closed over the SCADA system. We provide an impact analysis in case of a cyber-attack on the AGC signal. We first carry out a feasibility analysis based on reachability and optimal control theory, that provides an information regarding the existence of an attack pattern that can disturb the power system. We then deal with the problem of synthesizing an attack signal and treat it as a nonlinear control synthesis problem. Third, performance of our methodologies are illustrated by means of dynamic simulations on IEEE-118 bus network.

1 Introduction

A well-functioning society relies heavily on the proper operation of the electric power system. Large power outages may be difficult and time- consuming to restore and may also have devastating economic and humanitarian consequences. The importance of electric power delivery is illustrated, for instance, by the

Maria Vrakopoulou · Göran Andersson
Power System Laboratory, ETH Zurich, Physikstrasse 3, Zurich 8092, Switzerland
e-mail: {vrakopoulou,andersson}@eeh.ee.ethz.ch

Peyman Mohajerin Esfahani · John Lygeros
Automatic Control Laboratory, ETH Zurich, Physikstrasse 3, Zurich 8092 , Switzerland
e-mail: {mohajerin,lygeros}@control.ee.ethz.ch

Kostas Margellos
Department of Industrial Engineering and Operations Research, UC Berkeley, Hearst Avenue 2594, Berkeley CA 94720, US
e-mail: kostas.margellos@berkeley.edu

S.K. Khaitan, J.D. McCalley, and C.-C. Liu (eds.), *Cyber Physical Systems Approach to Smart Electric Power Grid*, Power Systems, DOI: 10.1007/978-3-662-45928-7_11

economic and social impacts of the 2003 northeast American blackout during which 50 million people were affected [1]. Therefore, in large electric power systems, an Information Technology (IT) infrastructure, referred to as Supervisory Control and Data Acquisition (SCADA) system, provides system-wide supervision and control [2]. The SCADA system measures data through remote devices installed throughout the grid and gathers the information at a control center through communication channels, where from, after computer processing, control commands are sent back to the power system. The dependence of the power system (physical) on the IT infrastructure (cyber) constitutes a cyber-physical interaction that despite the fact that it is designed to lead to a more efficient operation of the power system, it renders it more susceptible to operational errors and external attacks. Recent advancements in the design of cyber-physical systems are reviewed in [3,4].

The power system is typically divided in control areas, each of them monitored and controlled by a separate SCADA system. After gathering the measurements in the control center, state estimation is conducted so as to determine the most probable state of the system given that the measurements might be inaccurate or incomplete. Based on the estimated state, the SCADA system alerts the operator if control actions should be taken. The various power system points that are controlled by the SCADA system are the status of switches, generator voltage setpoints, generator active power setpoints, turns ratio of load tap changing transformers and other configuration settings. These control actions aim on a more efficient and secure operation of the power system and are typically manually driven. One of the few control loops that are closed over the SCADA system without human operator intervention is the Automatic Generation Control (AGC). This is a continuous[1] time control and involves the adjustment of the generator active power setpoints. All the aforementioned inputs and outputs of the control center (or the power system, respectively) constitute vulnerable points of the cyber-physical system that could be possibly manipulated as part of a cyber-attack to deteriorate the performance of the system.

Numerous analyses have described the systems potential vulnerabilities to cyber-attacks, while actual incidents have confirmed these vulnerabilities and underscored the importance of reducing them. The authors of [5] proposed a framework in order to clarify the interaction between the power system and the IT infrastructure and identify the vulnerabilities and the malfunctions of both that could lead to an abnormal operation of the power network. In [6–8], the vulnerabilities of a cyber-attack on the state estimation system were assessed. From another perspective the authors of [9] attempted to quantify the impact of a cyber-attack in a power market environment, while in [10–13] real examples of cyber-attacks were reported.

In this context, the VIKING research project [14] proposed a novel concept to address the challenges introduced by the interaction between the SCADA system and the power transmission and distribution systems. Tools and methodologies were developed to identify the vulnerabilities of these safety critical infrastructures [15], to determine the impact that possible failures or attacks might have [16, 17] and to develop strategies to mitigate these effects [18].

[1] Practically is applied every 2-4 seconds.

Motivated by the research carried out within the VIKING project, in this chapter we built on our earlier work [19, 20] and investigate the impact of a cyber-attack on the AGC signal in a power system. We focus on this control loop, since its automatic nature renders it more susceptible to external attacks. AGC actions are usually determined for each control area at the control center. The main objective is to regulate frequency to its nominal value and maintain the power exchange between the control areas at the scheduled level. To achieve this, measurements of the system frequency and the tie line power flows are sent to the dispatch center and then a feedback signal that regulates the generated power is sent back to the generators, participating in the AGC, through the SCADA system.

In this chapter we assume that an attacker has gained access in one of the areas of the power system. We first provide a feasibility analysis and show whether there exists an attack signal that could irreversibly disturb the system. Our methodology employs tools from reachability theory and optimal control for nonlinear systems [21,22]. We next focus on the problem of synthesizing an attack signal; we treat it as a controller synthesis problem where the objective is to drive the system outside the safety margins. Different alternatives are provided ranging from open loop approaches, based on Markov Chain Monte Carlo (MCMC) optimization [23, 24], to close loop schemes based on feedback linearization and gain scheduling [25,26]. Due to the complexity (large-scale, nonlinear) of the models that describe power systems, for our analysis and synthesis investigation we use a two-machine frequency model where each machine represents a different control area. To evaluate the performance of our methodology, we apply the attack signal that is constructed based on the aforementioned abstraction to the detailed power system model.

In Section 2 the physical description and the mathematical model of the two-machine power system is provided. Section 3 provides a feasibility analysis that provides intuition regarding how plausible it is for the system to be disturb by an attack signal. Section 4 provides different attack signal synthesis alternatives and illustrates their efficacy on a detailed simulation enviroment. Finally, Section 5 provides some concluding remarks.

2 Power System Modeling

The fact that power systems are generally exposed to disturbances originating from the uncertainty and variability of the loads, unpredictable line outages etc., has deemed necessary the integration of many control systems. These control systems aim to keep the power system within acceptable operating limits maintaining the security and the quality of supply in satisfactory levels. Due to the various time constants of the processes, the system is controlled in an hierarchical way. Some quantities are rapidly controlled locally and other, with a relatively slower response, via the SCADA system. The nonlinear nature of the power flow equations and the various control schemes that power systems are equipped with result in a very complex model characterized by large-scale nonlinear continuous and discrete dynamics. Such complex models cannot be used efficiently in the development of advanced control strategies thus we have to rely on different levels of abstraction that simplify certain elements of the initial model and/or take advantage of the possible de-coupling between control loops.

One example of a possible control loop de-coupling involves the two main control loops of power system. These are responsible for the regulation of the voltage magnitudes and the frequency of the system so as not to exceed certain specified limits. The time constants of the local Automatic Voltage Regulator (AVR) are quite smaller than the ones of the frequency control loops and hence in load-frequency studies one can use a quasi-state model that considers only the steady state point of the voltage control loop ignoring its fast dynamics.

In this chapter we investigate the impact of a cyber-attack on the AGC in one control area. To facilitate the needs of this study, we divide a network into two independent control areas, and consider the case where an attacker has gained access to the AGC signal of one of them being able to inject an undesirable input. Since we are dealing with load-frequency studies, we consider only frequency dynamics. For this purpose, a simplified nonlinear frequency model that includes governor and AGC dynamics is developed. We represent each control area by a single generator to apply and illustrate better the control design that will be presented in Section 4.

In the following subsections, we first describe the basic principles of the model abstraction we employ and the frequency dynamics and then present the two-area power system model.

2.1 Frequency Dynamics in One Area

In this subsection the frequency dynamics of a single control area as driven by different frequency control levels are described. The analysis is mainly based on [27,28]. As mentioned above, to simplify the dynamics of each area, we condense them into one single generating unit by considering aggregated quantities based on the center of inertia of the area. For that purpose we consider the following lumped quantities:

$$
\begin{aligned}
f &= \frac{\sum H_i f_i}{\sum H_i} && \text{Centre of inertia frequency}(Hz),\\
S_B &= \sum S_{B_i} && \text{Total rating (MVA)},\\
H &= \frac{\sum H_i S_{B_i}}{\sum S_{B_i}} && \text{Total inertia constant (sec)},\\
P_m &= \sum P_{m_i} && \text{Total mechanical power (MW)},\\
P_e &= \sum P_{e_i} && \text{Total electrical power (MW)},\\
\frac{1}{R} &= \sum \frac{1}{R_i} && \text{Equivalent droop constant (Hz/MW)},\\
\frac{1}{D_l} &= \sum \frac{1}{D_{l_i}} && \text{Equivalent damping coefficient (Hz/MW)},
\end{aligned}
$$

where $i \in G$ and G is the set of the generators that belong to control area i.

The aggregated principal frequency dynamics of each area can be described by

$$\Delta \dot{f} = \frac{f_0}{2HS_B}(\Delta P_m - \Delta P_e), \tag{1}$$

where operator Δ returns the deviation of its arguments from their reference values. The frequency of the rotor is denoted by f whereas P_m and P_e represent the generated (mechanical) power and consumed (electrical) power, respectively. A brief description of additional dynamics due to the term of the generated power (ΔP_m) and the consumed power (ΔP_e) follows.

2.1.1 Generated Power

The frequency of the system can be controlled adjusting properly the generated power. Every change in the setpoint of the generated power is first filtered by the turbine dynamics where it is converted in mechanical power. For simplicity we ignore here the turbine dynamics and hence for the setpoint of the generated power we can refer directly to ΔP_m. This setpoint depends on the output of the frequency control loops and on manual interventions. This can be expressed by

$$\Delta P_m = \Delta P_{m,p} + \Delta P_{m,AGC} + \Delta P_{m,set}, \tag{2}$$

where $\Delta P_{m,p}$ represents the change in the produced power due to the primary frequency control (governor) action, $\Delta P_{m,AGC}$ the change due to the secondary frequency control (AGC) action and $\Delta P_{m,set}$ a scheduled step change. In the following, modeling details for the primary frequency control and the AGC loop are presented.

- **Primary Frequency Control**

Primary frequency control refers to control actions that are done locally at every plant. The governor adjusts the setpoint of the produced power to bring the frequency close to its nominal value. The response should be in a scale of a couple of seconds. According to a simplified model of a governor with speed droop characteristic the rotor measured frequency is compared with the nominal one and the error signal is amplified to produce the control signal ΔP_p. Specifically the control law is given by $\Delta P_p = -\frac{1}{S}\Delta f$, where the quantity S at the proportional gain is referred to as droop or speed regulator.

However, every generator is set to have a specific reserve amount of power that is able to offer according to its availability, the optimal performance of the system, but also market rules. Hence, there are upper and down limits at the produced power deviation that primary and secondary frequency controllers can impose.

Therefore the final change of the produced power due to the primary control action is

$$\Delta P_{m,p} = \begin{cases} \Delta P_p^{min} & \text{if} \quad \Delta P_p \leq \Delta P_p^{min}, \\ \Delta P_p & \text{if} \quad \Delta P_p^{min} < \Delta P_p < \Delta P_p^{max}, \\ \Delta P_p^{max} & \text{if} \quad \Delta P_p \geq \Delta P_p^{max}. \end{cases} \tag{3}$$

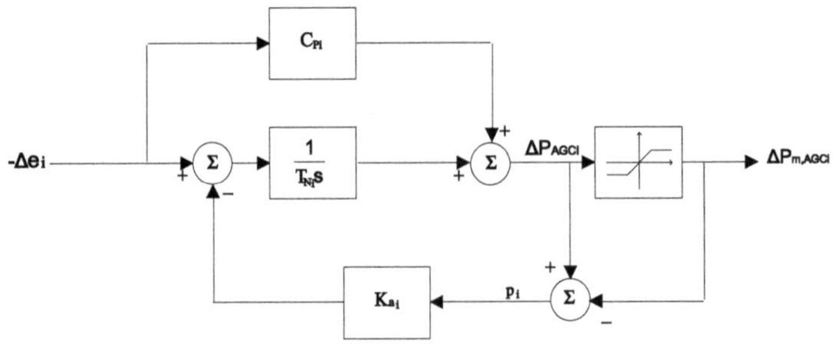

Fig. 1 PI controller with anti-wind up for the AGC loop

- **Automatic Generation Control**

As already discussed, the main objectives of the AGC are to regulate frequency to the specified nominal value and maintain the interchanged power between the controlled areas to the scheduled values by adjusting the generated power of specific generators in the area. It consist the secondary frequency control loop acting in a scale of a couple of minutes.

AGC actions are usually determined for each control area at the control center via the SCADA system. Measured system frequency and tie line flows are sent to this center, where computer processing takes place, and finally a feedback signal that regulates the generated power is sent back to the generators.

However, as mentioned above, there is saturation at the imposed control signal. Therefore the final change of the produced power due to AGC signal in area *i* is

$$\Delta P_{m,AGC_i} = \begin{cases} \Delta P^{min}_{AGC_i} \text{ if } & \Delta P_{AGC_i} \leq \Delta P^{min}_{AGC_i}, \\ \Delta P_{AGC_i} \text{ if } & \Delta P^{min}_{AGC_i} < \Delta P_{AGC_i} < \Delta P^{max}_{AGC_i}, \\ \Delta P^{max}_{AGC_i} \text{ if } & \Delta P_{AGC_i} \geq \Delta P^{max}_{AGC_i}, \end{cases} \quad (4)$$

where ΔP_{AGC_i} is the AGC control signal before saturation, and $\Delta P_{m,AGC_i}$ is the AGC control signal that finally affects the produced power.

The secondary control of area *i* is typically a proportional-integral (PI) controller. To avoid wind up in case of saturation, an anti-wind up circuit is also used [29]. The overall block diagram for the AGC of a single area is shown in Fig. 1 where has as input an error signal and output the control signal that will adjust the production.

The error signal Δe_i, considering an interconnected system of N-areas each of them equipped with its own AGC controller, is:

$$\Delta e_i = \sum_{j \in \Omega_i} \Delta P_{ij} + B_i \Delta f_i, \quad (5)$$

where $\Delta f_i = f_i - f_0$, $\Delta P_{ij} = P_{ij} - P_{0_{ij}}$. Quantity P_{ij} is the power transmitted from area *i* to area *j*, $P_{0_{ij}}$ the scheduled transmitted power from area *i* to area *j*, Ω_i the set of indices corresponding to the areas connected to area *i*, f_i the frequency of area *i* and f_0 the nominal frequency of the system (same for all areas in steady state).

Parameter B_i is the so called frequency bias factor and its value is given by $B_i = \frac{1}{S_i}$ (based on the non interactive control), where S_i represents the equivalent total droop of area i.

The output of the AGC controller of area i is

$$\Delta P_{AGC_i} = -(C_{P_i} + \frac{1}{sT_{N_i}})(\sum_{j\in\Omega_i} \Delta P_{ij} + \frac{1}{S_i}\Delta f_i) - \frac{K_{a_i}}{T_{N_i}s} p_i, \tag{6}$$

or in the time domain

$$\Delta \dot{P}_{AGC_i} = -C_{p_i}(\frac{\Delta \dot{f}_i}{S_i} + \sum_{j\in\Omega_i} \Delta \dot{P}_{ij}) - \frac{1}{T_{N_i}}(\frac{\Delta f_i}{S_i} + \sum_{j\in\Omega_i} \Delta P_{ij}) - \frac{K_{a_i}}{T_{N_i}} p_i, \tag{7}$$

where C_{p_i} is the proportional factor of the AGC controller, T_{N_i} the integration time constant of AGC controller and

$$p_i = \begin{cases} 0 & \text{if} \quad \Delta P_{AGC_i}^{min} < \Delta P_{AGC_i} < \Delta P_{AGC_i}^{max}, \\ \Delta P_{AGC_i} - \Delta P_{m,AGC_i} & \text{else}, \end{cases} \tag{8}$$

where p_i is defined in (8). The power of the tie lines depends on the state of the system and will be specified later in the modeling of the two-area power system.

2.1.2 Consumed Power

A change in the consumed power can be due to a change in the actual load or due to frequency dependency of the load. For instance, the amount of power that motor loads consume differs with frequency since their speed changes also. Moreover, the fact that kinetic energy can be stored in rotating masses of large motors causes an additional contribution depending on $\dot{f}$. Considering an area i in an interconnected system the load that the whole area has to compensate depends also on the power that is transmitted through the tie lines to other areas. Thus a change in the consumed power is expressed through

$$\Delta P_{e_i} = \Delta P_{L_i} + \Delta P_{L,f_i} + \sum_{j\in\Omega_i} (\Delta P_{ij}), \tag{9}$$

where ΔP_{L_i} is the actual deviation of the load, and $\Delta P_{L,f_i}$ is the deviation due to the frequency dependence of the load that is given by

$$\Delta P_{L,f_i} = \frac{1}{D_{l_i}}\Delta f_i + 2\frac{W_{0_i}}{f_0}\Delta \dot{f}_i. \tag{10}$$

where the first term directly depends on the frequency, whereas the second one represents the fact that kinetic energy can be stored in the rotating masses of large motors.

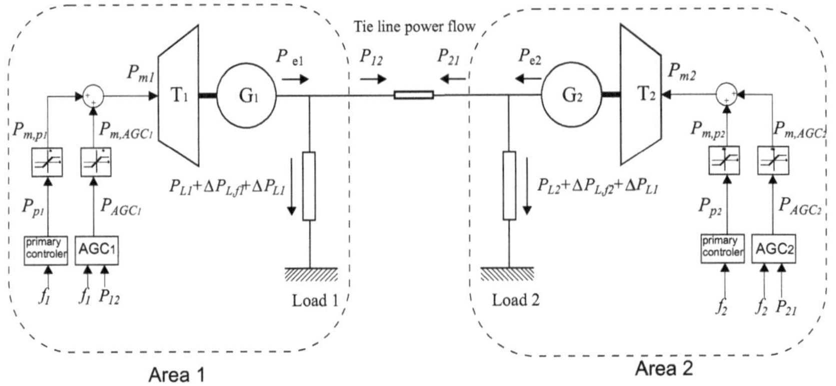

Fig. 2 Two-Area Power System. One generator model for each area equipped with primary control and AGC.

2.2 Two-Area Power System Model

Consider now the system of Fig. 2 that consists of only two control areas, each one equipped with its own AGC, connected by a tie line of reactance X. Each area is represented by an equivalent generating unit equipped also with an equivalent primary frequency control.

Based on the discussion of the previous subsection the frequency dynamics for area i that is connected with area j composing the two-are system are given by

$$\Delta \dot{f}_i = \frac{f_0}{2(H_i S_{B_i} + W_{0_i})} \left(\Delta P_{m_i} - \Delta P_{L_i} - \frac{1}{D_{l_i}} \Delta f_i - \Delta P_{ij} \right), \tag{11}$$

All considered quantities are defined in the previous subsection except the power flow on the tie-line. The power flow from area i to area j is described by (12), where P_{ij} is positive when area i sends active power to area j. Also, since the active power losses on the line are neglected $P_{ji} = -P_{ij}$.

$$P_{ij} = \frac{V_i V_j}{X} \sin(\delta_i - \delta_j) = P_T \sin(\delta_i - \delta_j), \tag{12}$$

where $P_T = \frac{V_i V_j}{X}$ and X is the reactance of the tie line, V_i, V_j the voltage magnitude at the ends of the line, δ_i, δ_j the voltage angles at the ends of the line. Assuming the steady state point of the voltage controllers, we consider constant voltage magnitudes at the ends of the line during load deviation.

We set $\phi_{ij} = \delta_i - \delta_j$ and then define the variables according to the deviation from their initial (scheduled) value, here highlighted by the '0' subscript (i.e $\phi_{ij} = \Delta\phi_{ij} + \phi_{0_{ij}}$). Since $\dot{\delta}_i = 2\pi\Delta f_i$, then $\Delta\dot{\phi}_{ij} = 2\pi(\Delta f_i - \Delta f_j)$ and P_{ij} and its derivative results in

$$\begin{aligned} \Delta P_{ij} &= P_T \sin(\Delta\phi_{ij} + \phi_{0_{ij}}) + P_{0_{12}}, \\ \Delta \dot{P}_{ij} &= 2\pi P_T (\Delta f_i - \Delta f_j) \cos(\Delta\phi_{ij} + \phi_{0_{ij}}). \end{aligned} \tag{13}$$

Based also on the discussion in the previous subsection we get the following equations for the dynamics of the two-area system for $(i,j) \in (1,2),(2,1)$:

$$
\begin{aligned}
\Delta \dot{f}_i &= \tfrac{f_0}{2(H_i S_{B_i}+W_{0_i})}\Big(\Delta P_{m_i} - \Delta P_{L_i} - \tfrac{1}{D_{l_i}}\Delta f_i - P_T \sin(\Delta\phi_{ij}+\phi_{0_{ij}}) + P_{0_{ij}}\Big),\\
\Delta \dot{\phi_{12}} &= 2\pi(\Delta f_1 - \Delta f_2),\\
\Delta \dot{P}_{AGC_i} &= \Big(\tfrac{1}{D_{l_i}}\tfrac{C_{p_i} f_0}{2S_i(H_i S_{B_i}+W_{0_i})} - \tfrac{1}{S_i}\tfrac{1}{T_{N_i}}\Big)\Delta f_i - \tfrac{C_{p_i} f_0}{2S_i(H_1 S_{B_i}+W_{0_i})}\Delta P_{m_i}\\
&\quad -\Big(\tfrac{1}{T_{N_i}} - \tfrac{C_{p_i} f_0}{2S_i(H_i S_{B_i}+W_{0_i})}\Big)(P_T \sin(\Delta\phi_{ij}+\phi_{0_{ij}}) - P_{0_{ij}})\\
&\quad -2\pi C_{p_i} P_T(\Delta f_i - \Delta f_j)\cos(\Delta\phi_{12}+\phi_{12}) + \tfrac{C_{p_i} f_0}{2S_1(H_i S_{B_i}+W_{0_i})}\Delta P_{L_i} - \tfrac{K_{a_i}}{T_{N_i}} p_i,\\
\Delta P_{m,p_i} &= \begin{cases} \Delta P_{p_i}^{min} \text{ if } & \Delta P_{p_i} \le \Delta P_{p_i}^{min},\\ \Delta P_{p_i} \text{ if } & \Delta P_{p_i}^{min} < \Delta P_{p_i} < \Delta P_{p_i}^{max},\\ \Delta P_{p_i}^{max} \text{ if } & \Delta P_{p_i} \ge \Delta P_{p_i}^{max},\end{cases}\\
\Delta P_{m,AGC_i} &= \begin{cases} \Delta P_{AGC_i}^{min} \text{ if } & \Delta P_{AGC_i} \le \Delta P_{AGC_i}^{min},\\ \Delta P_{AGC_i} \text{ if } & \Delta P_{AGC_i}^{min} < \Delta P_{AGC_i} < \Delta P_{AGC_i}^{max},\\ \Delta P_{AGC_i}^{max} \text{ if } & \Delta P_{AGC_i} \ge \Delta P_{AGC_i}^{max},\end{cases}
\end{aligned}
\tag{14}
$$

$$
\begin{aligned}
\Delta P_{p_i} &= -\tfrac{1}{S_i}\Delta f_i,\\
\Delta \dot{P}_{AGC_i} &= -C_{p_i}\big(\tfrac{\Delta \dot{f}_i}{S_i} + \Delta \dot{P}_{ij}\big) - \tfrac{1}{T_{N_i}}\big(\tfrac{\Delta f_i}{S_i} + \Delta P_{ij}\big) - \tfrac{K_{a_i}}{T_{N_i}} p_i,\\
p_i &= \begin{cases} 0 & \text{if } \Delta P_{AGC_i}^{min} < \Delta P_{AGC_i} < \Delta P_{AGC_i}^{max},\\ \Delta P_{AGC_i} - \Delta P_{m,AGC_i} & \text{else.}\end{cases}
\end{aligned}
\tag{15}
$$

where $\Delta\phi_{ij} = -\Delta\phi_{ji}$.

For the analysis of the following sections we consider the model in (15) and assume that an attacker has disabled the AGC signal in the second control area and applies an arbitrary input $u \in U \subseteq \mathbb{R}$. Under this assumption and using a compact notation (15) is transformed in a continuous time, non nonlinear control system of the form

$$\dot{x} = f(x,w) + g(x,w)u, \tag{16}$$

where $x = [x_1,\ x_2,\ x_3,\ x_4]^T = [\Delta f_1,\ \Delta f_2,\ \Delta\phi_{12},\ \Delta P_{AGC_1}]^T \in \mathbb{R}^4$, $u \in U \subseteq \mathbb{R}$ is the attack input, and w is a vector containing all constants parameter in (11).

Moreover, Let $\mathscr{U}_{[t,t']}$ denote the sets of Lebesgue measurable functions from the interval $[t,t']$ to U. Following [21], if U is compact, f is Lipschitz in x and continuous in u, and $T \ge 0$ is an arbitrary time horizon, then this system with initial condition $x(t) = x \in \mathbb{R}^4$ admits a unique solution $x(\cdot) : [t,T] \to \mathbb{R}^4$ for all $t \in [0,T], x \in \mathbb{R}^4$, $u(\cdot) \in \mathscr{U}_{[t,T]}$. For $\tau \in [t,T]$ we will use $\sigma(\tau,t,x,u(\cdot)) = x(\tau)$ to denote this solution.

3 Feasibility of AGC Attack

3.1 Safety Considerations

The AGC scheme outlined in the previous section is vital to the satisfactory performance of the power system, since it tries to keep the system frequency to its nominal value because too large deviations could damage the power system devices. This action may in the end jeopardize the stability of the entire system and in the worst case lead to a system blackout. In normal operation the frequency deviation of each area should not exceed 1.5Hz.

The amount of power that a line can transfer is also limited to maintain reliability and stability in the system. The limiting value for the permissible power transfer is influenced, according to the line length, by three factors: the thermal limit, the voltage drop and the stability limits. In the case-study of the two-area system, the amount of power that can be transferred is considered to be limited only by the steady state stability limit. This limit is a percentage of the maximal power P_T. We consider a minimum allowable steady state margin of 30% [30] which implies that $\Delta P_{12} \in [-70\% P_T, +70\% P_T]$. Since P_T is assumed constant, the aforementioned limits are translated into a bound $x_3 \in [-44°, 44°]$ in the phase angle difference.

In summary we consider the system to be safe when the state trajectories of (16) lie inside the following safe set of the state space:

$$x_1 \in [-1.5, +1.5], \ x_2 \in [-1.5, +1.5], \ x_3 \in [-44°, 44°] \tag{17}$$

We consider the model in (16) and investigate whether there exists a policy $u(\cdot)$ for the attacker, that can drive the system trajectories $\sigma(\cdot, t, x, u(\cdot))$ outside the safety margins in (17), and/or lead to unstable swinging in the power exchanged between the two control areas by exceeding the limits of x_3 for a sufficiently large amount of time. It should be noted that power swinging results in large power oscillations in the tie-line which are undesirable and can lead to triggering out-of-step protection relays that trip generating units in order to avoid potential damaging and mechanical vibrations [30].

3.2 Violating the Safety Margins

We first examine if the attacker, selecting a suitable policy, can lead the system trajectory outside the safe region defined in (17). Define $K_1 \subset \mathbb{R}^4$ by

$$K_1 := \{x \in \mathbb{R}^4 \mid |x_1| \leq 1.5, |x_2| \leq 1.5, |x_3| \leq 44°\}, \tag{18}$$

and let $l_1(\cdot) : \mathbb{R}^4 \to \mathbb{R}$ be the signed distance to the set K_1, defined by $l_1(x) = \min\{x_1 + 1.5, 1.5 - x_1, x_2 + 1.5, 1.5 - x_2, x_3 + 44°, 44° - x_3\}$, for any $x \in \mathbb{R}^4$. Clearly, $K_1 = \{x \in \mathbb{R}^4 \mid l_1(x) \geq 0\}$. Note that the last state x_4, which corresponds to the AGC signal in the first area is restricted indirectly due to the line saturation.

The problem of interest can be though of as a reachability problem where the objective is to compute the set of states at some initial time $t < T$ for which there exists a control policy that can drive (at least for some time instance) the system trajectories in K_1^c, i.e. outside the safe region (17). The desired set can be encoded by

$$Reach(t, K_1) = \{x \in \mathbb{R}^4 \mid \exists u(\cdot) \in \mathscr{U}_{[t,T]} \\ \exists \tau \in [t, T]\ \sigma(\tau, t, x, u(\cdot)) \notin K_1\}. \tag{19}$$

It is shown in [21, 22] that $Reach(t, K_1)$ can be related to the zero sub-level set of

$$V(x,t) = \inf_{u(\cdot) \in \mathscr{U}_{[t,T]}} \min_{\tau \in [t,T]} l_1(\sigma(\tau, t, x, u(\cdot))). \tag{20}$$

In particular, $Reach(t, K_1) = \{x \in \mathbb{R}^4 \mid V(x,t) < 0\}$ and $V(x,t)$ is the unique, bounded and uniformly continuous viscosity solution to the Hamilton-Jacobi equation

$$\frac{\partial V}{\partial t}(x,t) + \min\{0, \inf_{u \in U} \frac{\partial V}{\partial x}(x,t) f(x,u)\} = 0, \tag{21}$$

with terminal condition $V(x,T) = l_1(x)$.

Therefore, to compute $Reach(t, K_1)$ it suffices to solve the partial differential equation in (21). The latter can be achieved using standard numerical tools for such problems based on Level Set Methods [31].

For the analysis of this section, we performed a series of reachability computations for different bounds of the attack input. Fig. 3 shows a family of curves that correspond to the different bounds of the attack signal. These curves quantify how the volume of the safe set changes in time. The safe set is defined as the complement of $Reach(t, K_1)$ since it includes all states from which the system trajectories can remain state for the entire horizon. By inspecting this figure, since the volume of the safe set vanishes for an attack authority greater than or equal to $200MW$, we can conclude that the attacker would need a signal at least $200MW$ to disturb the system starting from the nominal operating point.

We also depict the case that the attacker is able to inject an arbitrary signal up to 100MW, i.e. $|u| \leq 100$MW. In Fig. 4, as also expected from Fig. 3, it is clear that the safe set has saturated after approximately 10 seconds, which means that despite the attack, there are still some states, including the nominal point, that system trajectories can start and remain in the safe region K_1 of (18). The complement of the red surfaces correspond to $Reach(t, K_1)$.

3.3 Power Swinging between Two Areas

Next we consider the possibility of keeping the angle x_3 outside $[-44°, 44°]$ for a sufficiently large amount of time, thus leading to an unstable power swinging between the two areas. To this end we define the set $K_2 \subset \mathbb{R}^4$ by

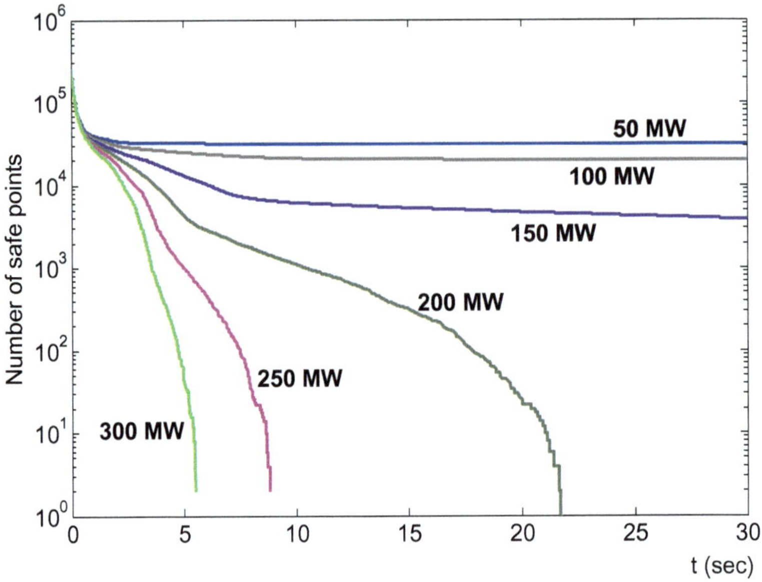

Fig. 3 Volume of the safe set (complement of $Reach(t, K_1)$) for different bounds of the attack signal

$$K_2 := \{x \in \mathbb{R}^4 \mid |x_3| > 44^\circ\}, \tag{22}$$

and a function $l_2(\cdot) : \mathbb{R}^4 \to \mathbb{R}$ to be the signed distance to the set K_2, defined by $l_2(x) = \min\{-x_3 - 44^\circ, x_3 - 44^\circ\}$, for any $x \in \mathbb{R}^4$. Clearly, $K_2 = \{x \in \mathbb{R}^4 \mid l_2(x) \geq 0\}$.

We first perform a so called viability computation and determine the set of states for which there exists an attack policy such that the emanating trajectories remain in K_2 for the entire horizon. The desired set can be encoded by

$$\begin{aligned} Viab(t, K_2) = \{x \in \mathbb{R}^4 \mid \exists u(\cdot) \in \mathscr{U}_{[t,T]} \\ \forall \tau \in [t, T]\ \sigma(\tau, t, x, u(\cdot)) \in K_2\}. \end{aligned} \tag{23}$$

It is shown in [21] that $Viab(t, K_2)$ can be related to the zero sub-level set of

$$\widetilde{V}(x,t) = \sup_{u(\cdot) \in \mathscr{U}_{[t,T]}} \min_{\tau \in [t,T]} l_2(\sigma(\tau, t, x, u(\cdot))). \tag{24}$$

In particular, $Viab(t, K_2) = \{x \in \mathbb{R}^4 \mid \widetilde{V}(x,t) \geq 0\}$ and $\widetilde{V}(x,t)$ is the unique, bounded and uniformly continuous viscosity solution to the Hamilton-Jacobi equation

$$\frac{\partial \widetilde{V}}{\partial t}(x,t) + \min\{0, \sup_{u \in U} \frac{\partial \widetilde{V}}{\partial x}(x,t) f(x,u)\} = 0, \tag{25}$$

with terminal condition $\widetilde{V}(x,T) = l_2(x)$.

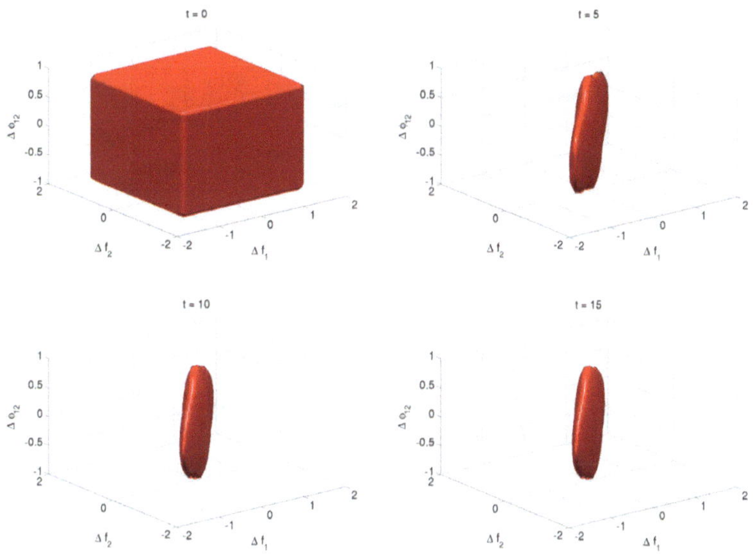

Fig. 4 Safe set for the case where $u \in [-100, +100]MW$ and $x_3 \in [-44^\circ, 44^\circ]$

The result of this calculation is shown in Fig. 5, where it was assumed that the attack signal is bounded in $[-350\ 350]$ MW due to the AGC saturation. It can be observed that the viability set $Viab(t, K_2)$ is saturated in approximately 7 seconds; namely, there exists a non-empty set such that if the system starts from that set, the attacker can construct an input sequence to keep the angle above or below 44° for the specified time horizon. Notice that since the other states (except x_3) are free in this case, the constraint ($|x_3| > 44^\circ$) in the definition of K_2 divides the state space to two parts; one part between the two surfaces of Fig. 5, and one outside. The latter is the set where the attacker is trying to steer the system trajectories.

Having computed the set $Viab(t, K_2)$ it remains to verify whether the attacker is able to force the system to that set. If so, then once reaching the viability set, the attacker could change his control policy and keep the angle deviation in the unsafe region for sufficiently large amount of time. The latter can cause power swinging and its undesirable consequences. For this purpose, we define the set K_4 by

$$K_3 := \{x \in \mathbb{R}^4 \mid \widetilde{V}(x, 0) > 0\}, \tag{26}$$

where $\widetilde{V}(x, 0)$ is the value function that characterizes the set $Viab(0, K_2)$ obtained from the viability computation. We then compute the set $Reach(t, K_3)$ defined as in (19) with K_3 in place of K_1.

As shown in Fig.6, since the safe set (complement of $Reach(t, K_3)$) is empty, for every initial condition, there exists at least one control policy for the attacker so as to reach the viability set $Viab(t, K_2)$ in 8.5 seconds. Then, the attacker could switch policy and keep the state trajectory in K_2.

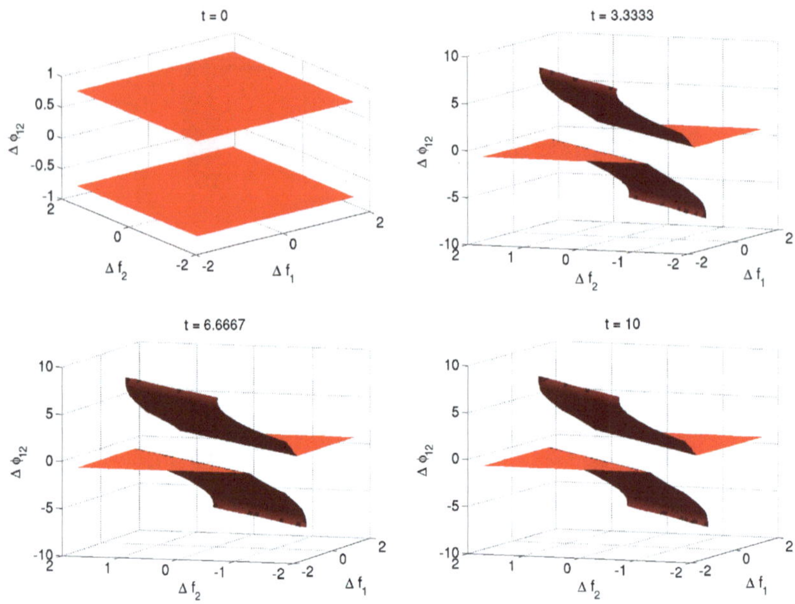

Fig. 5 Computation of the viable set $Viab(t,K_2)$

Fig. 7 summarizes the previous analysis, which comprises of two stages; the first stage provides a way to compute the viability set $Viab(t,K_2)$ of K_2, whereas the second stage describes the computation of $Reach(t,K_3)$. One can see how the volume of the safe part of the state space changes. Following the definition of the reachable and the viability set, at the first stage the safe set coincides with the viable set, and in the second one it corresponds to the complement of the reachable set. At the 7th second the viability set is saturated, and the attacker could change policy so as to keep the angle increasing. That way, the power will start swinging and this in turn might lead to activation of the out-of-step protection relays.

4 Attack Signal Synthesis

Using the frequency model of Section 2 it was shown in Section 3 that if an attacker gains access to the AGC signal in one control area, then she can cause undesirable effects to the network. Using tools from optimal control and reachability theory the existence of such an attack policy was verified. However, to construct such an attack policy is a difficult task since it is based on the spatial derivatives of the value functions V, $\widetilde{V}$, whose computation is affected by discretization errors. To overcome this difficulty we present here different alternatives for a synthesis of an attack signal.

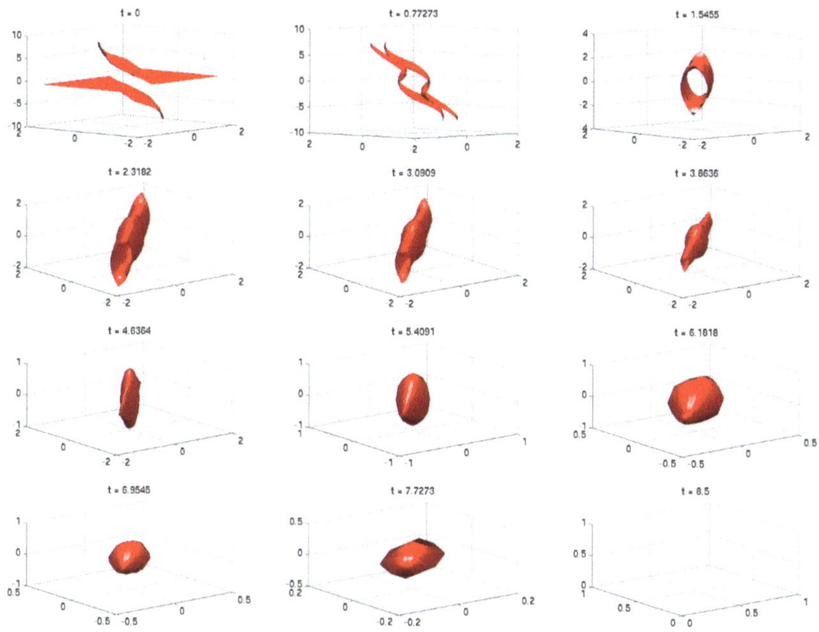

Fig. 6 Computation of the reachable set $Reach(t, K_3)$

4.1 *Open Loop*

4.1.1 Naive Attack Signal

The optimal attack input generated based on the analysis of the previous section is shown to be a bang-bang signal. Motivated by this fact, we show here that constructing a bang-bang input sequence with arbitrarily selected switching instances is not sufficient for the attacker to disturb the system.

We select the naive, bang-bang attack signal to be a pulse sequence as the one shown in Fig. 8(a). By inspection of Fig. 8(b), such an attack signal leads only to minor deviations in the frequency of each area from its nominal value, and to affordable oscillations in the power exchanged between the two areas. Other random bang-bang signals have been tested as well, leading to similar performance. This reveals the need to resort to more sophisticated attack synthesis techniques.

4.1.2 MCMC Based Attack Signal

It was shown that applying a naive, open loop bang-bang signal is not sufficient for the objectives of an attacker. Here we construct an open loop signal of similar type, but selecting this time the switching instances by means of an optimization problem. Specifically, we assume that the nominal parameter values are available to the attacker, i.e. $w = w_0$ for the model (16) considered in the proposed design.

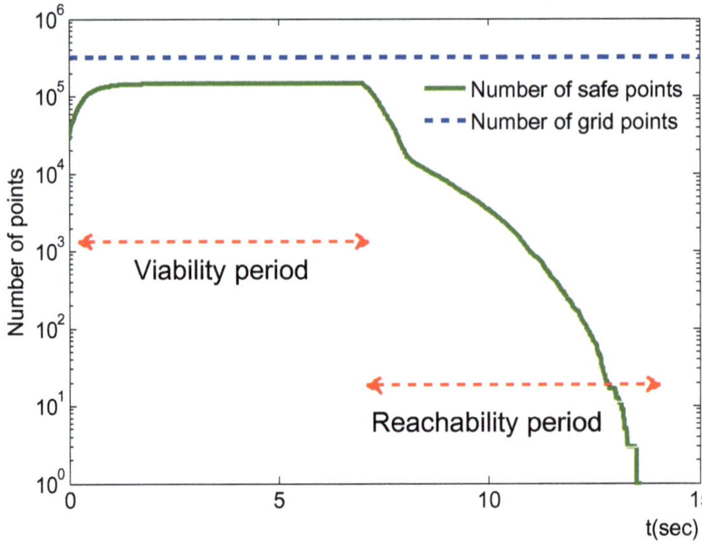

Fig. 7 The dash line is the total number of grid points; the solid line indicates the volume of the safe set over time

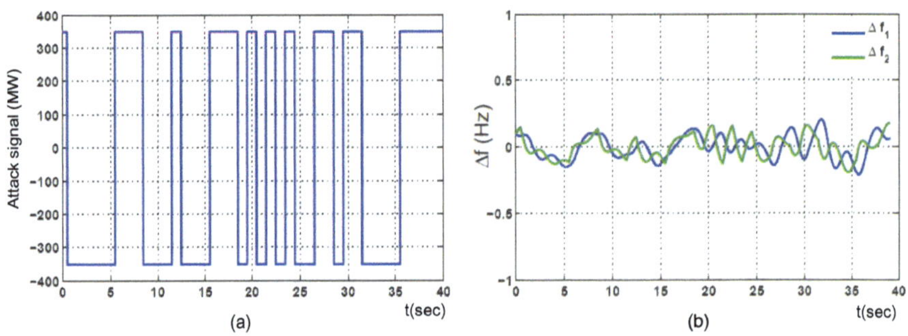

Fig. 8 (a)Naive bang-bang attack signal, (b)frequency deviation of the two areas

We consider input sequences of the form

$$\mathbf{u}(t) = u_\kappa, \ \frac{T}{N}\kappa \leq t < \frac{T}{N}(\kappa+1), \text{for} \kappa = 0, \ldots, N-1, \tag{27}$$

where $u_\kappa \in \{-350, 350\}$, T denotes the optimization horizon and T/N is the time discretization step. Identifying an optimal control policy for the attacker that leads the system trajectories outside the safety margins in (17) can be thought of as an optimization problem. We seek for a vector $\theta = (u_1, \ldots, u_N) \in \{-350, 350\}^N$ that maximizes the objective function $J = e^{\int_0^T x_2^2 dt}$, subject to the system dynamics (16). By maximizing the criterion J we implicitly maximize the deviation of the frequency

Algorithm 1. MCMC algorithm

1: Let θ_0 denote an initial choice for θ.
2: Define as N the total number of iterations.
3: **For** $i = 0, \ldots, N$
4: Fix $\alpha > 0$ and extract $\theta_{i+1} \sim p_\theta(\cdot|\theta_i)$

$$\rho = \min\{1, \frac{p_\theta(\theta_i|\theta_{i+1})}{p_\theta(\theta_{i+1}|\theta_i)} \frac{J(\theta_{i+1})^\alpha}{J(\theta_i)^\alpha}\},$$

$$\theta_{i+1} = \begin{cases} \theta_{i+1} & \text{with probability } \rho, \\ \theta_i & \text{with probability } 1-\rho, \end{cases}$$

5: **end**

from its nominal value and hence force the system trajectories outside the safe region in (17).

This is a nonlinear optimization problem over a discrete domain; to solve it we use the Markov Chain Monte Carlo (MCMC) method. This is a randomized optimization technique, which explores the search space via a Markov chain. We approximate the maximizer of the problem by extracting candidate solutions from a proposal distributions of our choice. The algorithm involves then a sophisticated accept-reject mechanism, which involves a trade-off between the objective value of the samples solution and the rareness.

At each step i of the algorithm, the extracted variable θ_i is accepted with a probability ρ, as this is defined in the algorithm above. Otherwise, it is rejected and the previous state of the chain is replicated. At the end of the algorithm, the extracted points are concentrated at different regions, and based on the peakedness, the optimal value for θ is determined. The probability density $p_\theta(\cdot|\theta_i)$ denotes the proposal distribution, which at the first step is chosen to be a uniform distribution, so as to search evenly the decision space. At a next step the entire process is repeated, this time sampling from gaussian distributions centered at the accepted samples of the first run. That way, a local search is performed and a more accurate maximizer is identified. To investigate the performance of the attack policy constructed according to the MCMC algorithm we considered two different case studies. In the first set-up the obtained solution, which is based on a model with $w = w_0$, is applied to a perfect model that has the nominal set of parameter values, i.e. $w = w_0$. The second study involves the application of the obtained solution on a model with $w \neq w_0$. For the MCMC algorithm we selected $T = 40$ sec and $N = 40$. We performed in total 82306 iterations until the accepted states of the chain were 50000. The ratio between accepted and total states of the chain is 0.61.

Perfect Model: We consider a set-up with $w = w_0$.Fig. 9(a) depicts an open loop policy for the attacker, obtained via the MCMC optimization method. Fig 9(b) shows the frequency response in the two areas. Clearly, the impact of the attack signal is extremely severe. The swings of the transferred power on the tie line will result in triggering the out-of-step protection relays. If the system is not equipped

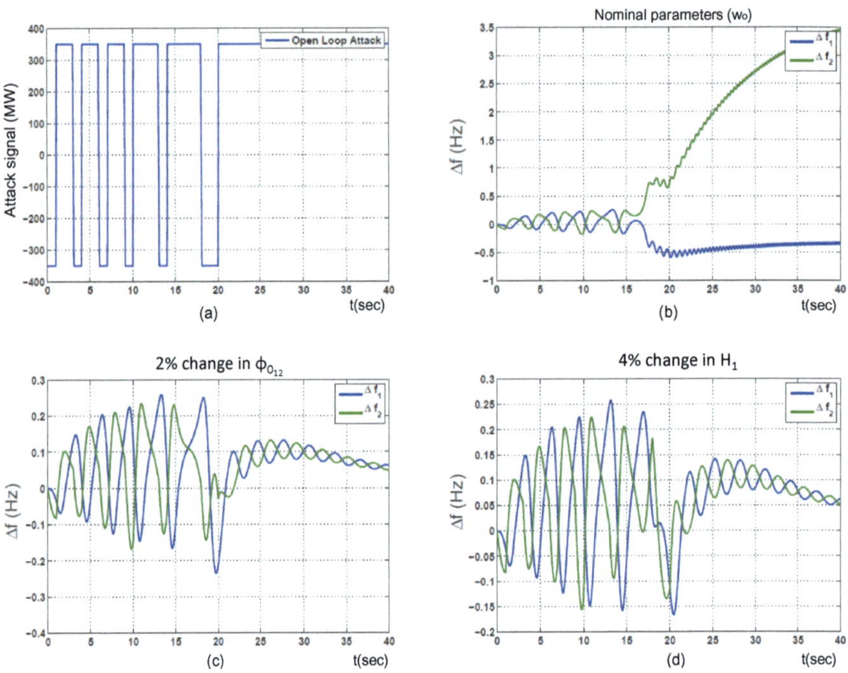

Fig. 9 (*a*) Open loop policy generated by the MCMC algorithm. (*b*) Frequency trajectories for perfect model with $w = w_0$. (*c*) Frequency trajectories for imperfect model with $w \neq w_0$ (2% mismatch in $\phi_{0_{12}}$). (*d*) Frequency trajectories for imperfect model with $w \neq w_0$ (2% mismatch in H_1).

with such a protection scheme, the generators of the second area would start to trip by the time that the frequency of that area would exceed the safety margins. The latter may lead to cascading failures and even to a wide-area blackout.

Imperfect Model: In Fig. 9(c) and 9(d), we assume that the attacker does not have perfect information of the system. Specifically we consider the case where the angle $\phi_{0_{12}}$ and the inertia H_1 in the first area that the attacker considers in her design are 2% and 4%, respectively, higher than true parameter values. It is clear that the open loop strategy is extremely sensitive to such a model mismatch and hence the open loop policy does not serve practically as an efficient solution.

4.2 *Closed Loop*

The poor performance of the naive attack signal and the sensitivity to parameter uncertainty of the MCMC based signal motivate the synthesis of a feedback attack policy. Two alternatives are proposed, one based on feedback linearization and one based on gain scheduling. For simplicity we assume perfect state information for the attacker. We refer to [20] for the case of partial state information, where a nonlinear observer is constructed.

4.2.1 Feedback Linearization Based Attack Signal

We consider an attack scheme that is based on feedback linearization and the MCMC algorithm presented in the previous subsection. Feedback linearization is based on applying a nonlinear coordinate transformation and a nonlinear feedback to transform a nonlinear input affine system as the one in (16) to a system that is linear in the new coordinates.

The feedback linearization procedure is based on the notion of relative degree γ. Specifically, for the nonlinear system (16) with output $y = l(x)$, for some $l(\cdot) : \mathbb{R}^4 \rightarrow \mathbb{R}$, is said to have relative relative degree γ with $1 \leq \gamma \leq n$, in a region $D \subset \mathbb{R}^4$ if $L_g L_f^{i-1} l(x) = 0$ for $i = 1, 2, \cdots, \gamma - 1$, and $L_g L_f^{\gamma-1} l(x) \neq 0$ for all $x \in D$. Note that $L_f^1 l(x) = \frac{\partial l}{\partial x} f$ is called the Lie derivative of l with respect to f, whereas higher order Lie derivatives are defined recursively [25]. It should be also noted that for the relative degree to be well-defined $L_f^i l(x)$ needs to be differentiable and hence $f(\cdot, w)$ should be smooth. Due to the saturation of primary and secondary loop control this might not be the case; however, we assume that that none of the dynamic saturations is activated and $f(\cdot, w)$ is sufficiently smooth. The saturations will be explicitly taken into account in the design of the attack signal.

It can be easily seen that by choosing $y = l(x) = x_3$, (16) has relative degree $\gamma = 2$. It is then shown in [25] that, for every $x_0 \in D$, there exists a nonlinear transformation $T(\cdot, \cdot) : \mathbb{R}^4 \times \mathbb{R}^3 \rightarrow \mathbb{R}^4$ such that $[\eta,\ \xi]^T = T(x, w)$, $\eta, \xi \in \mathbb{R}^2$, and a nonlinear feedback $v = \alpha(x, w) + \beta(x, w)u$ with $\alpha(x, w) = L_f^{\rho} l(x)$, $\beta(x, w) = L_g L_f^{\gamma-1} l(x)$, that results in a dynamical subsystem that is linear in the ξ coordinates.

Upon using the linearizing transformation T and the associated functions α and β, (16) is transformed to

$$\begin{aligned} \dot{\eta} &= f_0(\eta, \xi), \\ \dot{\xi} &= A_c \xi + B_c v, \\ y &= C_c \xi, \end{aligned} \tag{28}$$

where $A_c \in \mathbb{R}^{2\times 2}$, $B_c \in \mathbb{R}^2$ and $C_c \in \mathbb{R}^{1\times 4}$ are canonical controllability matrices [26] and $f_0(\cdot, \cdot) : \mathbb{R}^2 \times \mathbb{R}^2$ is a nonlinear function referred to as zero dynamics. This form decomposes the system into a linear subsystem in the ξ coordinates and an internal nonlinear subsystem in the η coordinates. Here our main goal is to push the system trajectories to the unsafe region in contrast to the usual stabilization idea. Hence, unstable behavior of the internal dynamics would be a benefit for our objectives, i.e. destabilize the system.

In the linearized subsystem we can apply the state feedback $v = K\xi$, which results in the feedback law

$$u(x, w, K) = \frac{K\xi - \alpha(x, w)}{\beta(x, w)} = \frac{K[0\ \ I]T(x, w) - \alpha(x, w)}{\beta(x, w)}, \tag{29}$$

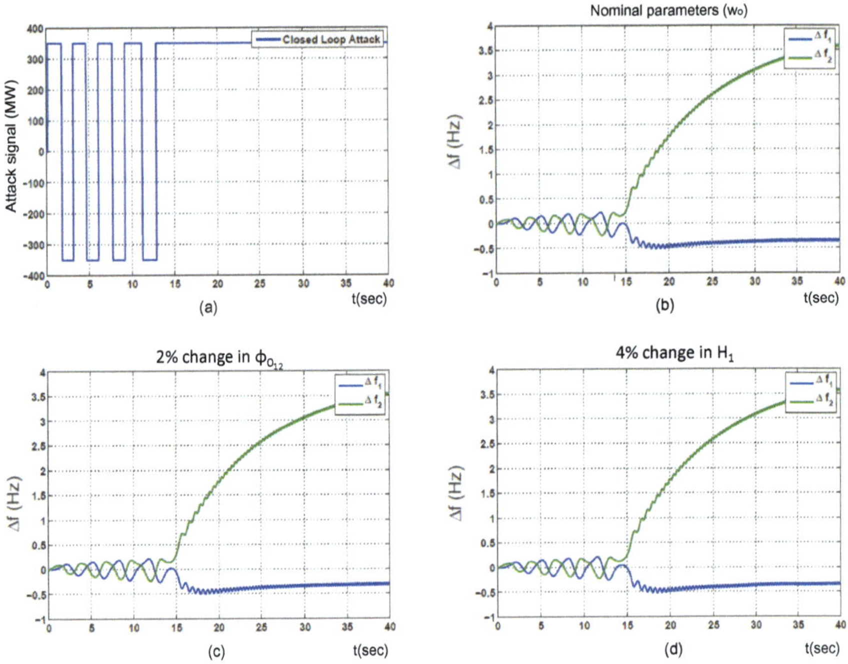

Fig. 10 (*a*) Closed loop policy generated using feedback linearization. (*b*) Frequency trajectories for perfect model with $w = w_0$. (*c*) Frequency trajectories for imperfect model with $w \neq w_0$ (2% mismatch in $\phi_{0_{12}}$). (*d*) Frequency trajectories for imperfect model with $w \neq w_0$ (4% mismatch in H_1).

in the original coordinates. The feedback gain $K \in \mathbb{R}^{1\times\gamma}$ is a constant vector. To consider the saturation limits of AGC, $|u(x,w,K)| \leq U_0 = 350$ MW, we pass the control law through a saturation operator as

$$\bar{u}(x,w,K) = \mathrm{sat}(u(x,w,K),U_0) = \begin{cases} u(x,w,K) & \text{if } |u(x,w,K)| \leq U_0, \\ U_0 \,\mathrm{sign}(u(x,w,K)) & \text{if } |u(x,w,K)| > U_0, \end{cases} \tag{30}$$

where the sign$(\cdot)$ operator returns the sign of its argument.

The attack signal would be then given by (30), where the gain K remains to be defined. To determine K we seek to maximize $J = \max(\|x_1\|_\infty, \|x_2\|_\infty)$ subject to the dynamics in (16) when the feedback law in (30) is applied. That way the gain K chosen by an attacker would result in the maximum possible deviation of the frequencies from their nominal values and lead the system trajectories outside (17). To solve this maximization problems we use the MCMC algorithm described in the previous subsection. Note that J is different from the objective function used in the previous subsection since it resulted in this case to a better performance.

As in the case of the MCMC based attack design we investigated two cases according to whether the attacker has perfect model knowledge (i.e. $w = w_0$) or not. For the MCMC algorithm employed to determine the gain K we used $T = 40$ sec and $N = 40$. We performed in total 10000 iterations and the ratio between accepted and total states of the chain is 0.37.

Perfect Model: We considered a scenario with $w = w_0$. Fig. 10(a) shows the feedback policy of the attacker and Fig. 10(b) shows the frequency trajectory of each area and proves the severe impact that a suboptimal attack signal could have on the system.

Imperfect Model: Similarly to the open loop simulations (Fig. 10(c), 10(d)), we assume that the attacker has an imperfect knowledge of the system, and the angle $\phi_{0_{12}}$ and the inertia H_1 in the first area that she considers in her design are 2% and 4%, respectively, higher than true parameter values. In contrast to the open loop strategy, the feedback policy is considerably robust to such a model mismatching and consequently it provides an effective and practical solution to construct an attack signal.

4.2.2 Gain Scheduling Based Attack Signal

If the system dynamics in (16) were described by a linear dynamical system, a natural choice for the attacker would be to choose her signal among the class of linear feedback policies. The feedback gain would be then selected so that the eigenvalues of the linear system have positive real part, resulting in an unstable behavior.

In our case, however, (16) is nonlinear and in fact, due to the saturation limits of the primary frequency controller and the AGC, it involves multiple modes of operation. Motivated by control design techniques based on gain scheduling [26], we apply the following procedure [32]. We first linearize (16) around a nominal operating point at every mode of operation. In total 27 modes are distinguished due to the saturation limits of the primary frequency controller of each area and the AGC of the second area (the attacker has gained access of the AGC of the second area). We then have a family of linear systems. For each one we design a linear feedback so that the eigenvalues of the corresponding system have positive real parts.

The attack signal is then a switched linear feedback of the state, since the feedback gain changes according to the mode of operation. Moreover, to ensure that the attack signal satisfies the saturation limits of the AGC, we pass it through a saturation function as in (30).

Fig. 11(a) illustrates that by applying the attack signal generated by the aforementioned procedure, unacceptable deviations in the frequencies are obtained. It should be also noted that if the feedback gain is not updated according to the mode of operation then the effect of the corresponding attack signal is very different. Fig. 11(b) shows the negligible deviations in the frequency response for each area, to highlight the necessity of the gain scheduling scheme.

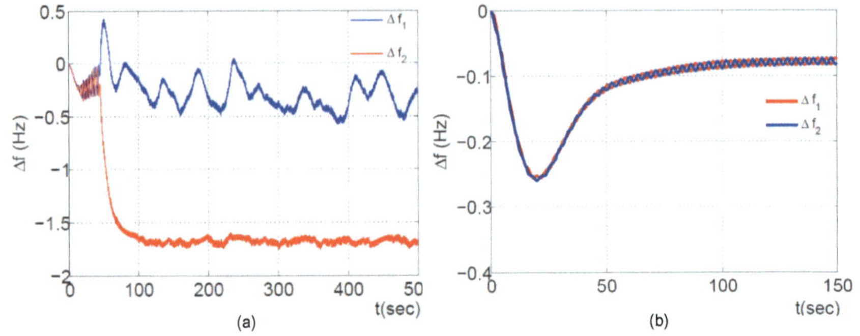

Fig. 11 Frequency response at each area as an effect of the gain scheduling based attack signal

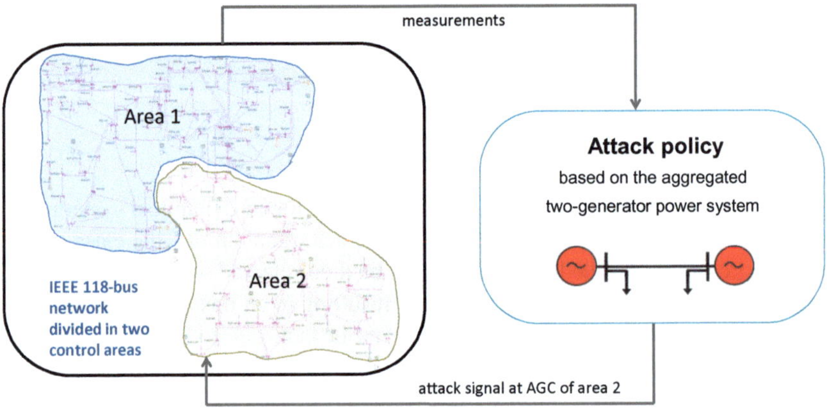

Fig. 12 Interaction between the power system and the cyber-attack policy. The abstracted model of Section 2 is used for synthesizing the attack signal that is applied to the two-area detailed model.

5 Evaluation on a Detailed Simulation Environment

In the previous sections, for the synthesis and the evaluation of the constructed attack signal, the two-generator power system model of Section 2 was employed. Here we investigate the performance of the constructed signals when they are applied to a detailed model of the network. For this study, we used the IEEE 118-bus network, and a detailed power system simulation environment implemented in MATLAB by [33]. All generators are represented by the classical model and are also equipped with primary frequency control, Automatic Voltage Regulator (AVR) and Power System Stabilizer (PSS). Moreover, we divide the network into two control areas, each one equipped with each own AGC loop. The data of the model are

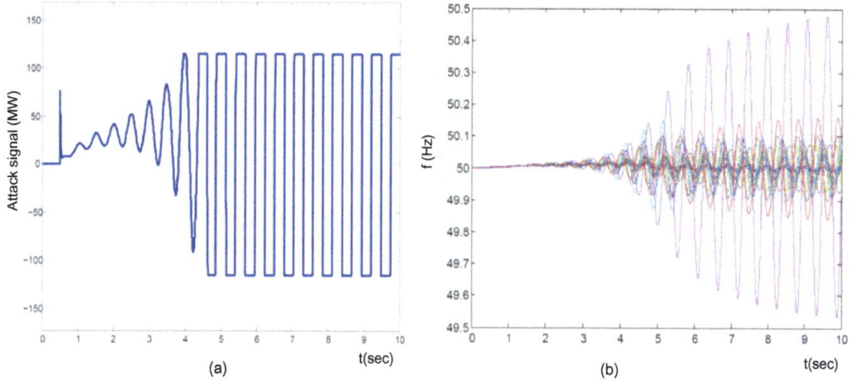

Fig. 13 (*a*)Attack signal, (*b*) Frequency response of the generating units

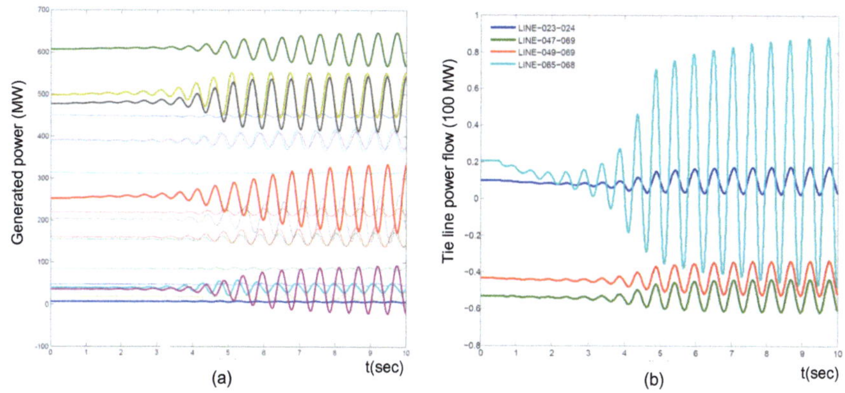

Fig. 14 (*a*) Swinging on the produced power of the generators and (*b*) power swinging between the two areas as an effect of the application of the attack signal

retrieved from a snapshot available at [34]. Since there were no dynamic data available, typical values provided by [35] are used for the simulations.

For the control synthesis, we use the abstracted two-generator model by aggregating each area into one generator based on the center of inertia as discussed in Section 2. In Fig. 12, we show the interaction between the power system and the cyber-attack policy. The model abstraction is only used for the attack signal synthesis and serves as feedback to the detailed simulation environment.

Fig. 13 shows the effect in the detailed system once the feedback linearization based attack signal is applied. In contrast to the results of the previous section, applying the attack signal to the detailed system does not lead to significant frequency deviations. However, as shown in Fig. 14, swings on the generator output and the power flows across the tie lines connecting the two areas are observed. The swinging behavior can be dangerous for the system since they may trigger out-of-step protection relays and cause a cascade of undesirable effects. It should be noted that the qualitatively different effect of the application of the attack signal to the detailed model compared with those obtained when it is applied to the model of Section 2, are due to the mismatch between the two models (abstraction error).

6 Conclusions

In this chapter we investigated the impact that a cyber-attack on the AGC loop may have in the power system. We employed an abstraction of the detailed power system model and carried out an feasibility analysis based on reachability and optimal control theory. This analysis offered us intuition on whether there exist an attack pattern that can disturb the power system. We also investigated the problem of synthesizing an attack signal by using open and closed loop nonlinear control approaches. The efficancy of the proposed methods was investigated by means of simulations on the IEEE-118 bus network.

The fact that our results show that the power system can indeed be disturbed by an AGC attack, highlight the necessity of devising an attack detection scheme. A complimentary study towards this direction can be found in [36, 37] where a detection algorithm and an intuitive mitigation strategy is proposed.

Acknowledgements. Research was supported by the European Commission under the project VIKING, FP7-ICT-SEC-2007-1. The authors would also like to thank the coordinator of the VIKING project Gunnar Björkman for his valuable input and insight in the current work. We would also like to thank Dr. Turhan Demiray for his support and for generously providing the Matlab-based detailed simulation environment.

References

1. Andersson, G., Donalek, P., Farmer, R., Hatziargyriou, N., Kamwa, I., Kundur, P., Martins, N., Paserba, J., Pourbeik, P., Sanchez-Gasca, J., Schulz, R., Stankovic, A., Taylor, C., Vittal, V.: Causes of the 2003 Major Grid Blackouts in North America and Europe, and Recommended Means to Improve System Dynamic Performance. IEEE Transactions on Power Systems 20(4), 1922–1928 (2005)
2. Zima, M., Bockarjova, M.: Operation, Monitoring and Control Technology of Power Systems. Lecture Notes, ETH Zurich (2007)
3. Khaitan, S., McCalley, J.: Cyber physical system approach for design of power grids: A survey. In: IEEE PES GM 2013, July 21-15, pp. 1–5 (2013)
4. Khaitan, S., McCalley, J.: Design techniques and applications of cyber physical systems: A survey. IEEE Systems Journal PP, 1–16 (2014)
5. Kirschen, D., Bouffard, F.: Keep the Lights On and the Information Flowing. IEEE Power and Energy Magazine 7(1), 50–60 (2009)
6. Teixeira, A., Amin, S., Sandberg, H., Johansson, K., Sastry, S.: Cyber security analysis of state estimators in electric power systems. In: 2010 49th IEEE Conference on Decision and Control (CDC), pp. 5991–5998 (December 2010)

7. Vukovic, O., Sou, K.C., Dan, G., Sandberg, H.: Network-aware mitigation of data integrity attacks on power system state estimation. IEEE Journal on Selected Areas in Communications 30(6), 1108–1118 (2012)
8. Hug, G., Giampapa, J.: Vulnerability assessment of ac state estimation with respect to data injection cyber-attacks. IEEE Transactions on Smart Grid (2012)
9. Negrete-Pincetic, M., Yoshida, F., Gross, G.: Towards Quantifying the Impacts of Cyber Attacks in the Competitive Electricity Market Environment. In: IEEE Power Tech Conference (2009)
10. Forbes, Congress Alarmed at Cyber-Vulnerability of Power Grid, http://www.forbes.com/2008/05/22/cyberwar-breach-government-tech-security_cx_ag_0521cyber.html
11. CNN, Sources: Staged cyber attack reveals vulnerability in power grid, http://www.cnn.com/2007//US/09/26/power.at.risk/index.html
12. Comptuterworld, DHS to review report on vulnerability in West Coast power grid, http://www.computerworld.com/s/article/9138017
13. Wang, J.-W., Ronga, L.-L.: Cascade-based attack vulnerability on the US power grid. Elsevier, Safety science 47(10), 1332–1336 (2009)
14. VIKING Project, http://www.vikingproject.eu
15. Vulnerability assessment of scada systems. Deliverable D3.1, VIKING project (2011)
16. Impact analysis of adverse events. Deliverable D3.2, VIKING project (2011)
17. Consequence and cost analysis of scada system vulnerabilities. Deliverable D3.3, VIKING project (2011)
18. Mitigation and protection strategies. Deliverable D4.3, VIKING project (2011)
19. Mohajerin Esfahani, P., Vrakopoulou, M., Margellos, K., Lygeros, J., Andersson, G.: Cyber attack in a two-area power system: Impact identification using reachability. In: American Control Conference, pp. 962–967 (2010)
20. Mohajerin Esfahani, P., Vrakopoulou, M., Margellos, K., Lygeros, J., Andersson, G.: A robust policy for automatic generation control cyber attack in two area power network. In: 49th IEEE Conference Decision and Control, pp. 5973–5978 (2010)
21. Lygeros, J.: On reachability and minimum cost optimal control. Automatica 40(6), 917–927 (1999)
22. Mitchell, I., Bayen, A.M., Tomlin, C.: A time-dependent Hamilton-Jacobi formulation of reachable sets for continuous dynamic games. IEEE transactions on Automatic Control 50
23. Lecchini-Visintini, A., Lygeros, J., Maciejowski, J.: Stochastic optimization on continuous domains with finite-time guarantees by markov chain monte carlo methods. IEEE Transactions on Automatic Control 55(12), 2858–2863 (2010)
24. Robert, C., Casella, G.: Monto Carlo Statistical Methods. Springer
25. Sastry, S.: Nonlinear systems: analysis, stability and control. Springer, New York (1999)
26. Khalil, H.: Nonlinear Systems, 3rd edn. Prentice-Hall, NJ (2002)
27. Kundur, P.: Power System Stability and Control. McGraw-Hill (1993)
28. Andersson, G.: Dynamics and Control of Electric Power Systems. ETH Zürich (2009)
29. Franklin, G.F., Powell, J.D., Emami-Naeini, A.: Feedback Control of Dynamic Systems. Prentice Hall (2002)
30. Kundur, P.: Power System Stability and Control. McGraw-Hill Inc. (1994)
31. Mitchell, I.: Application of level set methods to control and reachability problems in continuous and hybrid systems. Stanford University, PhD thesis (2002)
32. Panagou, A.: Cyber-security issues in the Automatic Generation Control. Semester thesis, Power System Laboratory. ETH Zurich, Switzerland (2013)

33. Demiray, T.: Simulation of Power System Dynamics using Dynamic Phasor Models. PhD thesis, Diss. ETH No.17607, ETH Zurich, Switzerland (2008)
34. Power Systems Test Case Archive. College of Engineering, University of Washington, `http://www.ee.washington.edu/research/pstca/`
35. Anderson, P.M., Fouad, A.A.: Power System Control and Stability. IEEE Computer Society Press (2002)
36. Mohajerin Esfahani, P., Vrakopoulou, M., Andersson, G., Lygeros, J.: A tractable nonlinear fault detection and isolation technique with application to the cyber-physical security of power systems. In: 2012 IEEE 51st Annual Conference on Decision and Control (CDC), pp. 3433–3438 (December 2012)
37. Mohajerin Esfahani, P., Lygeros, J.: A tractable fault detection and isolation approach for nonlinear systems with probabilistic performance. IEEE Transaction of Automatic Control (TAC) (conditionally accepted, November 2014), `http://arxiv.org/abs/1408.1767`

Intrusion Detection for CPS Real-Time Controllers*

Christopher Zimmer, Balasubramany Bhat, Frank Mueller, and Sibin Mohan

Abstract. Security in CPS-based real-time embedded systems controlling the power grid has been an afterthought, but it is becoming a critical issue as CPS systems are networked and inter-dependent. This work presents a set of mechanisms for time-based intrusion detection, i.e., the execution of unauthorized instructions in real-time CPS environments. The novelty is the utilization of information obtained by static timing analysis for intrusion detection. Real-time CPS systems are unique in that timing bounds on code sections are readily available since they are required for schedulability analysis. We demonstrate how micro-timings can be exploited for multiple granularity levels of application code to track execution progress. Through bounds checking of these micro-timings, we develop techniques to detect intrusions (1) in a self-checking manner by the application and (2) through the operating system scheduler, which are novel contributions to the real-time/embedded systems domain to the best of our knowledge.

1 Introduction

The presence of embedded systems is altering today's life in many facets, and often in a subtle way that may go unnoticed — unless system failure impacts our lives.

Christopher Zimmer · Balasubramany Bhat · Frank Mueller
North Carolina State University, Raleigh, NC 27695-8206
e-mail: mueller@cs.ncsu.edu

Sibin Mohan
University of Illinois at Urbana-Champaign, Urbana IL 61801
e-mail: sibin@illinois.edu

* This work was supported in part by NSF grants 1329780, 1239246, 0812121 and U.S. Army Research Office (ARO) grant W911NF-08-1-0105 managed by NCSU Secure Open Systems Initiative (SOSI). This is an extended version of a prior conference paper [50].

S.K. Khaitan, J.D. McCalley, and C.-C. Liu (eds.), *Cyber Physical Systems Approach to Smart Electric Power Grid*, Power Systems, DOI: 10.1007/978-3-662-45928-7_12

Examples range from non-critical systems (televisions, toasters), moderately critical systems (HVAC control systems, PHEV charging stations, traffic lights) to highly critical ones (power grid control, anti-lock brakes, and flight control systems). The latter two categories are examples of cyber-physical systems (CPS) where system control affects human lives or interacts with the environment. Most of these systems have real-time constraints, and ensuring that such systems are secure from intrusion and tampering is a design challenge of utmost importance. Securing CPSs dramatically deviates from security in general-purpose computing systems. In the latter, attacks may result in slower response or no execution at all. Imminent system failures, if detected, can be mitigated by rebooting or re-installation with a temporary lapse of services to users.

In safety critical real-time systems, in contrast, slower response or failure could result in significant environmental damage or even in loss of life. System restarts often cannot be instant due to an unstable physical system state, e.g., when an aircraft is in flight or a car is subject to slick roads requiring break control.

In practice, real-time software may have stringent requirements for CPS control. However, this still leaves vulnerabilities exposed by libraries and specific embedded domain device software. Attackers may exploit these by eventually executing arbitrary code that they have injected. Such code injection attacks have been common for several years in the general-purpose domain. As more embedded applications utilize networks they become more susceptible to such attacks, a problem particularly for CPS applications due to their increasing network connectivity.

One critical observation for this work lies in how embedded real-time systems are designed today. Their unique requirements lend themselves well to security methodologies that simply do not apply to general-purpose computing. The key idea of this work is to rely on static analysis of application code that yields detailed timing bounds, which can subsequently be exploited to raise the protection of CPS systems in terms of cyber security.

During system design, timing analysis of embedded real-time tasks provides so-called worst case execution time (WCET) bounds. These bounds lend themselves naturally to security analysis. As WCET safely bounds the upper execution times for specific code sections, execution times above these bounds provide indications of a system compromise due to intrusion. We have designed a technique for embedded real-time systems where general-purpose domain protection may prove ineffective: Techniques such as address-space layout randomization [37] and StackGuard [13], designed for a 64-bit address space, can be defeated more easily in embedded 8/16/32-bit processors with brute-force attacks. Instruction Set Randomization [19] and other hardware enhancements [38, 20] impose high-overhead due to binary rewriting or require additional hardware (with limitations given their static buffer constraints), the cost and overhead of which simply cannot be accommodated in lower-end CPS platforms.

Contributions: This work contributes three mechanisms utilizing both instrumentation of and analysis from within real-time applications to detect timing perturbations resulting from the execution of unauthorized code. The approaches are demonstrated to be effective both under simulation and on a hardware platform. Using timing metrics and comparing them with worst-case bounds allows the detection of security breaches due to system intrusion. In addition, prior to an actual deadline miss, one can detect that an application is about to exceed its timing requirements, which allows one to still trigger appropriate actions in a timely manner before the deadline. The three mechanisms are:

1. We first introduce T-Rex, which utilizes timing bounds to detect intrusion at a fine-grained level through instrumentation of return paths. This method allows the detection of code injections due to smallest timing dilations, i.e., depending on system parameters as small as 5-22 cycles.
2. The second method, T-ProT, validates intra-task checkpoints via synchronous scheduler invocations to uncover coarser-grain injections between 9 and 5k cycles.
3. The third approach, T-AxT, exploits asynchronous scheduler-triggered timing validations of application code sections. It does so without requiring the application code to be instrumented.

These security checks can be strategically scheduled to utilize otherwise idle time in the schedule. By offering different levels of granularity through these schemes, sufficient time is given to transition to a fail-safe state after intrusion detection. If properly designed, evasive actions can still be accommodated within real-time deadlines.

2 Attack Model and Scenario

In this section, we discuss the attack and adversary models that are the premise for our contributions. We then demonstrate a sample attack under these constraints.

There are a number of scenarios for attacks on embedded systems with or without real-time constraints. Past security work predominantly focused on wireless networks in the domain of embedded systems, such as [45]. Models range from passive packet sniffing to various active attacks, such as network traffic disruption (*e.g.*, jamming, spoofing) and packet data tampering/rewriting. Our approach complements network-centric protection with application-level intrusion detection.

We assume that one or more network nodes have been compromised or an attacker has successfully authenticated a node under our adversary model. Node authentication may provide adversaries with control to the local (wired, wireless or ad-hoc) network. Such nodes can be embedded or general-purpose systems, they may be mobile or stationary. We assume that hardware parameters are not modified during an attack, *i.e.*, memory latencies and processor frequencies are not modified by the initial attack code. In contrast to network-level security, we take an application-centric approach for protection. While past work has focused on the application-layer network interface for providing protection [48, 49, 47], we focus on application-intrinsic protection, which does not compete but rather complements the above schemes. This is based on the premise that attacks originate from applications before the operating system is compromised. Our work focuses on early intrusion detection at the application level before other system or hardware parameters can be manipulated, *i.e.*, on the detection of intrusion on uncompromised nodes *via* code injection. Data injection attacks are beyond the scope of this work. We assume that the user data space is unsafe (partially or fully compromised) at the time of detection but the operating system space is still trusted as it has not been penetrated (yet).

In this work, we seek to protect embedded control software by enhancing it with sanity checks to uncover execution of unauthorized code in addition to regular application code. Consider the example in Figure 1 that obtains input data (via fscanf) from an array of input sensors (*e.g.*, temperatures) that are aggregated and later analyzed to drive feedback-control of an actuator valve. In our attack scenario, a network packet supplies the sensor data from a spoofed or compromised node, which we implemented on a MIPS ISA platform.

```
void Sum() {
    char localcpy[MAXSIZE];
    fscanf(input,"%s\n",&localcpy);
    for (i = 0; i < MAXSIZE; i++) {
        // Search for data, increment counter, ...
    }
    // Checkpoint 1 instr. in assembly
}
void read_data() {
    input = fopen("SomeNetworkDevice","r+");
    Sum();
    // Checkpoint 2 instr. in assembly
}
```

Fig. 1 Sample Code Vulnerability

A buffer overflow is caused by supplying an initial input string that exceeds the bound of the localcpy array. It overwrites both frame pointer and return address. When returning from the function after the loop, control is subsequently transferred to the first instruction in the Sum function (see Figure 2). Upon the second execution of Sum, a second input corrects both frame pointer and return address to resume execution as normal. Without ever causing a program fault, this attack results in 2 × `MAXSIZE` aggregations of legitimate sensor data within thresholds, yet the result would be averaged incorrectly over just `MAXSIZE` elements (code omitted). This may lead to an incorrect overall value that would usually go undetected.

In embedded systems, general-purpose and network-level protection methods are insufficient for such attacks for a number of reasons.

1. While this attack exploited a common library routine to trigger a buffer overflow, constraining analysis to a subset of vulnerable routines is insufficient in

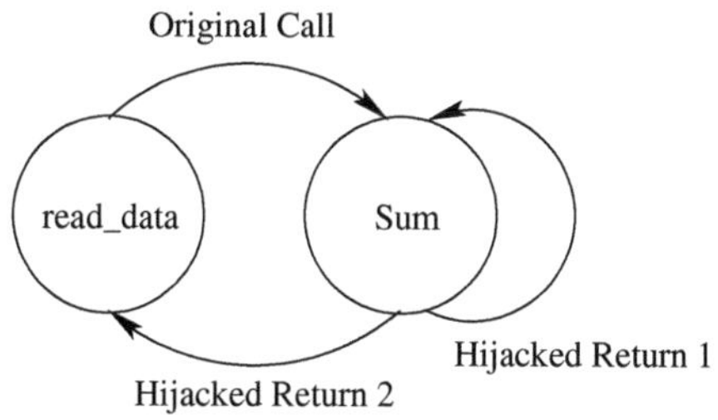

Fig. 2 Diverted Control Flow

embedded systems where custom hardware devices expose non-standard input routines beyond POSIX library routines that may have exploits.
2. Statistical detection methods [22] can be defeated in such a scenario by adaptively changing sensory input over time, which requires multiple repetitions of attacks if they can be detected at all.
3. Signature-based methods can be defeated through spoofing as embedded systems have limited computational capabilities that allow only symmetric signatures/encryption to be employed. Stronger public/private key pair signatures or encryption typically cannot be accommodated in given utilization bounds of lower-end embedded real-time systems [41].

Since our focus is on real-time systems, we follow an approach that differs significantly. In real-time systems, statically analyzable timing bounds are calculated at multiple granularity levels. We exploit time-bound checking as means to detect intrusions. For the attack in Figure 1, the time from the initially diverted return to the second return from Sum accounts for 14K additional cycles on the MIPS ISA. We have developed a number of application-centric techniques that can detect timing dilations as small as 5-22 cycles. With only minimal runtime overhead in the order of 1% of the application's execution time, the above intrusion was instantly detected. Our method detects not only this injection attack but also a variety of others. The approach is orthogonal to methods that protect against other attacks, such as data injection, timing, and denial of service attacks. Each of these attacks may require separate approaches for prevention or detection, *i. e.*, it is not realistic to expect a *single* method to secure against all of types of adversary approaches. Overall, time-based security can *complement* other security mechanisms. While it does not categorically prevent all attacks, it will raise the bar for code injection attacks.

3 Establishing Execution Time Bounds

In hard real-time systems, a priori determination of execution time bounds is a strict requirement. After all, a missed deadline may render the entire application incorrect. Timing analysis determines an application's best-case and worst-case execution time bound (BCET and WCET). This allows verification if a task's deadline can always be met. Timing analysis can be performed via dynamic [8, 42], static techniques [44, 31] or hybrids of them [5, 30, 43].

Dynamic timing analysis determines the effect of different inputs on execution time to approximate the WCET, *e.g.*, to determine that an inversely sorted list maximizes bubblesort's computational complexity. Static analysis bounds aggregate costs of instructions in blocks and then compound the costs of paths throughout the program taking architectural timing effects into account to derive a safe WCET bound at compile time. Static timing analysis has been shown to provide *safe* WCET bounds [42], much in contrast to the dynamic approach.

There are two reasons for deficiencies of dynamic analysis. (1) Due to explosion of the input space for just moderately complex software, it quickly becomes infeasible to determine worst-case inputs or exhaust all inputs during testing. (2) Even if worst-case inputs were known, hardware complexity no longer guarantees that worst-case timing occurs for the algorithmic worst-case input but may rather occur on other inputs, e.g., cache misses or branch mispredictions.

In this work, we utilize WCET bounds obtained from static timing analysis. While the objective of traditional timing analysis is to determine WCET bounds along the *longest* execution path, our work capitalizes on the ability to exploit timing results along *arbitrary* paths. Our work relies on WCET bounds for such paths but for *security* reasons and not for schedulability. We utilize the tool chain [18, 34, 32] depicted in Figure 3 to conduct our study. This enables us to accurately gauge the WCET bounds of an application (macro view) as well as small groups of instructions (micro view). A compiler translates the application to annotated Portable Instruction Set Architecture (PISA) assembly, which is a MIPS-like ISA [10]. This intermediate code along with loop bounds information is then fed into a control-flow analysis tool. Subsequently, control-flow analysis and static cache analysis are performed. The respective outputs are then consumed by a timing analyzer. It derives safe WCET bounds based on the annotated assembly and loop bounds.

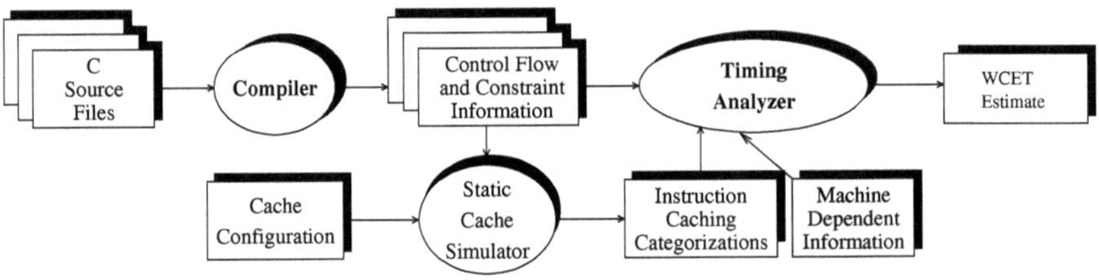

Fig. 3 Timing Analysis Tools

To support real-time security, we modified the timing analysis toolset in Figure 3. The original toolset provided timing feedback at the functional and loop level. We enhanced the toolset to supply timing bounds for a series of smaller ranges during the same analysis run including aggregate values of WCET bounds for sequential instructions plus the cost of branch mispredictions. The resulting bounds are tight and enable us to determine, within a reasonable margin, if a security breach has occurred, *e.g.*, through code injection.

4 Time-Based Intrusion Detection

There may be a variety of motivations for attackers to intrude systems, ranging from changing data for personal benefit to causing potentially catastrophic damage to the CPS environment, *e.g.*, to overload a power transformer by changing safety bounds data resulting in irreversible physical damage requiring components to be replaced (e.g., costly substation transformers). The common idea of our approach is not to prevent but rather detect intrusions, namely by verifying timing bounds at checkpoints during application execution. Our approach is generalized by a common methodology and systematic placement of checks within multiple system components as described in the following. We distinguish two checkpoint placement strategies, one that instruments the application and one where the real-time scheduler triggers checks called T-AxT. For application-side checkpoints, we promote what we term *macro* and *micro* checks of timing bounds. T-ProT competes with scheduler-triggered T-AxT checking at the macro level while T-Rex complements the other two schemes at the micro level.

Checkpoints are realized as synchronous system calls for application instrumentation or reside in the scheduler at preemptions. It is necessary to use system calls because user space provides insufficient data protection. Thus, we are using the real-time operating system as our trusted computing base. Critical security data, such as timing bounds, reside in a different address space than application code to decrease their vulnerability due to tampering.

Overall, the primary goal of this work is to design and assess methodologies that provide real-time CPS applications with an intrusion detection security mechanism. We next present several novel methods that work independently of one another or in a concerted fashion to provide elevated levels of protection within CPS applications.

4.1 Timed Return Execution (T-Rex)

Our first method, T-Rex, employs application-level checkpoints to detect code injections resulting in buffer overflow attacks. Typically, such attacks overwrite the return address of a routine whose frames are stored on the stack. Upon return from a function, control is transferred to the location indicated by the overwritten return address. Attackers often divert execution to hand-written instructions intentionally placed in global/stack variables, or they may spawn new programs. T-Rex detects the former while T-AxT (see below) addresses the latter.

Our T-Rex method employs a pair of checkpoints to compare WCET timing bounds with actually elapsed wall-clock time along a return (from subroutine) path. Figure 4 depicts this scenario.

The first checkpoint sets a timer equal to the WCET, and the second checkpoint cancels this timer. Failure to cancel this timer (due to time overrun) would result in an interrupt indicating a compromised system. Notice that T-Rex is equally applicable to arbitrary control transfers, such as function pointers or large conditional switch/case statements resulting in indirect jumps.

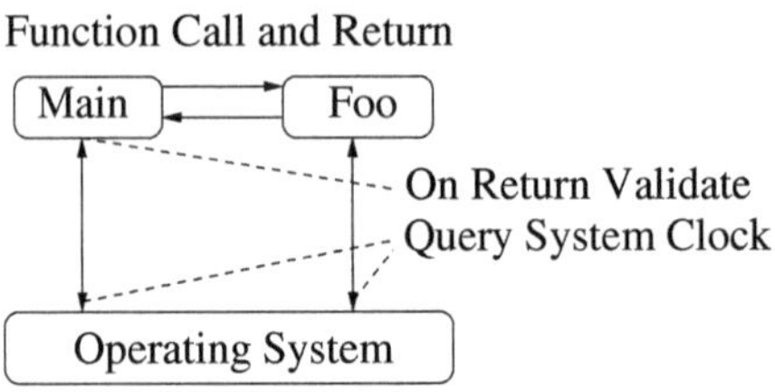

Fig. 4 Timed Return Execution (T-Rex)

In general, T-Rex stipulated that if the dynamically observed wall-clock delta between checkpoints exceeds the WCET bound then an excess amount of instructions must have been executed. Such a bound violation provides an indication of a security breach. In contrast to coarser code sections with conditional control flow, static timing analysis on these straight-line execution regions yields tight WCET bounds. Return-from-subroutine code comprises a series of loads and stores to restore prior processor state and unwind the stack. Figure 4 depicts the communication structure of this method. It shows the application interfacing with the system twice to obtain values from the system clocks before checking the time-stamp delta against WCET bounds. A region that exceeds the path-based WCET bound may not necessarily cause the entire program to exceed its overall WCET bound. This is due to conditional execution where shorter paths may be taken during the remaining of execution, which compensates for the injection overhead.

As such, T-Rex is well suited for detecting attacks that could not easily be detected at task-level granularity due to deadline misses. This is because violation of micro-path WCET bounds is a necessary but not a sufficient condition for violation of a task's deadline or WCET bound.

The design of T-Rex integrates a state machine into the operating system. T-Rex requires the use of two separate calls whose order is tracked. In the motivating example, the attack would cause the timer initiated at the first checkpoint to never be canceled as the second checkpoint is skipped. A potential system intrusion is indicated by the corresponding timeout interrupt.

We also check the addresses of the checkpoint to insure that they fall within the address range of instructions as part of the state machine. Thus, any attack would have to return back to the application code to shut off the timer using the second checkpoint. For tight WCET bounds, even the simple code from the attack to jump to the second checkpoint would be detected. An attacker could potentially disrupt the control flow of the application by jumping to a non-corresponding second checkpoint if slack was available. However, such illegal control flow transitions would be detected with the T-ProT technique described in the following section.

4.2 *Timed Progress Tracking (T-ProT)*

Our second mechanism, *Timed Progress Tracking* (T-ProT), is depicted in Figure 5. T-ProT utilizes synchronous calls at security checkpoints to the scheduler and validates WCET bounds of longer code sections than T-Rex. The scheduler assumes the job of checking these bounds against actual elapsed time to provide separation between protected application and corresponding timing data as the latter resides within the operating system, *i.e.*, at a higher privilege level and in an address domain disjoint from the application's domain. Hence, our timed security does not rely on data / knowledge embedded within an application. Since such data can potentially be compromised, separation is a critical design decision.

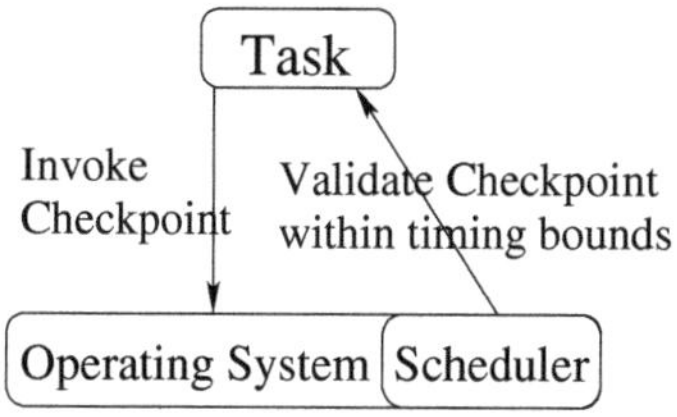

Fig. 5 Timed Progress Tracking (T-ProT)

Consider a scenario where the program diverts from the expected control flow. T-ProT detects several such intrusion scenarios, such as large sections of application code that are skipped or failure to return control to the base application, *e.g.*, by replacing the executable of a real-time task (through "exec" system calls). Upon encountering a timing checkpoint, instrumentation forces a synchronous scheduler call. The scheduler subsequently checks timing bounds for the code section between the previous and this checkpoint. It then activates a timeout equal to the WCET distance until the next checkpoint. An intrusion is flagged if no checkpoint is encountered before this timer elapses, i.e., when the respective timer interrupt is triggered (instead of being canceled when encountering the next checkpoint). Assuming that the application was not aborted prematurely due to an attack, we ensure that these checkpoints are always traversed when a job completes or its deadline expires. This is ensured by placing checkpoints in control-flow blocks guaranteed to be traversed during execution (*e.g.*, using post-dominator information [1]).

We controls the sensitivity of protection by determining the instrumentation points (checkpoints). In some algorithms, the best-case execution time may deviate significantly from the worst-case execution time. For instance, the insertion sort algorithm has a best-/worst-case complexities of $O(n)$ and $O(n^2)$, respectively. The difference between these bounds provides a substantial margin to orchestrate code injection. To overcome this problem, checkpoints need to be inserted such that time distribution is divided in a (uniform) manner to minimize the time between two consecutive checkpoints. An example of this would be checkpoints within the loops of the insertion sort that fire every k iterations. Here, the choice of k determines the strength in protection. We assure that there is sufficient slack in the task schedule to accommodate the timing checks, where scheduler invocations provide a call-back interface to trigger these checks. This also meshes well with code obfuscation techniques employing multi-version binaries: One can instrument at disjoint points for otherwise functionally equivalent binaries of the same application. Any attempt to systematically defeat our timed security approach becomes increasingly more difficulty for attackers by doing so.

A combined approach that uses these two methods bears additional benefits. By themselves, each approach can detect certain types of attacks. In combination, they become far stronger. T-Rex provides more fine-grained views of the internals of application timings thus allowing for targeted detection thresholds for code sections. However, this fine-grained approach has the shortcoming that detection is constrained to localized code sections. An attack may remain undetected by method one if the compromised code never returns to the original application code at all or only returns to locations that bypass these checks. In another approach to counter detection, WCET bounds stored within the application could be tampered with. In such cases, checks would fail to indicate bounds violations. This is precisely where T-ProT complements T-Rex.

T-ProT provides an "outside-of-the-application" mechanism to ensure that specific security checks, placed strategically on the critical path of the application, are actually executed. These checks occur either when a job has completed or when a deadline expires, whichever happens first.

Checks allow the operating system to determine if injected attack code caused a job to bypass our checks or if a return never reverted back to the job's code at all. T-ProT, though operating within looser timing bounds at the macro level, uses timing data in a safer manner. It is protected from application-side buffer overflows because the data is stored inside the operating system scheduler, an address space in a different protection domain than that of the application. In combining the benefit of the two methods into a single system, we enable a more secure real-time environment suitable for the CPS domain.

4.3 *Timed Address Execution Tracking (T-AxT)*

Our approaches so far, T-Rex and T-ProT, both require application instrumentation for checkpoint placement. An attacker could exploit this fact through application-specific checkpoint bypass techniques, even though such bypasses are non-trivial to construct within given timeout bounds. To overcome this weakness, we designed T-AxT as an asynchronous checkpoint technique coexisting with unmodified application code. T-AxT exclusively utilizes the scheduler and timing bounds information provided at system start to maintain timed security. In T-AxT, the scheduler preempts the application upon timeouts. It then probes the PC value of the preempted application and compares execution progress to WCET bounds associated with the code section between the previous and current PC values of consecutive preemptions.

As T-AxT operates without synchronous calls, it presents an alternative to T-ProT. But with this technique, bounding the WCET of loops presents a challenge. As PC values are agnostic towards the progress of loops, the current iteration point within nests of loops needs to be known. We probe actual values of induction variables whose locations (registers/memory) are obtained via static analysis (offline, prior to system start). The scheduler dynamically evaluates polynomial functions parametrized by actual iteration points to determine if the WCET bound of a code section has been exceeded. Such sections may span multiple loop nests and iterations. Codes are systematically supplemented during static code analysis with an induction variable should any loops lack induction variables altogether.

We determine the WCET comparison bounds in either absolute or relative time in our experiments. We utilize WCET bounds *relative* to task activation when multiple execution paths exist. This allows us to eliminate path-aggregate over-estimations of WCET bounds due to conservative static timing analysis. In contrast, we utilize absolute WCET bounds for sequential straight-line code for finer granularity of timings. This duality is tailored to tighten WCET bounds checks in loops since scheduler preemption tends to occur in hot code regions, *i.e.*, predominantly within loop execution.

In practice, we mostly rely on checks of WCET bounds between two checkpoints at the highest nesting level. This interaction is depicted in Figure 6. The first check in the loop is calculated as an absolute checkpoint since no previous checkpoints exist. The second checkpoint is measured as a delta from the previous checkpoint. This strengthens timed security as a means of intrusion detection as bounds are tightened by this method.

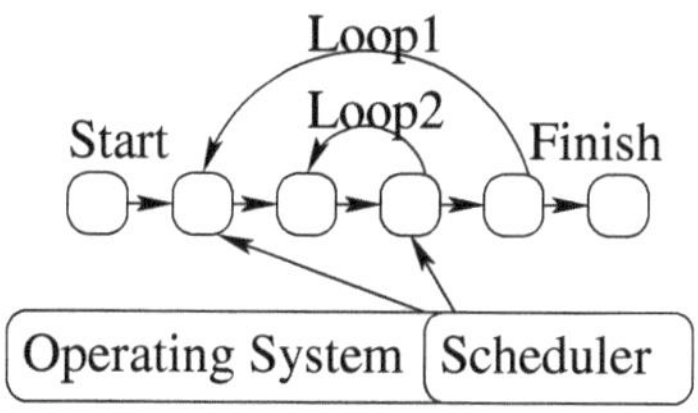

Fig. 6 Timed Address Execution Tracking (T-AxT)

Since we utilize application instrumentation for two of the timed security techniques, the overall real-time task set has to be reanalyzed after instrumentation. This ensures that WCET bounds include the instrumentation code. Timing checks by the scheduler have to be accounted for as well before the real-time schedulability is reassessed. To avoid that such overheads becomes excessive, which might render task sets infeasible in terms of real-time scheduling, checkpoints are selected based on profiled frequencies that are representative task executions in our experiments. Any detected timing bounds violation indicating intrusion further needs to result in evasive actions, such as transitioning to fail-safe states, *e.g.*, through a mode change that replaces all existing tasks with a new task set governing a shut-down sequence and network isolation. Such evasive actions are beyond the scope of this work, i.e., the focus of this work is on time-based intrusion detection.

Summary: Table 1 presents a high-level comparison between our novel techniques. T-Rex protects against buffer overflows commonly exercised on the return path of function calls, which requires fine-grained, cycle-level checks in conjunction with tight bounds on this return path. The overhead of such checks can be high if functions are called very frequently in tight loops, but could be lower when code is inlined instead of calling functions in these loops. Since protected code sections are sequential and bounds are tight, virtually all buffer overflows can be detected, and no source code changes are required. In contrast, T-Prot requires source code changes to insert checkpoint calls in between which code sections are timed at a medium grain. Due to variable loop bounds and conditionals in these code sections, bounds are moderately tight, and so is the overhead assuming that calls are inserted judiciously. T-AxT has comparable, if not lower, overhead than T-Prot. It tracks progress using the program counter and loop invariants, which allows coarser bounds checking, yet without requiring source code changes as checks are integrated into the scheduler at preemption points with low overhead.

Table 1 Comparison of Intrusion Detection Techniques

Property	T-Rex	T-Prot	T-AxT
Timing Method	return path	add checkpoints	at scheduling points
Progress Tracking	cycles/instr.	time of a task	inspect loop counters
Granularity	fine: instructions	medium: blocks	coarse: interrupts
Bounds Tightness	very tight	moderate	loose
Cost/Overhead	high	moderate	low
Intrusion Detection	very strong	moderate/high	moderate/lower
Source code changes	none	insert calls	none

5 Implementation

We implemented the mechanisms of T-Rex, T-ProT and T-AxT in two different experimental frameworks. The first one that combines static timing analysis with architectural simulation yields simulation results in experiments detailed later. The second realizes dynamic timing analysis on a concrete embedded system hardware platform, where we subsequently obtain runtime results. We tested both of our implementations using a set of C-Lab benchmarks [11].

Table 2 C-Lab Benchmarks

C Benchmark	Function
adpcm	Adaptive Differential Pulse Code Modulation
cnt	Sum and count of positive and negative numbers in an array
lms	An LMS adaptive signal enhancement
srt	Bubble Sort
fft	Fast Fourier Transform

5.1 Simulation Framework

Figure 7 depicts the overall experimental framework. We enhanced a static analysis framework as discussed in Section 3 to support check-pointing instructions. These check-pointing instructions allow us to determine the worst case cycle time at which a particular instruction finishes execution. This information is essential to determining the WCETs between two consecutive checkpoints under T-ProT. We further utilize a custom SimpleScalar processor simulator [9] enhanced to support multitasking and scheduler threads / tasks, which we exploited to implement earliest deadline first (EDF) scheduling [31]. The instruction set architecture for this simulator is PISA. This matches the input assembly utilized by our timing analysis tools. For the purpose of this work, we assess benchmark results in SimpleScalar that match the configurations of the static timing analysis tools.

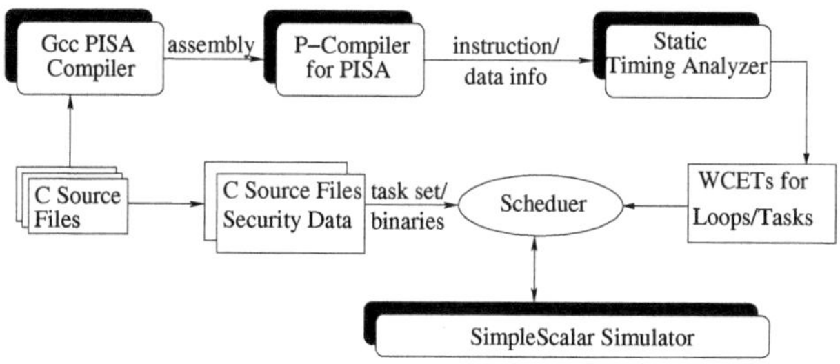

Fig. 7 Framework

As discussed before, these configurations provide a lower bound on the amount of code injection that may remain undetected. If we were to relax our configuration constraints, WCET bounds obtained by static analysis would become less tight implying that an attacker could potentially execute more instructions prior to being detected. To assess this trade-off, we also deployed T-Rex and T-ProT on a concrete hardware platform (see below).

The scheduler in the SimpleScalar framework supports multiple preemptive and non-preemptive scheduling algorithms. For the course of this work, we used a preemptive EDF schedule to most accurately show the side effect of our mechanisms on the scheduler itself. The scheduler is customized to support relative time for each thread aggregated during preemptions and at security checks of a task to most accurately track execution progress.

The cache configuration for both the static cache simulator and the timing analyzer were configured without data caches and with perfect instruction caches, *i.e.*, with an I-cache capacity exceeding that of our largest program sizes so that we only had to account for cold misses. The choice of the cache configuration parameters was intentional as our objective here was to assess a bound on detectable code injections. In other words, given the tightest possible timings on application code, we wanted to determine the largest number of cycles that would remain undetected by our security-enhancing mechanisms. For such a metric, the smaller this threshold, then stronger the protection will be by our mechanisms.

For T-Rex, SimpleScalar enhancements include two system calls to query timing information (a) before a return from a function / method, and (b) at the destination of a function / method return and compare the difference to static bounds. We utilize a timer and also verify correct sequential ordering of these calls. If call one was issued without the other, a control-flow violation (intrusion) is detected, that would result from a buffer overflow attack that returns control flow past the second call. Subsequently, a system-defined action, such as transitioning into a fail-safe state, can be initiated. In effect, the imposed call ordering represents a security side-check that provides the means to detect certain attacks missed if only execution cycles were checked. For example, if an attacker were to execute injected code and then transfer control to the instructions past our second system call in an attempt to bypass our imposed security, the absence of the second system call would be detected at the next return from a function when another instance of the first system call is issued. Call sites are identified by their call stack / PC and frame pointer signature so that calls from injected attack code are easily identified.

5.2 Embedded Hardware Framework

We also experimented with an actual embedded hardware platform, namely the DSK6713 kit from Spectrum Digital. These experiments combine dynamic timing analysis with implementations of T-Rex and T-ProT. The experimental board has a Texas Instruments C6 (TMS320C6713) DSP chip running at 150MHz featuring a 32-bit processor with Very Long Instruction Word (VLIW) architecture, eight independent functional units that can execute up to eight instructions per cycle, fixed and floating point arithmetic, 2 levels of caching and up to 256KB of on-chip SRAM programmed under Code Composer Studio v3.1. All programs were written in C and assembly.

This board is also utilized in a CPS project for controlling power devices (solid state transformers) in a renewable energy project (solar and wind power generation in microgrids). There, the TI DSP controls silicon-based solid state transformers during the DC/DC conversion from low to high voltage levels. These transformers represent the link between micro-grids and the regional power grid backbone. The project focuses on controlling renewable energy sources locally and feeding their power into the regional grid without disruptions. Due to the decentralized nature of micro-grids, software security is deemed critical in power grids as malicious attacks could potentially damage equipment upstream. Such damage would impact at a minimum entire suburbs and require manual maintenance. Hence, our work and the choice of platform are very much motivated by a concrete CPS scenario.

The effectiveness of our mechanisms depends on how accurately we can determine the WCET bounds and how tight these bounds are relative to average execution times. The objective of this study is to assess the lower bounds on tightness. In the experiments on the embedded platform, WCET bounds are determined by dynamically timing execution paths under worst-case scenarios while running the program on a cycle-accurate simulator from Texas Instruments that simulates the C6713 processor along with its on-chip peripherals. Executing the actual code segment repeatedly on this simulator using worst-case inputs and hardware settings provides the observed maximum number of CPU clock cycles for a given code segment. We then convert these dynamically determined WCET cycles into microseconds by considering the CPU clock speed. In addition, we tried to reduce the effect of any factors that adversely influence tightness of WCET bounds.

The following is a list of such factors on the given hardware platform configured for maximum predictability:

Caches: The TMS320C6713 has separate L1 instruction/data caches and a unified L2 cache. We chose to disable all caches resulting in tight WCET bounds relative to average timings. Enabling caches would considerably alter the WCET bounds to deviate more significantly from their average case, yet still preserve the safety and validity of upper WCET bounds.

SRAM *vs.* DRAM: In our experiments, the program code and data reside in static, non-volatile memory (SRAM), *i.e.*, we do not utilize dynamic, volatile memory (DRAM) at all. The TMS320C6713 processor has 256KB of on-chip SRAM. If, in contrast, DRAM were used, we would need to account for periodic self-refresh cycles. The DRAM controller refreshes row data in different banks of the DRAM in a row-cyclic manner. This issue is common to all embedded platforms utilizing dynamically buffered memory and refresh delays are known to present a challenge in real-time systems.

During such self-refresh cycles that last for a few microseconds, the CPU bus remains busy. Any attempt to read from the DRAM or other external devices would then stall the processor as long as the self-refresh cycles are in progress. These self-refresh cycles are asynchronous events as far as program execution is concerned and completely transparent. They would thus affect the timing calculations used in T-Rex. However, strategies exists for exactly measuring the duration of DRAM self-refresh cycles [3] and to treat DRAM refresh as a higher priority task [6, 7]. Since the refresh overhead challenge is orthogonal to our work and our aim was to assess how tight WCET bounds could become, we decided to eliminate these overheads in experiments by avoiding DRAM altogether and exploiting SRAM instead.

Compiler-Generated Runtime Overhead: In our current experiments we coded all tests and runtime / operating system code in C to reduce the amount of runtime overhead added by the compiler. Hence, instructions between the first and second instrumentation points around a function return of T-Rex are limited to stack unwinding operations and register restores. An object-oriented language, such as C++, would further add destructor overhead for objects locally declared within the method. Since destructors are user defined, providing tight WCET bounds for them presents a challenge.

Our implementation features a layered system architecture depicted in Figure 8. We ported a commonly used real-time operating system, MicroC OS II [21], which supports fixed-priority preemptive scheduling. We then implemented a scheduler based on rate-monotonic analysis (RMA) [25] on top of MicroC OS II. This scheduler supports threads of arbitrary periods imposing strict execution time control.

Failure to complete by a deadline results in preemption and rescheduling during the next period. Most hard real-time systems use similar schedulers in order to guarantee deadline constraints on periodic tasks. We also provide synchronous application checkpoint calls for implementing T-ProT and monitoring of aggregate execution time per task with a one microsecond precision, but we exclude the time spent inside interrupt service routines and scheduler overheads (due to the complexity of measuring these).

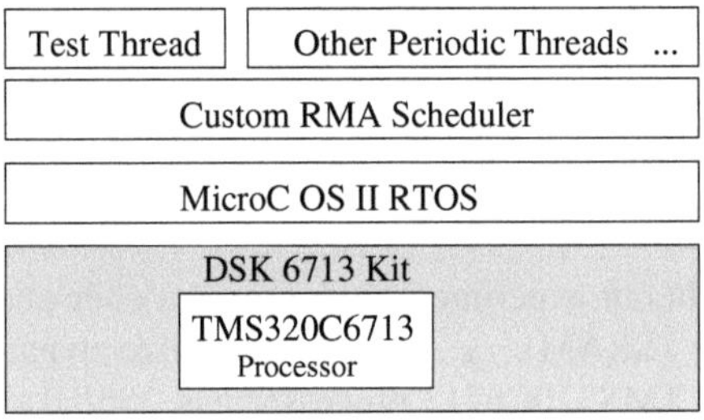

Fig. 8 System Architecture

6 Experiments

We first report the results of our simulation environment before discussing measurements obtained on the embedded hardware platform.

6.1 Common Attack Cycles

In the following, we first consider common shell codes used on Linux systems to determine typical attack scenarios. This is necessary since timing values of actual attacks for embedded systems are sparse in literature, at best. Metasploit, a repository for shell attacks, contains approximately 35 different Linux/Unix shell code examples of the same fundamental structure. A jump in the first line of the shell code transfers to another location within the shell code. This aids in determining the relative offset for addressing. An "exec" system call then invokes a command of the attacker's choice. The most common examples found on Metasploit are useradd, shell, and tcp open directives.

Figure 9 provides measured timing values for common portions of attack code. We measure the average cost of execution from the hijacked return to the first instruction in the shell code ("Start") and the average time of an execution system call ("Execpl") with null arguments. If actual values are passed, measurements are significantly larger. *E.g.,* passing "Chmod", a common attack to modify file permissions, dramatically increases the cycle overhead. The motivation here is to consider the effectiveness of our methods, and these examples of common shell code attacks provide realistic timings to this end.

Location	Cycles
Start	90
Execpl	2,800
Chmod	5,151,720

Fig. 9 Shell Code Timings

		No Caches		4KB I-Cache	
Program	**Function**	**WCET**	**Sensit.**	**WCET**	**Sensit.**
SRT	Initialize	39	5	21	13
SRT	BubbleSort	39	5	30	13
LMS	LMS	39	5	30	9
FFT	FFT	39	5	25	8
ADPCM	Encode	39	5	30	22
ADPCM	Decode	39	5	30	22

Fig. 10 T-Rex WCET and Sensitivity cycles

6.2 Simulation Experiments

In our implementation, T-Rex utilizes an *absolute* task timer to determine the total time since the simulation start. T-ProT and T-AxT are exercised in a modified preemptive real-time scheduler under the SimpleScalar environment developed elsewhere [31] to keep an *aggregate* timer for each of the executing jobs. This aggregate timer is compared against WCET bounds from static timing analysis. It is further saved in the scheduler-maintained thread control block at preemption and restored at reactivation. The value is reset at thread / task completion to prepare for the execution of the task's next periodic job.

Timed Return Execution (T-Rex) Results

The attack outlined in Figure 1 was successfully detected by T-Rex as a buffer overflow since the injected code accounts for 14k cycles, which far exceeds its detection granularity of 5-22 cycles. Under legitimate sensor inputs, the sample program produces the correct output with an additional 40 cycles relative to the application itself. Figure 10 shows the sensitivity results of T-Rex for varying benchmarks and their respective functions. In this experiment, the attack code, after executing its injected code, returns to the exact spot in the code that the original return for a call would have jumped to. The table then reports the WCET in cycles for the return sequence as reported by timing analysis (WCET in column 3) and the number of slack cycles that would remain undetected (sensitivity in column 4), first without considering caches and in next two columns with a 4KB instruction cache.

The slack amounts to the difference between WCET and actual execution time, the latter of which is observed from SimpleScalar simulation. The WCET bound is extremely tight since T-Rex assesses time on a straight-line path of the control flow. Hence, the window of vulnerability is restricted to a sensitivity of 5 cycles without and 8-22 with caches. If an attack was to go undetected, it would have to be constrained to such a small amount of code as an injection. These results provide a lower bound. The upper bound for undetectable injections is given by the T-ProT or T-AxT methods, which address larger injections and omission of code sections in favor of injected code. However, it would be non-trivial to disguising the side effects of polluting stacks and registers.

The timing bounds and subsequent security checks for straight-line code are very precise as results in Figure 10 illustrate. Instruction cache effects loosen these bounds proportionally to the cache miss penalty of 10 cycles (as seen for ADPCM). Overall, this leaves little room for injected code to go undetected.

Timed Progress Tracking (T-ProT) Results

T-ProT relies on synchronous scheduler checkpoints to dynamically detect intrusions by WCET bounds violations. Its effectiveness is assessed by the results in Table 3, which reports checkpoints between adjacent instrumentation points in the control flow for each application. For example, checkpoint 0-1 denotes execution from entry of main() to a later basic block in CNT, 2-3 and 3-4 denote loop entry and exit, respectively, while 3-2 denotes a back-edge within the outer and inner loops, respectively (see Figure 11). For these code sections, Corresponding WCET bounds (column 3) and sensitivities (column 4) are reported in cycles.

Table 3 T-ProT WCET and Sensitivity cycles

		No Caches		4KB I-Cache	
Program	**Checkpoint**	**WCET**	**Sensit.**	**WCET**	**Sensit.**
LMS	0 - 1	1,500	44	844	173
LMS	1 - 2	5975	65	3279	774
LMS	2 - 2	17199	259	8699	2120
LMS	2 - 3	11330	210	5549	1430
FFT	0 - 1	1,600	195	846	228
FFT	1 - 2	950	54	697	220
FFT	2 - 2	19,283	2,787	13,955	5,334
FFT	2 - 3	12,709	1,997	9,451	3,831
FFT	3 - 3	5,084	460	3,150	659
FFT	3 - 4	208	48	120	49
CNT	0 - 1	1814	120	786	147
CNT	1 - 2	69	9	46	14
CNT	2 - 3	14083	283	4341	1493
CNT	3 - 2	13599	239	4199	1481
CNT	3 - 4	13726	266	2760	1534

Several checkpoints were instrumented in benchmarks as illustrated for CNT in Figure 11:

1. immediately after the original variable declarations but prior to the invocation of loop 1;
2. within the outer loop just prior to the inner loop invocation;
3. in the inner loop with logic surrounding it to only perform the check during half way through the total iterations of the inner loop; and
4. in the final block of the application just prior to exiting.

T-ProT has a coarser granularity for the reported bounds on undetectable injections as indicated by the results in Table 3. These bounds, while smaller in some case, range up to nearly 5k cycles on the upper end. Hence, scheduler callbacks result in less sensitivity than return path instrumentation. The more complex control flow (than just straight-line code as in T-Rex) causes this lower sensitivity.

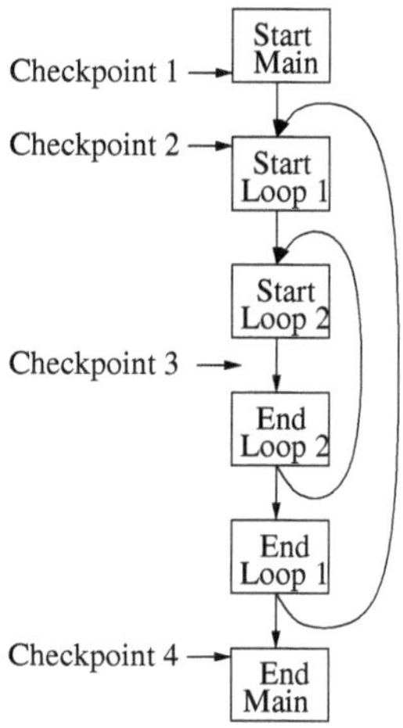

Fig. 11 CNT Control Flow

Checkpoints are scattered throughout the application as they may cross loop levels, as indicated by Table 4. This reduces the tightness of WCET bounds. WCET bounds of a loop iteration are generally less tight than straight-line code due to fluctuations in the number of iterations or conditionals inside the loop body. To obtain safe worst-case results, we have to conservatively calculate the worst case scenario (upper bound on loop iterations, longer path for conditional execution) in our static analysis.

The utilization of instruction caches, as depicted in the last two columns of Table 3, has an impact on the overestimation. This is due to the fact that relative checkpoints tend to not incur cache misses as most cold misses occur prior to the first checkpoint hit.

These scheduler checks result in strengthened support for security. Moreover, T-ProT is quite versatile in that it may be used to instrument code sections at arbitrary points in the application. This makes T-ProT suitable to detect compromised subroutines in a targeted manner.

There are additional security benefits to using T-ProT. Timing bounds on preemption require a look-up of the previous checkpoint and a comparison of the current timing values with the corresponding WCET bounds. When factored into the application execution, this cost is hardly noticeable and requires only insignificant additional slack in the real-time schedule of the task set at the benefit of *more secure cyber-physical systems* (see Section 8).

Table 4 T-ProT Checkpoint Hits

Program	Total Checkpoints	Total Hits
LMS	3	203
FFT	4	114
CNT	4	132

Timed Address Execution Tracking (T-AxT) Results

The coarsest granularity of our mechanisms is provided by T-AxT. It is also the most difficult to attack directly because it resides within the kernel and is not triggered by checkpoints from tasks. The periodic timer for these results was set at 20k cycles on a 100 MHz processor clock in simulation. This value was chosen to balance overhead, *e.g.*, SRT required 2051 checkpoints during job execution (see Table 5).

The coarser granularity of T-AxT is due to aggregation of conservative bounds during static timing analysis and approximate matching of PC values with WCET bounds. WCET values were associated with the next-smaller blocks of code relative to a PC value to conserve storage overhead for WCET bounds. The LMS benchmark generally retained the highest difference in cycle measurements *vs.* actual time. This is due to the complexity and size of multiple inner loops within LMS. The overestimation of WCET could be decreased using a finer granular configuration but at a larger storage cost. The benefit of T-AxT is its ability to bound the WCET of PC-constrained code sections within or across loops and to verify that the job's execution meets these bounds. For a given code section, bounds violations are a sufficient indication of intrusion.

Table 5 Timed Address Execution Tracking

Program	Period	WCET	Sensit.
CNT	20,000	21,225	1,225
CNT	20,000	28,200	8,200
CNT	20,000	27,750	7,750
CNT	20,000	27,225	7,225
CNT	20,000	26,775	6,775
LMS	20,000	30,991	10,991
LMS	20,000	28,434	8,434
LMS	20,000	33,473	13,473
LMS	20,000	28,918	8,918
LMS	20,000	32,597	12,597
SRT	20,000	23,400	3,400
SRT	20,000	24,128	4,128
SRT	20,000	22,701	2,701
SRT	20,000	22,372	2,372
SRT	20,000	22,701	2,701

6.3 Experiments on an Embedded Hardware Platform

The DSP hardware provides a platform for the next set of experiments, where both T-Rex and T-ProT were implemented. The first experiment features the benchmark ADPCM deployed as a single periodic task. The code of this task is enhanced by T-Rex to provide timed security. The single-task constraint allows us to control the experiment by eliminating additional preemptions between first and second calls that obtain clock values. We determined that the calls themselves add only negligible overhead. We used "assert" statements at checkpoints to check timing bounds. The tested assertion here is given by the comparison of the actual time elapsed since obtaining the first clock value and the expected WCET bound.

Figure 12(a) depicts the output of assertions that were added for trace visualization purposes. The first word in every output line indicates the ADPCM function instrumented, followed by the result of the assertion indicating if it passed or failed. The number before '>' indicates the WCET bound in microseconds for the corresponding function return and the number after '>' indicates the actually measured time for the same in microseconds.

Assertions compare these times with a predetermined WCET bound, which in this case is determined to be about 3.1 μsecs (rounded up conservatively to 4) for all functions using the C6713 device cycle-accurate simulator. The output shows that all timed return path values are within a range of 1-2 μsecs. Hence, all the assertions pass, *i.e.*, no timing violations were detected implying that no intrusion was seen.

scalel: ASSERT PASSED 4 > 1	scalel: ASSERT PASSED 4 > 1
dh: ASSERT PASSED 4 > 2	dh: ASSERT PASSED 4 > 1
uppol2: ASSERTPASSED 4 > 2	uppol2: ASSERT PASSED 4 > 1
uppol1: ASSERT PASSED 4 > 2	uppol1: ASSERTPASSED 4 > 2
encode: ASSERT PASSED 4 > 2	encode: ASSERT FAILED 4 > 16
filtez: ASSERT PASSED 4 > 2	filtez: ASSERT PASSED 4 > 1
filtep: ASSERT PASSED 4 > 1	filtep: ASSERT PASSED 4 > 1

Fig. 12 (a) All Asserts Pass (b) Some Asserts Fail

In the second experiment, calls to a dummy function are issued after obtaining the first clock value but before a return from a function. In other words, we created a code injection scenario. The dummy function simply executes an empty loop (no-op) for 100 iterations before returning to the caller. This simulates code injection that returns to the original control flow without harming stack values, *i.e.*, the only noticeable effect is time dilation. Results of this experiment are depicted in Figure 12(b). As illustrated by the results, code injection through the dummy function resulted in a large deviation in elapsed time between obtaining clock values on the return path. Notice that even ten iterations accounting for 1.4 μsecs would suffice for detection as $2.0 + 1.4 > 3.1$, which gives an attacker little room for devising malicious code. The next experiment features a set of periodic tasks with mixed periodicities (containing smaller and larger periods than ADPCM) to co-exist with the ADPCM task. We further experimented with explicit sleep statements prior to obtaining the first and second clock values in order to force preemptions. As expected, assertions indicated intrusions in all these cases. These results are omitted here, since they resemble those reported in the previous figures.

Finally, T-ProT was implemented on the embedded hardware platform. As before, the WCET bounds between various checkpoints are obtained as the maximum cycle count for executing the program in a loop on the C6713 cycle-accurate simulator under worst-case conditions and inputs plus complete path coverage. This cycle bound is then converted into execution *time* by adjusting for the CPU clock speed before comparing with measured time on the hardware at a checkpoint. Our RMA scheduler provides a built-in mechanism to remember the previous checkpoint and assert the validity of the latest checkpoint. Table 6 shows the calculated WCET bounds and observed runtimes for FFT on the embedded TI DSP hardware platform. All checkpoints pass in this experiment indicating a safe execution in the absence of code injection (columns 2-4).

Table 6 Checkpoints of T-ProT for FFT on TI DSP

	No Injection			Code Injection		
Chkpt. #	**WCET**	**Actual**	**Chkpt**	**WCET**	**Actual**	**Chkpt**
Chkpt 0 - 1	3	2	pass	3	2	pass
Chkpt 1 - 1	5	3	pass	5	3	pass
Chkpt 1 - 2	7	5	pass	7	5	pass
Chkpt 2 - 2	4	3	pass	4	3	pass
Chkpt 2 - 3	3	2	pass	3	16	fail

We next injected code that executes between checkpoints 2 and 3 (depicted in columns 5-7 of Table 6). A small loop is introduced between these two checkpoints to simulate code injection. Results of Table 6 indicate that all tests between checkpoints 2 and 3 fail implying a detected intrusion. Overall, we have shown that our mechanisms facilitate intrusion detection in both preemptive and non-preemptive multi-tasking real-time environments. Thus, CPS applications can universally benefit from these approaches.

7 Trading off Security against Timeliness

The aim of our work is to increase the level of protection against attacks in systems at the cost of executing additional routines that monitor and check the system behavior. In cyber-physical systems with real-time constraints, these instrumentation and time validation checks affect system utilization and thus real-time schedulability.

Our sample attack in Section 2 shows that embedded systems with network connections, such as CPSs, are vulnerable to cyber attacks. Reports in practice reinforce this fact. Most notably, worms have entered monitoring equipment and disabled a safety system at a nuclear power plant [24]. In another incident, a virus reportedly spread past firewalls into the accounting system of the main Australian power company, which did not implement proper physical network separation between accounting and power control subsystems [33]. Further damage was only contained by reconfiguring servers between the two subsystems to prevent the virus from spreading uncontrolled into the power control subsystem.

These are just two examples illustrating the urgency of providing guards against cyber attacks in the CPS realm. Our timed security is one such technique readily deployable to complement existing intrusion detection techniques. The rationale of such deployment is to further strengthen security as a single protection mechanism can often be defeated by itself, yet a set of mechanisms is much harder to circumvent. In practice, the inherent costs of security are well justified. We also observe that many real-time systems provide sufficient slack in a task schedule so that security mechanisms could be accommodated under feasible schedulability. After all, real-time systems only have to ensure timeliness in the sense that deadlines are met. As long as deployed security methods, such as timed security, impose overhead within deadline bounds, correctness is guaranteed.

Conversely, systems with tight slack may limit the level of security that can be realized. Depending on vulnerability and criticality assessment, such networked systems may need to be redesigned for more powerful hardware targets, or a paradigm is needed to provide the ability to selectively augment code with security measures. Selectivity amounts to a tradeoff between availability of slack to meet deadlines and safety and vulnerability considerations of code sections. T-Rex, for instance, increases the execution time of an application due to its inherent instrumentation. This overhead is assessed in the results of Section 8.

Return-path instrumentation results in the invocation of only few checking instances at execution time in many embedded applications since the bulk of the work is performed in loops whose bodies do not contain function calls, thus resulting in negligible timing overhead. In codes containing hot spots in tight inner loops with function calls, in contrast, security checks impose a significant overhead that may easily exceed the available slack. In such cases, application code should be refactored based on transformation techniques such as inlining, single caller function specialization, which avoids allocating a new stack frame in place (commonly performed by the Intel compiler), or reduction of function call frequencies through restructuring. In future work, the balance between such transformations and security overhead of T-Rex to target given slack margins should be studied. Overhead is imposed by synchronous upcalls and timeout preemptions under T-ProT. This results in scheduler activations to subsequently check if the application operates within expected timing bounds. The overhead of the former (upcalls) is more significant than that of the latter (timeouts) as timeouts are only triggered upon an intrusion but otherwise canceled. This method should be used in conjunction with selective placement of checkpoints using strategic and statistical means (*e.g.*, random placement and random activation). Attacks would also become more difficult as random activations strengthen security.

The overhead of T-AxT can easily be controlled through its scheduler activations. Should frequent checks be required, timer interrupts would have to be triggered in shorter intervals adding to the overhead of interrupt service routines. The overall objective is to provide adequate coverage of checkpoints to maximize overall security within the given timing constraints. All methods are designed to allow selective instrumentation, but the details of such placements and their trade-offs are beyond the scope of this chapter.

Overall, we developed three security-enhancing methods based on timing information already inherent to CPS real-time control systems. Their overheads have acceptable costs when properly tuned for providing security without compromising timeliness. By adjusting the frequency of dynamic checks, particularly for less critical sections, one can trade off overheads for an increase in the vulnerability level. The trade-off between overhead and level of security is common in general-purpose computing, yet the implications on timeliness add another equation to this trade-off. Our techniques target real-time CPS where system criticality outweighs performance concerns making security a mandate rather than an option. A future direction of research might investigate the viability of additional security measurements. Some of them are quite feasible, such as exploiting average case execution times for checks on timing outliers. Such methods are probabilistic and may result in large numbers of false positives. More accurate results with lower false positives should be expected based on parametric models of execution time that take actual loop trip counts of dynamic execution into account, both for BCET/WCET bounds and average times [31, 29]. Early warning indicators could be dynamically triggered to activate stringent security checks that bare higher costs or to reduce system functionality in order to limit potential damage to the *physical* side of the CPS application.

8 Instrumentation Overhead

We also designed experiments to assess the cost of instrumentation relative to the performance costs of each of our methods. Table 7 depicts these overheads in percent relative to the application's base execution time without the security methods. We distinguish the "default overhead" and "scaled overhead". The former corresponds to the experiments of Section 6 while under the latter, variations on the frequency of intrusion checks are featured.

Overheads (default) range from 0.22% to 1.54% for three of the four benchmarks under T-Rex. Such overheads are negligible assuming just minimal slack in a real-time task schedule. The higher overhead of 18.71% for ADPCM is due to its modular structure compared to other benchmarks. It consists of several small functions that are called within a loop. Thus, T-Rex checks are invoked more frequently at a deeper nesting level than in other benchmarks. Code restructuring, such as inlining, reduces this overhead to that of the other benchmarks. For example, after inlining calls at the inner-most loop levels for ADPCM, the T-Rex scaled overhead was reduced to just 0.32%, as depicted in the last column of table 7. For the remaining benchmarks, default overheads did not justify any inlining so no scaled overheads are reported for T-Rex. Occasional code restructuring only imposes an insignificant performance cost.

Depending on the application instrumentation frequency overheads for T-ProT vary. The default overhead for the experiments in Section 6 ranges between about 7% and 16%. Such instrumentation with a high level of coverage incurs a sizable performance penalty in performing finer grain security checks. The scaled overheads in last column of Table 7 of about 3%-8% correspond to a reduction in the number of instrumentation checkpoints by half relative to the default method. This is accomplished by selective activation of instrumentation checkpoints but selective placement would be a valid alternative as well.

Tunable performance overhead is provided by T-AxT depending on the frequency of the periodic wake up that initiates the intrusion check. We used a periodic wake up of 20,000 cycles, which provides a reasonably frequent security check at a dynamic overhead comparable to that of T-ProT with a constant default overhead of approximately 16%. The scaled overhead amounts to about 8% for a 40,000 cycle instrumentation period (see last column of the table).

Overall, overhead is observed to scale linearly with instrumentation frequency for all of our techniques. Such scaling is easily controlled (a) for T-AxT through selection of periods, (b) for T-ProT through rate control and (c) for T-Rex through inlining, rate control or a combination of both.

Table 7 Dynamic Performance Overheads

Method #	Benchmark	Default Overheads	Scaled Overheads
T-Rex	SRT	0.22%	N/A
	LMS	1.54%	N/A
	ADPCM	18.71%	0.32%
	FFT	0.021%	N/A
T-ProT	LMS	7.55%	3.68%
	FFT	16.17%	7.92%
	CNT	10.05%	4.92%
T-AxT	LMS	15.89%	7.94%
	SRT	15.89%	7.94%
	CNT	15.89%	7.94%

9 Related Work

Generic security features have been considered in the context of scheduling of real-time application tasks in past work. Often, certain out-of-the-box security mechanisms are applied at the cost of ensuring timeliness while arguing that security is improved [40, 46].

Past work on embedded systems security has focused on sensor networks including remote memory verification and network-related anomaly detection at the packet or application level [36, 48, 49, 47, 45]. Timing analysis is considered in literature as a means to reverse-engineer encryption techniques [35] instead of utilizing it for protection. The emphasis of this work is on utilizing timing analysis bounds to detect code injection attacks [28].

Shao et al. use a hardware/software combination to detect attacks [38], which is closely related to our work. The first technique adds a new stage to the processor pipeline to check on an address before data is written to it. If the value is greater than that of a special register delimiting vulnerable stack regions then write is denied. The second technique uses a new "sjmp" instruction to XOR the write address with the value stored in the special register to assess validity of the jump target. Other approaches rely on hardware buffers to store return addresses [20] when buffer space is available. These techniques do provide security with negligible performance overhead but at the cost of specialized modifications to hardware. Our work does not require special hardware support.

Significant work has been performed in the area of security of general-purpose and server environments in which attacks are more prevalent. These systems are generally much larger and more difficult to impose restrictions on due to their general-purpose nature. Efforts at reducing opportunities for code injection in these environments has resulted in concepts such as canary value placement. Buffer overflow may be detected in general-purpose systems by placing canaries adjacent to the return address on the stack, which may be overwritten in an attack [13]. If a tampered canary is detected prior to transferring control at a return, the program aborts itself. Canaries are typically pseudo-randomized at compile time to increase the difficulty of success during buffer overflow attacks. Thus, simply placing the canary value onto the stack next to the return address, which avoids detection for known canaries, becomes challenging [13]. Yet, even pseudo-randomized canaries can be exploited in systematic repeated attacks.

In general-purpose systems, another protection mechanism employed is to utilize address-space layout randomization (ASLR) [37]. The stack is placed in a hard-to-guess location in the memory. If an attacker attempts to jump to code placed on the stack, it becomes difficult to infer absolute stack addresses where attack code may have been injected. This method is best suited for systems that employ 64-bit addressing spaces, *i.e.*, where ample room for stack placement exists such that repeated brute-force attempts are statistically ineffective. However, such techniques may be circumvented by repeated attacks in a space-constrained embedded real-time system with 8/16/32-bit address spaces [37].

Dan et al. [15] discuss power grid challenges while Mitchell and Chen [27] provide a survey of CPS intrusion detection approaches of which we mention a few but otherwise refer the reader to the survey. Different detection techniques (knowledge vs. behavior) and deployment scenarios (host vs. network) are discussed. Many systems employ behavior-based techniques [23, 2], optionally using domain-specific knowledge [17] and are often targeted at wireless communication [39]. Several approaches follow a model-based approach utilizing varying methods ranging from regular expressions [12] over Petri nets [26] to neural nets [16].

This work extends our prior publication [50] by the following contributions: It contains more detailed explanations of our technical approach, tightness of bounds for detecting intrusions, motivational scenarios, more illustrative discussions of examples, a discussion of deploying hybrids of the proposed methods in a mutually complementing manner, consideration of scheduler interactions, discussions on resorting to fail-safe modes, measures to ensure tight WCET bounds, future early-warning enhancements, and a discussion of future work on cyber security specific to CPS.

10 Conclusion

Our work contributes three novel software methodologies that provide enhanced security in deeply embedded real-time systems, such as Power Grid control devices. We attain elevated security assurance through two levels of instrumentation that enable us to detect anomalies, such as timing dilations exceeding WCET bounds. (1) T-Rex: Tight timing bounds of selected code sections are obtained during static timing analysis at no extra cost during the required schedulability analysis and are subsequently utilized to monitor execution during run-time. Buffer overflow attacks are detected due to exceeded WCET bounds upon return path instrumentation for code injections as small as 5-22 cycles. (2) T-ProT: Application instrumentation issues synchronous scheduler calls to assess timing bounds validity for precisely delimited sections of code. T-ProT by itself uncovers coarser-grain injections between 9 and 5k cycles at controllable overhead and complements T-Rex. (3) T-AxT: Asynchronous scheduler-triggered validations of timing bounds are performed for approximated sections of code, which, compared to T-ProT, obviates application instrumentation, results in low overhead and complements T-Rex. Attacks uncovered by T-AxT alone are consequently the coarsest grained. These security checks can be strategically scheduled to utilize otherwise idle time in the schedule. Upon validation of timing bounds, no action is taken. Conversely, upon violation of bounds, an alert is raised that provides an opportunity to reduce system functionality, revert to a fail-safe state or shut down the system altogether pending further investigation/assessment. To the best of our knowledge, such detection of system compromises through micro-timing information is a novel contribution to real-time systems.

Within the realm of this work, overheads on performance and tightened security should become more balanced or tunable by a "dial". This may also include the exploitation of average-case execution time in statistical sanity checks or probabilistic timing analysis-based systems or parametric models [5, 29]. More gradual warning systems might provide several steps of reduced functionality while raising the bar for intrusions if threats are detected, similar to the Simplex approach for reliability [4, 14].

References

1. Aho, A.V., Sethi, R., Ullman, J.D.: Compilers – Principles, Techniques, and Tools. Addison-Wesley (1986)
2. Asfaw, B., Bekele, D., Eshete, B., Villafiorita, A., Weldemariam, K.: Host-based anomaly detection for pervasive medical systems. In: 2010 Fifth International Conference on Risks and Security of Internet and Systems, CRiSIS (2010)

3. Atanassov, P., Puschner, P.: Impact of dram refresh on the execution time of real-time tasks. In: Proc. IEEE International Workshop on Application of Reliable Computing and Communication, pp. 29–34 (2001)
4. Bak, S., Chivukula, D., Adekunle, O., Sun, M., Caccamo, M., Sha, L.: The system-level simplex architecture for improved real-time embedded system safety. In: IEEE Real-Time Embedded Technology and Applications Symposium, pp. 99–107 (2009)
5. Bernat, G., Colin, A., Petters, S.: Wcet analysis of probabilistic hard real-time systems. In: IEEE Real-Time Systems Symposium (2002)
6. Bhat, B., Mueller, F.: Making dram refresh predictable. In: Euromicro Conference on Real-Time Systems, pp. 145–154 (2010)
7. Bhat, B., Mueller, F.: Making dram refresh predictable. Real-Time Systems 47(5), 430–453 (2011)
8. Braberman, V., Felder, M., Marre, M.: Testing timing behavior of real-time software. International Software Quality Week (1997), http://citeseer.ist.psu.edu/braberman97testing.html
9. Burger, D., Austin, T., Bennett, S.: Evaluating future microprocessors: The simplescalar toolset. Tech. Rep. CS-TR-96-1308, University of Wisconsin - Madison, CS Dept. (1996)
10. Burger, D., Austin, T.M., Bennett, S.: Evaluating future microprocessors: The simplescalar tool set. Technical Report CS-TR-1996-1308, University of Wisconsin, Madison (1996)
11. C-Lab: Wcet benchmarks, http://www.c-lab.de/home/en/download.html
12. Chana, S.K., Karale, S.J.: Analysis of Intrusion Detection Response System (IDRS) In Cyber Physical Systems (Cps) Using Regular Expression (Regexp). IOSR Journal of Computer Engineering, IOSR-JCE (2014), http://dx.doi.org/10.6084/m9.figshare.1109874
13. Cowan, C., Beattie, S., Johansen, J., Wagle, P.: Pointguardtm: protecting pointers from buffer overflow vulnerabilities. In: SSYM 2003: Proceedings of the 12th Conference on USENIX Security Symposium, p. 7 (2003)
14. Crenshaw, T., Gunter, E., Robinson, C., Sha, L., Kumar, P.: The simplex reference model: Limiting fault-propagation due to unreliable components in cyber-physical system architectures. In: IEEE Real-Time Systems Symposium, pp. 400–412 (2007)
15. Dán, G., Sandberg, H., Ekstedt, M., Björkman, G.: Challenges in power system information security. IEEE Security Privacy 10(4), 62–70 (2012)
16. Gao, W., Morris, T., Reaves, B., Richey, D.: On scada control system command and response injection and intrusion detection. In: eCrime Researchers Summit (eCrime), pp. 1–9 (2010)
17. Hadeli, H., Schierholz, R., Braendle, M., Tuduce, C.: Leveraging determinism in industrial control systems for advanced anomaly detection and reliable security configuration. In: IEEE Conference on Emerging Technologies Factory Automation, ETFA 2009, pp. 1–8 (2009)
18. Healy, C.A., Arnold, R.D., Mueller, F., Whalley, D., Harmon, M.G.: Bounding pipeline and instruction cache performance. IEEE Transactions on Computers 48(1), 53–70 (1999)
19. Kc, G.S., Keromytis, A.D., Prevelakis, V.: Countering code-injection attacks with instruction-set randomization. In: CCS 2003: Proceedings of the 10th ACM Conference on Computer and Communications Security, pp. 272–280 (2003)
20. Kuperman, B., Brodley, C., Ozdoganoglu, H., Vijaykumar, T., Jalote, A.: Detection and prevention of stack buffer overflow attacks. Commun. ACM 48(11), 50–56 (2005)
21. Labrosse, J.: Micro C/OS-II. R & D Books (1998)

22. Lauf, A., Peters, R., Robinson, W.: Intelligent intrusion detection: A behavior-based approach. In: 21st Advanced Information Networking and Applications: Symposium for Embedded Computing (2007)
23. Lauf, A.P., Peters, R.A., Robinson, W.H.: A distributed intrusion detection system for resource-constrained devices in ad-hoc networks. Ad Hoc Netw. 8(3), 253–266 (2010)
24. Levy, E.: Crossover: Online pests plaguing the offline world. IEEE Security and Privacy 1(6), 71–73 (2003)
25. Liu, C., Layland, J.: Scheduling algorithms for multiprogramming in a hard-real-time environment. J. of the Association for Computing Machinery 20(1), 46–61 (1973)
26. Mitchell, R., Chen, I.: Effect of intrusion detection and response on reliability of cyber physical systems. IEEE Transactions on Reliability 62(1), 199–210 (2013)
27. Mitchell, R., Chen, I.R.: A survey of intrusion detection techniques for cyber-physical systems. ACM Comput. Surv. 46(4), 55:1–55:29 (2014)
28. Mohan, S.: Worst-case execution time analysis of security policies for deeply embedded real-time systems. SIGBED Rev. 5(1), 1–2 (2008)
29. Mohan, S., Hawkins, F.M.W., Root, M., Whalley, D., Healy, C.: Parametric timing analysis and its application to dynamic voltage scaling. ACM Transactions on Embedded Computing Systems (2007) (accepted)
30. Mohan, S., Mueller, F.: Preserving timing anomalies in pipelines of high-end processors. Tech. Rep. TR 2007-13, Dept. of Computer Science, North Carolina State University (2008)
31. Mohan, S., Mueller, F., Hawkins, W., Root, M., Healy, C., Whalley, D.: Parascale: Expoliting parametric timing analysis for real-time schedulers and dynamic voltage scaling. In: IEEE Real-Time Systems Symposium, pp. 233–242 (2005)
32. Mohan, S., Mueller, F., Whalley, D., Healy, C.: Timing analysis for sensor network nodes of the atmega processor family. In: IEEE Real-Time Embedded Technology and Applications Symposium, pp. 405–414 (2005)
33. Moses, A.: 'sinister' integral energy virus outbreak a threat to power grid (2009), `http://www.smh.com.au/technology/security/sinister-integral-energy-virus-outbreak-a-threat-to-power-grid-20091001-gdrx.html`
34. Mueller, F.: Timing analysis for instruction caches. Real-Time Systems 18(2/3), 209–239 (2000)
35. Ravi, S., Raghunathan, A., Kocher, P., Hattangady, S.: Security in embedded systems: Design challenges. ACM Trans. Embed. Comput. Syst. 3(3), 461–491 (2004)
36. Seshadri, A., Perrig, A., van Doorn, L., Khosla, P.: Swatt: Software-based attestation for embedded devices. In: IEEE Symposium on Security and Privacy, p. 272 (2004)
37. Shacham, H., Page, M., Pfaff, B., Goh-Jin, E.J., Modadugu, N., Boneh, D.: On the effectiveness of address-space randomization. In: CCS 2004: Proceedings of the 11th ACM Conference on Computer and Communications Security, pp. 298–307 (2004)
38. Shao, Z., Zhuge, Q., He, Y., Sha, E.H.M.: Defending embedded systems against buffer overflow via hardware/software. In: ACSAC 2003: Proceedings of the 19th Annual Computer Security Applications Conference, p. 352. IEEE Computer Society, Washington, DC (2003)
39. Shin, S., Kwon, T., Jo, G.Y., Park, Y., Rhy, H.: An experimental study of hierarchical intrusion detection for wireless industrial sensor networks. IEEE Transactions on Industrial Informatics 6(4), 744–757 (2010)
40. Son, S.H., Mukkamala, R., David, R.: Integrating security and real-time requirements using covert channel capacity. IEEE Transactions on Knowledge and Data Engineering 12, 865–879 (2000)

41. Venugopalan, R., Ganesan, P., Peddabachagari, P., Dean, A., Mueller, F., Sichitiu, M.: Encryption overhead for sensor networks and embedded systems: Modeling and analysis. In: Conference on Compilers, Architecture and Synthesis for Embedded Systems, pp. 188–197 (2003)
42. Wegener, J., Mueller, F.: A comparison of static analysis and evolutionary testing for the verification of timing constraints. Real-Time Systems 21(3), 241–268 (2001)
43. Whitham, J.: Real-time processor architectures for worst case execution time reduction. Ph.D. thesis, University of York (2008)
44. Wilhelm, R., Engblom, J., Ermedahl, A., Holsti, N., Thesing, S., Whalley, D., Bernat, G., Ferdinand, C., Heckmann, R., Mitra, T., Mueller, F., Puaut, I., Puschner, P., Staschulat, J., Stenstrom, P.: The worst-case execution time problem — overview of methods and survey of tools. ACM Transactions on Embedded Computing Systems 7(3), 1–53 (2008)
45. Wu, B., Chen, J., Wu, J., Cardei, M.: A survey of attacks and countermeasures in mobile ad hoc networks. Wireless Network Security 30(3), 103–135 (2007)
46. Xie, T., Qin, X., Lin, M.: Open issues and challenges in security-aware real-time scheduling for distributed systems. Journal of Information 6(9) (2006)
47. Zhang, L., White, G.B.: Analysis of payload based application level network anomaly detection. In: HICSS 2007: Proceedings of the 40th Annual Hawaii International Conference on System Sciences, p. 99 (2007)
48. Zhang, Y., Lee, W.: Intrusion detection in wireless ad-hoc networks. In: MobiCom 2000: Proceedings of the 6th Annual International Conference on Mobile Computing and Networking, pp. 275–283 (2000)
49. Zhang, Y., Lee, W., Huang, Y.A.: Intrusion detection techniques for mobile wireless networks. Wireless Networking 9(5), 545–556 (2003)
50. Zimmer, C., Bhat, B., Mueller, F., Mohan, S.: Time-based intrusion dectection in cyber-physical systems. In: International Conference on Cyber-Physical Systems, pp. 109–118 (2010)

Against Data Attacks on Smart Grid Operations: Attack Mechanisms and Security Measures*

Jinsub Kim and Lang Tong

Abstract. This chapter provides a survey and some highlights of recent developments on cyber security issues related to smart grid operations. In particular, we present data attack models and attack mechanisms on system state estimation, generation dispatch, and market operations. Security measures via sensor protection and data authentication are discussed. Although presented in the context of a smart grid, the main ideas are applicable to general cyber physical systems.

1 Introduction

A defining feature of a smart grid is the extensive use of measurement data and auxiliary information for real-time control and decision. To this end, a reliable and secure communications infrastructure is essential. While the evolution toward a smarter grid improves operation efficiency and resiliency against faults and natural failures, the heavier reliance on communications infrastructure also extends its vulnerability to threats of cyber attacks. See, *e.g.*, [16] on the vulnerabilities of smart grid. It is particularly relevant that errors in the cyber infrastructure contributed to the 2003 Northeast Blackout that affected 50 million people and caused the estimated loss of between 4 and 10 billion U.S. dollars [1]. The fact that accidental errors can cause economic loss of such scale begs the question of whether attacks orchestrated by

Jinsub Kim
School of Electrical Engineering and Computer Science, Oregon State University, Corvallis, OR 97330
e-mail: kimjinsu@eecs.oregonstate.edu

Lang Tong
School of Electrical and Computer Engineering, Cornell University, Ithaca, NY 14853, USA
e-mail: ltong@ece.cornell.edu

* This work is supported in part by the National Science Foundation under Grant CNS-1135844.

S.K. Khaitan, J.D. McCalley, and C.-C. Liu (eds.), *Cyber Physical Systems Approach to Smart Electric Power Grid*, Power Systems, DOI: 10.1007/978-3-662-45928-7_13

adversaries with backing of powerful international organizations would lead to even larger and more frequent blackouts.

Reports from the U.S. Department of Energy and the U.S. Department of Homeland Security point out that many legacy devices in current grids do not support advanced data authentication, integrity check, or encryption, and thus they are vulnerable to data attacks [2, 3]. In addition, even when data communications are secure, human factors present difficult challenges for the secure operation of a power grid that involves thousands of employees. Therefore, a starting point of developing effective and economic countermeasures of attacks is to gain an understanding of vulnerabilities of a large power grid and the impacts of data attacks assuming that adversaries are capable of penetrating the exterior cyber defense (*e.g.*, encryption, firewalls.)

In this chapter, we consider attacks on sensor measurements with the goal of misleading the control center with incorrect topology or state estimates. To distinguish them from attacks that physically harm the network, we refer to this kind of attacks as *data attacks*. We highlight some recent advances in the design of data attacks from the perspective of an adversary and the development of defense mechanisms from the perspective of a grid operator. In particular, we focus on attacks that affect power system state estimation, real-time dispatch, and real-time market operations. At the heart of real-time operation of a large power system is state estimation. Using streaming data from a large web of sensors that covers the entire network, state estimation not only plays a critical role of monitoring system operation but also is the key input to the real-time dispatch algorithm and is used in real-time contingency analysis and the computation of real-time electricity price.

We first focus on attacks aimed at affecting the state estimation of a power system, assuming a perspective of an attacker who has gained access to part of the measurements and is capable of altering data from a subset of sensors. For a successful attack, an attacker needs to know (or learn) the physical model and design the data modifications accordingly such that the altered data appear to be consistent to the control center. Our presentation starts by introducing a mathematical model of attack, analyzing feasibility of attacks, and reviewing various attack mechanisms that involve different costs and different ways of learning the physical system information.

Next, we take the perspective of a network operator, conscious about the presence of attacks. Because sensors and the communications systems are not always reliable, the data collected may be misread, missing, or erroneously aligned. Thus the power systems state estimation typically has a built-in detector of bad data. Although bad data detection can eliminate certain outliers and data anomalies, it does not provide an effective defense against malicious attacks. In particular, bad data detection does not provide the reasons why a data sample is "bad," nor does it provide any level of assurance that data used in state estimation have not been tampered. Missing in the standard practice of data collection are ways of authenticating data and means to alert the operator that the data used may not be secure. Because authenticating all sensor data is not practical, the challenge is to authenticate data from a small set of sensors such that any unauthorized change of data can be detected. In choosing the set of sensor data to authenticate, the analysis of the physical system model and the sensor measurement model plays a key role. We will review the existing security measures based on sensor data authentication and some approaches to facilitate fast detection of an attack.

Although this chapter presents the materials in the context of a smart grid, the main ideas of attack mechanisms and security measures are applicable to general cyber physical systems (CPSs) [19, 20]. In particular, the results presented in this chapter can be applied to a general CPS as far as its system measurement model can be approximated by a linear model.

1.1 Related Works

The vulnerability of a power grid to a data attack was first pointed out by Liu *et al.* in [30] where the authors consider an attacker who aims to perturb the state estimate by altering part of line flow and bus injection measurements. Under a linear (DC) model, it was shown that an attacker controlling only a few sensors can perturb the state estimate by an arbitrary degree without being detected by the control center. Such an attack aimed at perturbing the state estimate is referred to as a *state attack*, and especially the attacks that can avoid detection is referred to as *unobservable attack*.

Results following [30] have provided insights into the vulnerability of power grids to unobservable state attacks and potential attack mechanisms. The vulnerability to an unobservable state attack was characterized as the classical power systems observability condition [6, 28], which can be checked efficiently using graph-theoretic and algebraic techniques [4, 29, 32].

In an effort to assess the vulnerability of grids, security metrics are presented in [28, 39]. In [39], Sandberg *et al.* propose the security metric for each sensor, which is defined as the minimum number of sensors that the attacker needs to control in order to modify the sensor measurement using an unobservable state attack. Kosut *et al.* in [28] extend this definition and define the security index of a grid as the minimum number of sensors that an attacker needs to control to launch an unobservable state attack. Under the DC model, evaluating the security index is equivalent to finding the sparsest vector in the column space of a matrix, which is NP-complete for a general matrix [8, 43]. However, by exploiting the special structure of the DC measurement matrix, a polynomial time algorithm for the security index calculation is presented in [28].

From the perspective of an attacker with a limited power, cost-effective designs of unobservable state attacks were studied to minimize the necessary number of sensor data modifications under various scenarios [28, 30, 39]. These works assume an omniscient attacker knowing the system parameters and topology. For more practical attack scenarios, data-driven attacks that learn the necessary system information from partial sensor observations are proposed [11, 23, 26]. Section 3 reviews the detail of the vulnerability analysis and the attack mechanisms for unobservable state attacks.

Besides attacks on power flow measurements assumed in [30] and subsequent works, attacks on breaker sensor measurements should also be considered as such measurements are transmitted to the control center via the same communications infrastructure. If breaker sensor data are altered by an attacker in conjunction with proper modifications on line flow and bus injection measurements, the attack can mislead the control center with incorrect topology without being detected [22]. Kim and Tong [22] presented cost-effective attack mechanisms and local-data-driven attack mechanisms for unobservable topology attacks, which will be discussed in Section 4.

The topology estimate and the state estimate are key inputs to real-time grid operations and real-time electricity pricing. It has been demonstrated by several works [7,17,45] that attacks on topology and state estimates can potentially perturb the real-time economic dispatch signals and the real-time electricity prices such that certain load serving entities can benefit from the resulting changes. The details of the attack impact will be discussed in Section 5.

From the perspective of a network operator, a large number of security measures have been proposed [6,12,21,22,27,33]. A popular approach based on data authentication is to protect certain subset of sensors from adversarial data modification (by equipping them with data authentication protocols) such that any data modification on unprotected sensor data will cause inconsistency among the whole sensor data entries thereby leading to detection [6,12,21,22,27]. Because realizing a data authentication protocol on legacy devices may require costly hardware upgrades [3], many efforts have been made to minimize the number of sensors that require protection [6,21,22,27]. In addition, as the use of phasor measurement units (PMUs) is growing, strategic allocation of secure PMUs for protection against data attacks has been also studied [12,21,27]. To facilitate fast detection of an attack, the attack detection as a quickest detection problem was also studied in [9,14,15]. The security measures will be reviewed in Section 6.

Although there is an extensive literature focusing on the case when the adversary is able to control a sufficiently large number of sensors so that the attack launched is not detectable, the case when the adversary can alter only a small subset of sensors is less understood. For such cases, an attack is likely to introduce inconsistency among the sensor data thereby making it observable to the control center. However, bad data detection at the control center may falsely identify normal sensors as malfunctioning sensors while treating adversary-controlled sensors as valid. In [24], *observable attacks* that exploit such a flaw in bad data detection are considered. It is shown that the state estimates can be biased by such techniques. As such attacks require much less power compared to unobservable attacks, existing security measures against unobservable attacks are not able to prevent observable attacks. We will discuss open problems related to a security measure against observable attacks.

1.2 Organization and Notations

The rest of the chapter is organized as follows. In Section 2, we present the power system measurement model, topology and state estimation, and the attack model. The material here is standard with an emphasis on the graph-theoretic model of the network. Section 3 focuses on unobservable state attacks. Here we make explicit connection between feasibility of an unobservable state attack and the classical notion of network observability developed by [29, 32]. The connection leads to a graph-theoretic characterization of the grid vulnerability to unobservable state attacks. We review economic attack mechanisms and data-driven attack mechanisms for realizing unobservable state attacks with limited resource and system information. Section 4 assumes a more general framework where breaker state measurements and power flow measurements are jointly attacked. Here we develop the analysis for attacks on topology estimate. Feasibility of an unobservable topology attack is discussed, and economic and data-driven attack mechanisms are presented. The presentation here is based on the recent work in [22]. In Section 5, we discuss the potential attack impacts on real-time grid operations and real-time pricing. Then, security measures against unobservable attacks are reviewed in Section 6. We review security measures based on sensor data authentication that make any data attack detectable. Furthermore, from the perspective of a network operator with limited resource, optimal resource allocation to maximize the security level of the grid is discussed. Finally, Section 7 summarizes the works on unobservable attacks and introduces open problems related to observable attacks.

Throughout the chapter, we use a boldface lower case letter (*e.g.*, $\mathbf{z}$) to denote a vector and a boldface upper case letter (*e.g.*, $\mathbf{H}$) to denote a matrix. A set is denoted by a script upper case letter (*e.g.*, $\mathcal{A}$.) In addition, the hat is used to denote an estimate of certain object (*e.g.*, $\hat{\mathbf{x}}$ denotes an estimate of $\mathbf{x}$.) To enhance readability, Table 1 provides the notations used repeatedly throughout the chapter.

2 Power System State Estimation and Adversary Model

This section begins with a brief background on the power system measurement model and topology and state estimation. We will describe how the sensor measurements are related to the topology and the state and how they are used for estimating the topology and the state. The adversary model follows to give a mathematical model of a data attack.

2.1 Power System Measurement Model

A power grid is a network of buses connected by transmission lines, as illustrated by the IEEE 14-bus network in Fig. 1. The operating condition of a power grid is characterized by the topology and the system state. The *topology* of a grid is defined as an undirected graph $\mathcal{G} = (\mathcal{V}, \mathcal{E})$ where $\mathcal{V}$ is the set of buses, and $\mathcal{E}$ is the set of connected transmission lines: $\{i, j\}$ is in $\mathcal{E}$ if and only if there exists a

Table 1 Notations

$\mathbf{s}$	breaker sensor measurements
$\mathbf{z}$	line flow and bus injection sensor measurements
$\bar{\mathbf{s}}$	breaker sensor measurements corrupted by an attack
$\bar{\mathbf{z}}$	line flow and bus injection measurements corrupted by an attack
$\mathbf{b}$	an attack vector added to the breaker sensor measurements
$\mathbf{a}$	an attack vector added to the line flow and bus injection measurements
$\mathcal{B}$	the set of **b** that an attacker can use
$\mathcal{A}$	the set of **a** that an attacker can use
$\mathcal{S}_{\mathrm{B}}$	the set of breaker sensors controlled by the attacker
$\mathcal{S}_{\mathrm{A}}$	the set of line flow and bus injection sensors controlled by the attacker
$\mathcal{G}$	An undirected graph $(\mathcal{V}, \mathcal{E})$ that represents the grid topology
$\hat{\mathcal{G}}$	topology estimate
$\mathcal{V}$	the set of buses
$\mathcal{E}$	the set of connected transmission lines
$\mathbf{x}$	system state
$\hat{\mathbf{x}}$	state estimate
h	nonlinear power system measurement function
$\mathbf{H}$	DC measurement matrix
$\mathcal{R}(\mathbf{H})$	the column space of **H**
$\hat{\mathbf{H}}$	DC measurement matrix corresponding to $\hat{\mathcal{G}}$

transmission line between bus i and bus j, and the line is connected. Hence, the topology represents the connectivity of the grid. The system *state* $\mathbf{x}$ is defined as the vector of voltage magnitudes and phase angles at all buses except the voltage phase angle at the reference bus, which is set to zero.

The control center cannot directly observe the topology and the state; it estimates them based on real-time measurements from sensors deployed throughout the grid. There are two types of sensor measurements. The first are binary measurements from breaker sensors, denoted by $\mathbf{s} \in \{0,1\}^l$ where l is the number of breaker sensors. A breaker sensor measures whether certain line breaker is open or closed: '0' and '1' correspond to 'open' and 'closed' respectively. Since the connectivity of each line is determined by the statuses of line breakers on the line, each $\mathbf{s} \in \{0,1\}^l$ corresponds to a unique topology. The second type includes power measurements from line flow sensors and bus injection sensors[1]. These analog measurements, de-

[1] Other types of analog measurements (*e.g.*, voltage magnitudes, synchrophasor measurements) can also be considered. We restrict our attention to line flow and bus injection sensors merely for clearer presentation.

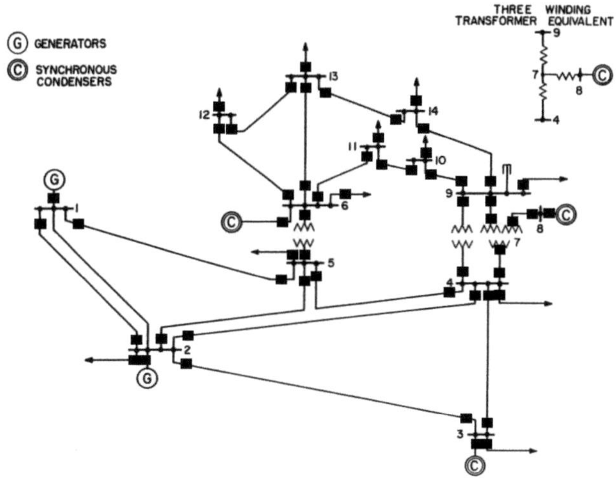

Fig. 1 IEEE 14-bus network. The black rectangle on bus i represents the bus injection sensor measuring the injection at bus i, and the rectangle on the line $\{i, j\}$, that is located closer to bus i, represents the line flow sensor that measures the line flow from bus i to bus j.

noted by a real vector $\mathbf{z}$, consist of real and imaginary parts of complex measurements of power flows through transmission lines and power injections at buses. The line flow and bus injection measurements are related to the state by the nonlinear AC model [4]:

$$\mathbf{z} = h(\mathbf{x}; \mathcal{G}) + \mathbf{e}, \tag{1}$$

where h is the nonlinear measurement function that depends on the topology $\mathcal{G}$, and $\mathbf{e}$ is the measurement noise, typically modeled as a Gaussian random vector with zero mean and covariance matrix $\boldsymbol{\Sigma}$.

In analyzing state estimation, a linear approximation of the AC model (1)—the so-called DC model—is often employed [4]. The DC model is obtained by linearizing the AC model at the nominal state. Specifically, the DC model relates real measurements $\mathbf{z}$ (line active power flows and bus active power injections) with bus voltage angles as states $\mathbf{x}$ in a linear form:

$$\mathbf{z} = \mathbf{H}\mathbf{x} + \mathbf{e}, \tag{2}$$

where the matrix $\mathbf{H}$ is obtained from the Jacobian of h evaluated at the nominal state. We should note that, while it is convenient and analytically tractable to use the DC model to construct attacks and analyze their properties, in evaluating the attack performance, it is necessary that the original AC model is used.

The measurement matrix $\mathbf{H}$ depends on the impedance of transmission lines and the topology. To describe the entries of $\mathbf{H}$, we consider a noiseless measurement $\mathbf{z} = \mathbf{H}\mathbf{x}$. Suppose that the kth entry of $\mathbf{z}$, denoted by z_k, corresponds to the line flow

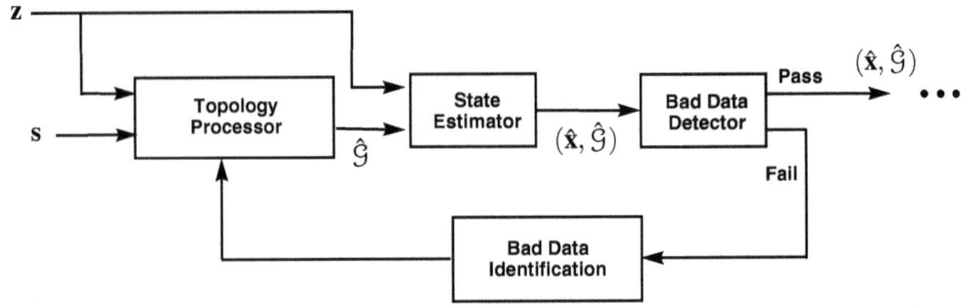

Fig. 2 Generalized state estimation

from bus i to bus j. If bus i and bus j are being connected by a transmission line (*i.e.*, $\{i, j\} \in \mathcal{E}$), z_k is equal to $B_{ij}(x_i - x_j)$ where B_{ij} is the susceptance of the line $\{i, j\}$, and x_i is the voltage phase angle at bus i. Otherwise, z_k is zero. If z_k corresponds to the bus injection measurement at bus i, z_k is equal to the sum of all outgoing line flows from bus i.

A grid is said to be *observable* if the state $\mathbf{x}$ can be uniquely identified from $\mathbf{Hx}$, or equivalently, $\mathbf{H}$ has full rank. In a practical power grid, a sufficient number of sensors are placed to guarantee observability. Krumpholz *et al.* in [29] show that observability can be verified merely based on the topology and the sensor locations, without knowing the exact $\mathbf{H}$. In particular, they presented the spanning-tree criterion, which is a graph-theoretical criterion to check observability. It is formally stated as follows.

Theorem 1 (Corollary 2, [29]). *A grid is observable if and only if there exists a way to assign each bus injection sensor to a line incident to the bus such that a spanning tree of $\mathcal{G}$ with at least one sensor on each edge of the tree exists.*

Based on the above relation, Krumpholz *et al.* in [29] provide a polynomial time algorithm for checking observability (polynomial with respect to the number of buses and lines.)

2.2 Topology and State Estimation

Once the control center receives the sensor measurements, it estimates the topology and the state. Fig. 2 illustrates the structure of the so-called generalized state estimation (GSE) [31], which looks for the pair of topology and state that fits the measurements best. In GSE, unless some breaker sensors are suspected faulty, the topology estimator first comes up with a topology estimate based on the sensor measurements. The topology estimator finds the topology estimate $\hat{\mathcal{G}}$ that corresponds to the breaker sensor measurements ($\mathbf{s}$). Then, the line flow and bus injection measurements ($\mathbf{z}$) are used to check whether $\hat{\mathcal{G}}$ is consistent with $\mathbf{z}$.

After the topology estimate $\hat{\mathcal{G}}$ is produced, the state estimate $\hat{\mathbf{x}}$ is obtained as a weighted least squares (WLS) estimate:

$$\hat{\mathbf{x}} = \arg\min_{\mathbf{x}} (\mathbf{z} - h(\mathbf{x};\hat{\mathcal{G}}))^T \boldsymbol{\Sigma}^{-1} (\mathbf{z} - h(\mathbf{x};\hat{\mathcal{G}})) \tag{3}$$

where the superscript T denotes the transpose operation. Note that the WLS estimate corresponds to the maximum likelihood estimate.

To validate the topology and state estimates, the control center conducts a bad data test to check whether the measurements used for estimation contain any bad data. For bad data detection, the tests based on the estimation residues are widely used [4]. In particular, the so-called $J(\hat{\mathbf{x}})$-test works as follows [4]:

$$\begin{cases} \text{Good data, if } (\mathbf{z} - h(\mathbf{x};\hat{\mathcal{G}}))^T \boldsymbol{\Sigma}^{-1} (\mathbf{z} - h(\mathbf{x};\hat{\mathcal{G}})) \leq \tau; \\ \text{Bad data, } \quad \text{otherwise.} \end{cases} \tag{4}$$

The threshold τ is set to satisfy the false alarm probability constraint. The $J(\hat{\mathbf{x}})$-test is widely used due to its simplicity and the fact that the test statistic approximately has a χ^2 distribution when the data are good. The latter is used to set the threshold τ.

If the bad data detector declares that the data are good, the topology and state estimates are used by other real-time operations. Otherwise, further investigation is made to correct or remove outliers in the measurements (see [4, 13] for the details.) Once the measurements are corrected, a new iteration begins with the corrected measurements. The iterations end only if the bad data detector declares that data are good.

Note that using the DC model (2), everything is the same as above except that $h(\mathbf{x};\hat{\mathcal{G}})$ is replaced with $\hat{\mathbf{H}}$, the measurement matrix corresponding to the topology estimate $\hat{\mathcal{G}}$. In this case, the state estimate is simply a linear WLS estimate:

$$\hat{\mathbf{x}} = \arg\min_{\mathbf{x}} (\mathbf{z} - \hat{\mathbf{H}}\mathbf{x})^T \boldsymbol{\Sigma}^{-1} (\mathbf{z} - \hat{\mathbf{H}}\mathbf{x}) = (\hat{\mathbf{H}}^T \boldsymbol{\Sigma}^{-1} \hat{\mathbf{H}})^{-1} \hat{\mathbf{H}}^T \boldsymbol{\Sigma}^{-1} \mathbf{z}. \tag{5}$$

2.3 Adversary Model

An adversary in a data attack is assumed to be capable of altering certain sensor measurements. The adversarial data modification is modeled as follows [22, 30].

$$\begin{aligned} \bar{\mathbf{s}} &= \mathbf{s} + \mathbf{b} \ (\text{mod } 2), \ \ \mathbf{b} \in \mathcal{B}; \\ \bar{\mathbf{z}} &= \mathbf{z} + \mathbf{a}, \ \ \mathbf{a} \in \mathcal{A}, \end{aligned} \tag{6}$$

where $\mathbf{b}$ and $\mathbf{a}$ are attack vectors for breaker status measurements and line flow and bus injection measurements, respectively. The sparsity patterns of the attack vectors are restricted by the set of sensors that the attacker can control. Let $\mathcal{S}_{\text{B}}$ and $\mathcal{S}_{\text{A}}$ denote the set of breaker sensors and the set of line flow and bus injection sensors that are being controlled by the attacker, respectively. Then, $\mathcal{B}$ and $\mathcal{A}$ are formally defined as follows:

$$\begin{aligned} \mathcal{B} &= \{\mathbf{b} \in \{0,1\}^l : \ b_i = 0, \ \forall \, i \notin \mathcal{S}_{\text{B}}\}; \\ \mathcal{A} &= \{\mathbf{a} \in \mathbb{R}^m : \ a_i = 0, \ \forall \, i \notin \mathcal{S}_{\text{A}}\}. \end{aligned} \tag{7}$$

In practice, an adversary can acquire the necessary ability for a data attack by launching man-in-the-middle attacks [35] or backdoor attacks [40] on proper links and devices in the supervisory control and data acquisition (SCADA) systems (see Hull *et al.* [16] for explanation of vulnerabilities of the SCADA system.) For instance, lack of a strong data authentication protocol may render the grid vulnerable to a man-in-the-middle attack. In a man-in-the-middle attack, an attacker sits between two devices, for instance a remote terminal unit (RTU) and a sensor [16]. The attacker impersonates the RTU when it communicates with the sensor, and it impersonates the sensor when it communicates with the RTU. In that way, the attacker can control the data packets communicated between the RTU and the sensor. Please refer to [35] and [40] for the detailed mechanisms of man-in-the-middle attacks and backdoor attacks, respectively.

This chapter focuses on the attacks, under which the corrupted measurements will pass the bad data detection. We refer to this type of attack as an *unobservable attack* as the attack is not visible to the control center. There also exists an attack approach that will be detected by the control center, but still perturb the state estimate [24]. We will discuss about such an *observable attack* at Section 7. Formally, an unobservable attack is defined as follows.

Definition 1. For $\mathbf{s}$ and $\mathbf{z}$ that correspond to a topology $\mathcal{G}$ and a state $\mathbf{x}$, *i.e.*, $\mathbf{z} = \mathbf{H}\mathbf{x} + \mathbf{e}$ where $\mathbf{H}$ is the measurement matrix corresponding to $\mathcal{G}$, an attack is *unobservable* if there exists $(\bar{\mathbf{x}}, \bar{\mathcal{G}}) \neq (\mathbf{x}, \mathcal{G})$ such that $\bar{\mathbf{s}}$ corresponds to the topology $\bar{\mathcal{G}}$, and $\bar{\mathbf{z}} = \bar{\mathbf{H}}\bar{\mathbf{x}} + \mathbf{e}$ where $\bar{\mathbf{H}}$ is the measurement matrix corresponding to $\bar{\mathcal{G}}$. In particular, an unobservable attack with $\bar{\mathcal{G}} \neq \mathcal{G}$ is called an *unobservable topology attack*, and an unobservable attack with $\bar{\mathcal{G}} = \mathcal{G}$ and $\bar{\mathbf{x}} \neq \mathbf{x}$ is called an *unobservable state attack*.

In other words, an unobservable attack modifies the measurements such that the corrupted measurements become consistent with a wrong pair of topology and state estimates. Note that an unobservable topology attack can perturb not only the topology estimate but also the state estimate. In contrast, an unobservable state attack focuses on perturbing only the state estimate.

3 Unobservable State Attack

This section reviews attack mechanisms for an unobservable state attack. We first review attack mechanisms for an adversary with the knowledge of the network topology and the line impedance values. Then, we introduce data-driven attacks that require little or no knowledge about the network topology and the line impedance values.

3.1 Unobservable State Attack: Structure and Feasibility

Liu *et al.* in [30] introduced an unobservable state attack, which was also the first data attack on a power grid. They assumed that breaker sensor data are intact from the adversary, *i.e.*, both $\mathcal{S}_\mathrm{B}$ and $\mathcal{B}$ are set to the empty set in (6).

In practice, breaker sensor measurements are also subject to attacks because they are sent to the control center via the same communications infrastructure as line flow and bus injection measurements. We will see in Section 4 that the attacker capable of perturbing some breaker sensor data may perturb the topology estimate without being detected, even when an unobservable state attack is not feasible. Nevertheless, Liu *et al.* provided insights into the structure of unobservable state attacks and revealed the possibility of data attacks for the first time.

If breaker sensor data are intact, the attack model (6) becomes

$$\bar{\mathbf{z}} = \mathbf{H}\mathbf{x} + \mathbf{e} + \mathbf{a}, \quad \mathbf{a} \in \mathcal{A}. \tag{8}$$

The main idea of an unobservable state attack is to set $\mathbf{a}$ to some nonzero element in $\mathcal{R}(\mathbf{H})$—the column space of $\mathbf{H}$. Suppose that there exists $\mathbf{a} \in \mathcal{A}$ such that $\mathbf{a} = \mathbf{H}\mathbf{y}$ for some nonzero $\mathbf{y} \in \mathbb{R}^n$. Then, the corrupt line flow and bus injection measurements become

$$\bar{\mathbf{z}} = \mathbf{H}(\mathbf{x} + \mathbf{y}) + \mathbf{e} \tag{9}$$

thereby leading to an unobservable state attack. Note that $\bar{\mathbf{z}}$ is not different from a normal measurement vector with the state $\mathbf{x} + \mathbf{y}$, and thus this attack will perturb the state estimate by $\mathbf{y}$ without being detected. Another powerful aspect of this attack is that by scaling up the attack vector, the degree of perturbation can grow arbitrarily large: for $\alpha \in \mathbb{R}$, if the attacker set $\mathbf{a}$ to $\alpha \cdot \mathbf{H}\mathbf{y}$, the resulting perturbation is $\alpha \cdot \mathbf{y}$.

On the other hand, for $\bar{\mathbf{z}}$ in (8) to be equal to $\mathbf{H}\bar{\mathbf{x}} + \mathbf{e}$ for some $\bar{\mathbf{x}} \in \mathbb{R}^n$, $\mathbf{a}$ has to be equal to $\mathbf{H}(\bar{\mathbf{x}} - \mathbf{x})$, *i.e.*, $\mathbf{a}$ should be in $\mathcal{R}(\mathbf{H})$. Therefore, we can have the following algebraic condition for existence of an unobservable state attack [30].

Theorem 2 (Theorem 1, [30]). *An unobservable state attack exists if and only if the dimension of the subspace* $\mathcal{A} \cap \mathcal{R}(\mathbf{H})$ *is nonzero.*

Note that $\mathcal{A}$ simply restricts $\mathbf{a}$ to have zero entries at the rows corresponding to the sensors that are not controlled by the attacker (*i.e.*, the sensors *not* in $\mathcal{S}_A$.) Let $\bar{\mathbf{H}}$ denote the submatrix of $\mathbf{H}$ obtained by removing the rows corresponding to the sensors in $\mathcal{S}_A$. Then, $\mathbf{a}$ is in $\mathcal{A} \cap \mathcal{R}(\mathbf{H})$ if and only if $\mathbf{a} = \mathbf{H}\mathbf{y}$ for some $\mathbf{y} \in \mathcal{N}(\bar{\mathbf{H}})$. This argument provides a way to construct an unobservable attack based on $\mathcal{N}(\bar{\mathbf{H}})$: choose a nonzero element $\mathbf{y}$ of $\mathcal{N}(\bar{\mathbf{H}})$ and set $\mathbf{a}$ to be $\mathbf{H}\mathbf{y}$. The larger the dimension of $\mathcal{N}(\bar{\mathbf{H}})$, the more flexible the direction of the resulting perturbation can be.

While Theorem 2 provides an algebraic condition for the grid vulnerability to an unobservable state attack, this condition can be transformed to the classical observability condition as stated in the following theorem [28].

Theorem 3 (Theorem 1, [28]). *An unobservable state attack exists if and only if removing the sensors in* $\mathcal{S}_A$ *from the grid makes the grid unobservable (i.e., removing the rows of* $\mathbf{H}$ *corresponding to the sensors in* $\mathcal{S}_A$ *makes* $\mathbf{H}$ *rank deficient.)*

As the power systems observability has been well studied for decades, Theorem 3 enables the control center to exploit a large number of existing results on observability analysis for analyzing the vulnerability of a grid to unobservable state attacks.

Specifically, Theorem 3, together with the spanning tree observability criterion in Theorem 1, provides a simple way to check the feasibility of an unobservable state attack:

Step 1. Remove the sensors in $\mathcal{S}_A$ from the grid.
Step 2. Apply the spanning tree criterion to check the observability of the grid with the remaining sensors.

Recall that the spanning tree criterion can be verified by the polynomial time algorithm in Krumpholz *et al.* [29].

The above feasibility check based on the spanning tree criterion directly leads to the following simple sufficient condition on $\mathcal{S}_A$ that guarantees the existence of an unobservable state attack.

Corollary 1. *Let $\{\mathcal{V}_1, \mathcal{V}_2\}$ be an arbitrary cut of $\mathcal{G}$, i.e., a partition of $\mathcal{V}$ consisting of two sets. Let $\mathcal{E}_c$ denote the corresponding cutset, i.e., a set of edges that connect one vertex in $\mathcal{V}_1$ with another vertex in $\mathcal{V}_2$. If $\mathcal{S}_A$ contains the set of the bus injection sensors on the endpoints of the edges in $\mathcal{E}_c$ and the line flow sensors on the edges in $\mathcal{E}_c$, there exists an unobservable state attack with $\mathcal{S}_A$.*

The intuition behind the above condition is that if we remove all the sensors adjacent to certain cut from the grid, then we cannot have a spanning tree of the topology with at least one sensor on each edge.

3.2 Cost-Effective Attack Strategy

The larger the size of $\mathcal{S}_A$, the more sensors the attacker should compromise. Hence, a smaller $\mathcal{S}_A$ is desirable for the attacker. A number of works addressed the problem of designing an unobservable state attack with a small $\mathcal{S}_A$ and characterizing the smallest cardinality of $\mathcal{S}_A$ that makes an unobservable state attack feasible, which is referred to as the *security index* of the grid. The security index not only quantifies the minimum effort needed for an unobservable state attack but also characterizes the robustness of the grid against state attacks.

Liu *et al.* in [30] consider the design of an unobservable state attack with a fixed amount of attack resource, represented by either a fixed $\mathcal{S}_A$ or a fixed size of $\mathcal{S}_A$. In particular, they present a *targeted* attack mechanism which aims to perturb certain subset of state variables. An adversary with a fixed target and a fixed amount of resource can utilize the attack mechanisms in [30]. The experimental results in [30] show that controlling data from only four sensors in the IEEE 300-bus network, where all bus injections and line flows are measured, an attacker can launch an unobservable state attack. The results demonstrate that even a grid with a small number of compromised sensors can be vulnerable to state attacks.

To characterize the minimum effort needed for a targeted attack, the security index was first defined by Sandberg *et al.* in [39]. In particular, the security index for the kth (line flow or bus injection) sensor, denoted by α_k, is defined as the minimum cardinality of $\mathcal{S}_A$ that enables an unobservable state attack to perturb the estimate of the power quantity measured by the kth sensor. It can be obtained as the optimal objective function value of

$$\begin{array}{ll} \min_{\mathbf{y}\in\mathbb{R}^n} & \|\mathbf{H}\mathbf{y}\|_0 \\ \text{subject to} & (\mathbf{H}\mathbf{y})_k = 1. \end{array} \tag{10}$$

The above optimization looks for the sparsest unobservable attack vector that has a nonzero value at the row corresponding to the kth sensor. Unfortunately, (10) is a non-convex problem, and thus it is difficult to solve (10) for a large system. Nevertheless, as pointed out in [39], a simple upper bound on α_k can be found by finding the sparsest column of $\mathbf{H}$ that has a nonzero entry at the kth row. In general, a grid topology is a sparse graph, and thus the columns of $\mathbf{H}$ are sparse vectors (recall the relation between $\mathbf{H}$ and the topology, described in Section 2.1.) Therefore, α_ks are expected to be small.

While Sandberg *et al.* in [39] focused on the security index for a targeted attack, Kosut *et al.* in [28] extended the definition to characterize the security index for an entire grid. The security index for an entire grid, denoted by α, is defined as the minimum cardinality of $\mathcal{S}_A$, for which an unobservable state attack exists. It can be obtained as the optimal objective function value of

$$\begin{array}{ll} \min_{\mathbf{y}\in\mathbb{R}^n} & \|\mathbf{H}\mathbf{y}\|_0 \\ \text{subject to} & \mathbf{H}\mathbf{y} \neq \mathbf{0}. \end{array} \tag{11}$$

Observing (10) and (11), it can be seen that α is equivalent to $\min_k \alpha_k$. The above optimization looks for the sparsest nonzero vector in $\mathcal{R}(\mathbf{H})$. For a general matrix $\mathbf{H}$, this problem has been known to be NP-complete (*e.g.*, see Corollary 1 in [43].) However, Kosut *et al.* show in [28] that the special structure of $\mathbf{H}$ can be exploited to have a polynomial time algorithm for solving (11). The solution of (11) provides the smallest $\mathcal{S}_A$ that enables an unobservable state attack and the sparsest attack vector.

As the security index characterizes the robustness of a grid against state attacks, it is also a useful metric to assess a security measure (*e.g.*, how much increase in the security index does the security measure result in?) We will revisit the security index in Section 6 when we review security measures.

3.3 Data-Driven Attack Mechanism

One common caveat of the aforementioned strategies of unobservable state attack is that they require the knowledge of $\mathbf{H}$ to construct an attack vector, or equivalently the knowledge of the topology and the line impedance values. The topology estimate and the line parameter information are kept secure at the control center, and thus it is difficult to acquire such information unless an attacker can penetrate the control center database. This section introduces some approaches that require little or no information about the topology and the line impedance values.

The attack strategies based on the partial topology and line parameter information are studied by Rahman and Mohsenian-Rad in [38]. They prove that an adversary can construct an unobservable state attack if he or she knows the impedance values of all transmission lines in a cutset associated with any cut of the topology. While their method significantly reduces the amount of required information, the attack is applicable only to few cases where the necessary parameter information is available.

From the attacker's perspective, a more preferable approach is to infer necessary system information based on available data, *esp.*, a subset of sensor data that the attacker can eavesdrop, and use it to construct an attack. The sensor data are more accessible than the topology or the line parameter information, because they can be acquired by hacking into local communications links or devices. Esmalifalak *et al.* in [11] made the first such attempt based on independent component analysis (ICA). They employ ICA to estimate a mixing matrix **HA** based on sensor measurements, where **A** is a linear operator that maps the load profile to the state vector. With **HA**, the attacker can find an attack vector in $\mathcal{R}(\mathbf{HA})$. One caveat of this approach is that its estimation technique depends on the assumption that loads are statistically independent and non-Gaussian. In addition, the method requires observations from all line flow and bus injection sensors in the grid.

Kim *et al.* in [23] propose a subspace approach to designing an unobservable state attack, which employs a subspace estimation technique to infer a basis matrix of $\mathcal{R}(\mathbf{H})$. The main idea of the subspace approach is that a basis matrix of $\mathcal{R}(\mathbf{H})$ is all that an adversary needs to know in finding a nonzero vector in $\mathcal{A} \cap \mathcal{R}(\mathbf{H})$. Estimation of a basis matrix from sensor measurements has been actively studied in the field of array signal processing [41]. It is further shown in [26] that under certain conditions, a basis matrix of the measurement space of a small fraction of sensors, *i.e.*, a basis matrix of $\mathcal{R}(\mathbf{H}_1)$ where $\mathbf{H}_1$ is a submatrix of **H** consisting of a small subset of rows, provides sufficient information for designing an unobservable state attack. This result leads to a data-driven attack based on observations from a small fraction of sensors. The experimental results with the IEEE 118-bus network in [26] show that observing only about 2 percent of sensors, an adversary can gain all necessary information to launch an unobservable state attack.

4 Unobservable Topology Attack

This section reviews vulnerability analysis and attack mechanisms for an unobservable topology attack. We first present a necessary and sufficient condition under which an unobservable topology attack is possible. Then, we present cost-effective attack mechanisms and the attack approaches that only require observations from few local sensors (no system information is necessary.)

In contrast to state attacks, the main objective of a topology attack is to perturb the topology estimate while the state estimate can of course be jointly perturbed. Kim and Tong in [22] first pointed out the possibility of a topology attack. They assume that $\mathcal{B} = \{0,1\}^l$, *i.e.*, all breaker sensor measurements are subject to adversarial modification, while adversarial modification on line flow and bus injection measurements is restricted by the set $\mathcal{A}$—the set of feasible attack vectors determined by the set of compromised sensors. Under this assumption, the attack model

(6) becomes

$$\begin{aligned} \bar{\mathbf{s}} &= \mathbf{s} + \mathbf{b} \ (\mathrm{mod}\ 2), \ \ \mathbf{b} \in \{0,1\}^l; \\ \bar{\mathbf{z}} &= \mathbf{H}\mathbf{x} + \mathbf{a}, \ \ \mathbf{a} \in \mathcal{A}, \end{aligned} \tag{12}$$

where the attack vector $\mathbf{b}$ is set to make $\bar{\mathbf{s}}$ correspond to certain target topology $\bar{\mathcal{G}}$. For the sake of simplicity, we consider the noiseless sensor measurements. Kim and Tong studied topology attacks in [22] for both noiseless and noisy measurements, and both cases led to the same condition for feasibility of an unobservable topology attack and very similar attack mechanisms. Readers interested in the analysis for the noisy measurement case can find the details in [22].

4.1 Feasibility of Unobservable Topology Attack

For the noiseless measurement model, the definition of an unobservable topology attack in Definition 1 can be simplified as follows.

Definition 2 (Definition 2.2, [22]). Given the measurements $\mathbf{z}$ from a topology $\mathcal{G}$ and a state $\mathbf{x}$, *i.e.*, $\mathbf{z} = \mathbf{H}\mathbf{x}$, $\mathbf{a}$ is an unobservable topology attack if there exists a topology $\bar{\mathcal{G}}$ different from $\mathcal{G}$ such that $\mathbf{H}\mathbf{x} + \mathbf{a} \in \mathcal{R}(\bar{\mathbf{H}})$ where $\bar{\mathbf{H}}$ denotes the measurement matrix for $\bar{\mathcal{G}}$.

Definition 2 provides an intuition about a condition on $\mathcal{A}$ (or equivalently, a condition on $\mathcal{S}_{\mathrm{A}}$) that guarantees feasibility of an unobservable topology attack. Suppose that $\mathcal{R}(\mathbf{H}) \subset \mathcal{R}(\bar{\mathbf{H}}) + \mathcal{A}$[2]. Then, for any $\mathbf{x} \in \mathbb{R}^n$, $\mathbf{H}\mathbf{x}$ is equal to $\bar{\mathbf{H}}\bar{\mathbf{x}} - \mathbf{a}$ for some $\bar{\mathbf{x}} \in \mathbb{R}^n$ and $\mathbf{a} \in \mathcal{A}$. Therefore, for any $\mathbf{x}$, there exists $\mathbf{a} \in \mathcal{A}$ such that $\mathbf{H}\mathbf{x} + \mathbf{a}$ will be in $\mathcal{R}(\bar{\mathbf{H}})$. Hence, $\mathcal{R}(\mathbf{H}) \subset \mathcal{R}(\bar{\mathbf{H}}) + \mathcal{A}$ is a sufficient condition for existence of an unobservable topology attack with the target $\bar{\mathcal{G}}$. Kim and Tong [22] showed that this condition is also a necessary condition, which is formally stated as follows.

Theorem 4 (Theorem 3.1-3.2, [22]). *If $\mathcal{R}(\mathbf{H}) \subset \mathcal{R}(\bar{\mathbf{H}}) + \mathcal{A}$, then for any $\mathbf{x}$, there exists an unobservable topology attack for the target $\bar{\mathcal{G}}$. Otherwise, for $\mathbf{x} \in \mathbb{R}^n$ almost everywhere, there does not exist an unobservable topology attack for the target $\bar{\mathcal{G}}$.*

4.2 State-Preserving Attack

Kim and Tong in [22] present an unobservable topology attack that guarantees the *minimum* size of $\mathcal{S}_{\mathrm{A}}$ under certain conditions. The attack design is based on the changes in the sensor measurements when the topology changes while the system state stays the same. The attack is referred to as the *state-preserving attack* because it does not perturb the state estimate. There is no incentive for an attacker to preserve the state estimate, but this way of attack construction provides a *sparse* attack vector, which means a small cardinality of $\mathcal{S}_{\mathrm{A}}$ and thus a small cost for the attacker.

[2] Here, the summation in $\mathcal{R}(\bar{\mathbf{H}}) + \mathcal{A}$ denotes the sum of two subspaces, which is defined as the set of elements that can be expressed as a sum of two vectors, one from $\mathcal{R}(\bar{\mathbf{H}})$ and the other from $\mathcal{A}$. In other words, $\mathcal{R}(\bar{\mathbf{H}}) + \mathcal{A} \triangleq \{\mathbf{c}_1 + \mathbf{c}_2 : \ \mathbf{c}_1 \in \mathcal{R}(\bar{\mathbf{H}}), \ \mathbf{c}_2 \in \mathcal{A}\}$.

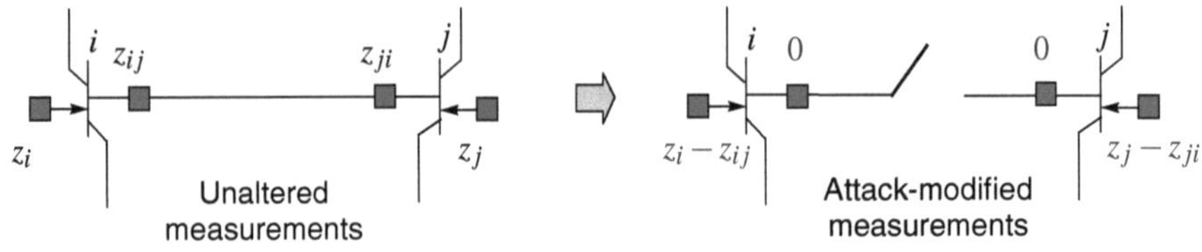

Fig. 3 The data-driven attack to remove the line $\{i, j\}$.

Given measurements $\mathbf{z}$ from a state $\mathbf{x}$, *i.e.*, $\mathbf{z} = \mathbf{H}\mathbf{x}$, the state preserving attack sets $\mathbf{a}$ to $(\bar{\mathbf{H}} - \mathbf{H})\mathbf{x}$. Then,

$$\bar{\mathbf{z}} = \mathbf{H}\mathbf{x} + \mathbf{a} = \bar{\mathbf{H}}\mathbf{x}, \tag{13}$$

and thus $\mathbf{a}$ is unobservable. Note that most rows of $\bar{\mathbf{H}} - \mathbf{H}$ are zero row vectors unless $\mathcal{G}$ and $\bar{\mathcal{G}}$ are drastically different. This means that $\mathbf{a}$ is sparse in general cases. To see this, note that the difference between $\bar{\mathbf{H}}$ and $\mathbf{H}$ exist in only the rows corresponding to the sensors that are located near the lines existing in only one of $\mathcal{G}$ and $\bar{\mathcal{G}}$ (see [22] for the details.) As far as the difference between $\mathcal{G}$ and $\bar{\mathcal{G}}$ involves only few lines, most rows of $\bar{\mathbf{H}} - \mathbf{H}$ will be zero row vectors thereby making $\mathbf{a}$ sparse.

4.3 Attack Based on Local Sensor Observations

While the state-preserving attack provides a sparse attack vector, the direct computation of $(\bar{\mathbf{H}} - \mathbf{H})\mathbf{x}$ seems to require an attacker to know $\bar{\mathbf{H}}$, $\mathbf{H}$, and the current state $\mathbf{x}$ (or equivalently, $\bar{\mathbf{H}}$, $\mathbf{H}$, and $\mathbf{z}$.) Surprisingly, if an attack aims to only *remove* some lines, the aforementioned information can be replaced with few local sensor measurements [22].

Fig. 3 illustrates the unobservable topology attack to remove the line $\{i, j\}$ based on the local sensor measurements. The attack observes the line flow measurement from bus i to bus j, denoted by z_{ij} in Fig. 3, and modifies the sensor data such that the line appears to be disconnected. Specifically, it modifies the breaker sensor data such that the breaker statuses indicate that the line is disconnected, and it sets the line flow measurements to zero and adjusts the bus injection measurements at bus i and bus j accordingly.

The above local data alteration is equivalent to the state-preserving attack and makes the altered measurements consistent with the topology without the line $\{i, j\}$ [22]. This can be easily seen by considering what change will happen to the line flows and the bus injections once a line gets disconnected while the state (*i.e.*, bus voltages) remains the same. An unobservable attack aimed at removing multiple lines can be constructed as a simple extension of this strategy.

5 Attack Impacts on Real-time Grid Operations

The state and topology estimates inform the control center about the real-time grid operating condition, and thus they are crucial inputs to many real-time functions at the control center. In particular, data attacks can have detrimental effects on the real-time economic dispatch and the real-time electricity pricing. This section introduces existing analyses of the potential attack impact on the real-time economic dispatch and the real-time electricity pricing.

The real-time economic dispatch (RTED) aims to find optimal generation adjustments to meet the demand of the next time period. A typical formulation of the RTED is as follows [7]:

$$\begin{aligned} \min_{\{g_i,\ i\in\mathscr{G}\}} \quad & \sum_{i\in\mathscr{G}} c_i(g_i) \\ \text{subject to} \quad & \sum_{i\in\mathscr{G}} g_i = \sum_{j\in\mathscr{D}} d_j. \\ & -\underline{\Delta g_i} \le g_i - \hat{g}_i \le \overline{\Delta g_i},\ \ \forall\, i\in\mathscr{G}. \\ & g_i^{min} \le g_i \le g_i^{max},\ \ \forall\, i\in\mathscr{G}. \\ & \left| \sum_{i\in\mathscr{G}} S_{ki}g_i - \sum_{j\in\mathscr{D}} S_{kj}d_j \right| \le F_k^{max},\ \ \forall \text{ line } k. \end{aligned} \tag{14}$$

The objective function is the sum of all generation costs, and $c_i(\cdot)$ and g_i denotes the cost function and the next-period generation level at bus i respectively. The sets $\mathscr{G}$ and $\mathscr{D}$ denote the set of generator buses and the set of load buses respectively. The first constraint is the balance constraint where d_j denotes the demand forecast at bus j for the next period. The second constraint is the ramping constraint which restricts the generation adjustment, $g_i - \hat{g}_i$, to be within a certain range where $\hat{g}_i$ denotes the estimate of the current generation at bus i obtained from the state estimate. The third constraint represents the generation capacity constraint. And, the fourth constraint is the line capacity constraint, where S_{ki} is the shift factor, which is the power flow change at line k due to a unit bus injection increase at bus i. The maximum power that the line k can deliver is denoted by F_k^{max}.

At the beginning of each period, the RTED sends out the optimal solution of (14), referred to as dispatch signals, to generation units such that they can adjust generations accordingly during the next period. Besides the demand forecast, the real-time topology and state estimates are key inputs to the RTED. The state estimate is used to obtain the estimates of the current generation levels, $\{\hat{g}_i,\ i\in\mathscr{G}\}$, which directly affect the ramping constraint. In addition, the topology estimate is used to obtain the shift factors in the line capacity constraints. Therefore, topology and state attacks can certainly affect the RTED result.

Another real-time function that significantly depends on the topology and state estimates is calculation of the real-time wholesale electricity prices. Nowadays, many independent system operators and regional transmission organizations (*e.g.*, PJM, ISO-NE, CAISO, ERCOT) offer the real-time wholesale electricity markets to reduce the peak electricity demand and resolve congestion in real time. While there exist a variety of real-time pricing mechanisms, PJM and ISO-NE employs the real-time ex-post locational marginal prices (LMPs). For the pth period, the real-time ex-post LMPs are calculated based on (i) the RTED solution for the pth period, (ii) the set of congested lines at the RTED solution, and (iii) the state estimate at the end of the pth period (see [34] for the detailed formulation.) Therefore, the ex-post LMPs are inevitably affected by topology and state attacks, and without a surprise, the impacts on the LMPs are closely related to the impacts on the RTED.

Several attack mechanisms have been presented to demonstrate the potential impacts of data attacks on the RTED and the real-time prices. Kim *et al.* in [25] present the dynamic data attacks that aim to make the feasible set of (14) empty such that the RTED cannot produce a generation schedule for the next period. Such lack of the RTED solution may signficantly increase the load and generation imbalance and necessitate the use of expensive fast-ramping generators. The dynamic attacks exploit the dependency of the RTED ramping constraint on the state estimate. In particular, the dynamic attacks perturb the ramping constraint throughout several RTED periods (by launching state attacks multiple times) such that the perturbed RTED solutions drive the system state toward a point where the RTED problem is infeasible.

Choi and Xie in [7] present an attack that aims to perturb the real-time LMPs. The proposed attack exploits both the dependency of the RTED ramping constraints on the state estimate and the dependency of the real-time LMPs on the RTED ramping constraints. They demonstrate that if an adversary deliberately alters the RTED ramping constraints by perturbing the state estimate, it can increase profits of certain load serving entities.

Xie *et al.* in [45] study a state attack that aims to maximize the attacker's profit by exploiting the virtual bidding mechanism. They assume that the attacker not only launches a state attack but also participates in the virtual bidding, *i.e.*, buys some energy from the day-ahead market and sells it in the real-time market or vice versa. Exploiting the dependency of the real-time LMPs on the state estimate, they present an optimization framework to design a state attack that will result in the maximum profit for the attacker.

Jia *et al.* in [17] study the worst case impact of data attacks on the real-time LMPs. They show that the state space can be partitioned into price regions, where each price region is a convex polytope, and each boundary between two price regions corresponds to a line capacity constraint. The real-time LMPs are determined purely based on which region the state estimate belongs to. Based on the state space partition, a state attack can move the state estimate into a different price region to perturb the real-time LMPs. In contrast to state attacks, a topology attack affects the real-time LMPs by restructuring the whole price regions. Jia *et al.* present numerical methods to find the maximum perturbation on the real-time LMPs that an adversary can create using either a state attack or a topology attack.

6 Security Measures against Unobservable Attacks

This section reviews the key ideas of the existing countermeasures. Most countermeasures are designed based on the feasibility conditions for unobservable attacks, presented in Theorem 2, Theorem 3, and Theorem 4. The main idea is to restrict the attacker's power (by protecting a subset of sensors) such that the feasibility conditions do not hold.

6.1 Security Measures against Unobservable State Attacks

Most popular approaches to preventing unobservable state attacks are based on *sensor data authentication*. Specifically, the control center protects certain subset of line flow and bus injection sensors such that their data cannot be altered by an adversary. The subset of sensors to protect is chosen such that the feasibility condition of an unobservable state attack in Theorem 2 (or Theorem 3) does not hold even when all unprotected sensor data are controlled by an attacker. Note that many devices in current grids are outdated and not able to support advanced authentication protocols [3]. Hence, implementing authentication for sensor measurements may incur a financial cost associated with hardware upgrades. Therefore, the number of the sensors to protect can be roughly considered as the cost of the protection strategy.

Bobba *et al.* in [6] propose protection of a set of *basic* sensors, which are defined as a minimum set of line flow and bus injection sensors that makes the gird observable. It is shown that protection of basic sensor measurements is necessary and sufficient for preventing unobservable state attacks, as formally stated in the following theorem.

Theorem 5 (Theorem 4.1, [6]). *An unobservable state attack does not exist if and only if a set of basic sensors are protected.*

Note that necessity and sufficiency of protecting basic sensors results directly from the feasibility condition in Theorem 3. Even when all unprotected sensors are removed, the grid is still observable based on the basic sensors. In addition, at least one set of basic sensors should be protected to guarantee the observability based on the protected sensors.

By definition, the cardinality of a set of basic sensors is equal to the dimension of the state vector, or equivalently one less the number of buses. Note that, in conjunction with Theorem 5, the spanning tree criterion in Theorem 1 provides a simple and optimal protection strategy: find a spanning tree of the topology, and protect one line flow sensor on each edge of the tree.

For the control center with limited amount of resource, even the spanning-tree protection might be costly. In such cases, the control center needs to prioritize protection of certain sensors such that the highest level of security can be achieved. Dán and Sandberg in [10] propose to find a set of protected sensors that maximizes the security index. Vuković *et al.* in [44] define a security metric that takes into account the SCADA communications infrastructure and the routing protocols, and a protection strategy is proposed to maximize the metric. It is shown that the routing protocols used by the SCADA network can be adjusted to increase the security level of the grid. Alternatively, Bi and Zhang in [5] propose to select a subset of state variables that are deemed important to the grid operations and prioritize sensor protection such that the estimates of the selected variables cannot be perturbed by an attack. Given the set of state variables to secure, they characterize a necessary and sufficient condition for protection and presented a protection strategy with the smallest set of sensors to protect.

The placement of protected phasor measurement units (PMUs) was also studied for protection against state attacks [12, 27]. PMUs provide more accurate measurements with much higher time resolution than legacy sensors, so they are considered as better alternatives to legacy sensors. Therefore, the control center has an incentive to protect PMUs rather than the legacy sensor devices. The experimental results in [27] suggest that placing protected PMUs to about a third of buses can prevent unobservable state attacks.

Once a protection strategy disables unobservable state attacks, any attack on state estimation will leave detectable traces in the corrupted measurements. A remaining important task is to figure out whether an unusual aspect of measurements is attributed to an attack or not. Therefore, a proper attack detector is needed to detect the presence of an attack quickly and accurately. Fast detection of an attack enables a timely response of the control center, and thus it is an important goal to meet. Huang *et al.* in [15] study the quickest detection of state attacks. They modify the classical CUSUM algorithm by Page [36] to handle the issue of an unknown statistical property of an attack vector. Similar approaches to the quickest attack detection can be found in [9, 14].

Besides protection based on sensor data authentication, there is an interesting alternative that can make an unobservable state attack detectable. Morrow *et al.* in [33] propose to change the impedance of certain lines by some degree and subsequently check whether the sensor data are consistent with the new impedance values. Assuming that an attacker is unaware of the impedance perturbation, an unobservable attack designed based on the old measurement matrix will not be consistent with the measurement matrix for the new impedance setting.

6.2 Security Measures against Unobservable Topology Attacks

Recall that Theorem 4 provides an algebraic feasibility condition with which we can check whether an unobservable topology attack for a certain target topology $\bar{\mathcal{G}}$ is feasible. If the control center restricts $\mathcal{S}_A$ (by installing data authentication protocols) such that the condition does not hold for any potential target topology, then the grid can be protected against unobservable topology attacks. Nevertheless, even verifying that the condition does not hold for every target topology is challenging, because the number of possible target topologies grows exponentially.

To avoid the computational burden, Kim and Tong in [22] present a simple graph-theoretical condition that guarantees protection against unobservable topology attacks.

Theorem 6 (Theorem 5.1, [22]). *Let $\mathcal{E}_0$ denote a set of lines such that $\{i, j\} \in \mathcal{E}_0$ if and only if there exists a transmission line, either connected or disconnected, between bus i and bus j. Let $\mathcal{T} = (\mathcal{V}, \mathcal{E}_T)$ denote a spanning tree of $(\mathcal{V}, \mathcal{E})$, and $\mathcal{B}$ denotes a vertex cover of the cotree[3] of $\mathcal{T}$ in $(\mathcal{V}, \mathcal{E}_0)$. If the control center protects a line flow sensor on each edge of $\mathcal{T}$ and bus injection sensors on the buses in $\mathcal{B}$, an unobservable topology attack does not exist.*

Note that finding a protection strategy satisfying the condition of Theorem 6 only involves finding a spanning tree and a vertex cover. Although the condition is not proven to be necessary, it provides a simple strategy that only requires additional protection of a vertex cover over the spanning tree protection; recall that the spanning tree protection is the minimum size protection to prevent unobservable state attacks, as implied by Theorem 5.

In the same spirit, placement of protected PMUs for preventing unobservable topology attacks was studied in [21]. In particular, the authors present a necessary and sufficient condition for protection, as stated in the following theorem.

Theorem 7 (Theorem 4.1, [21]). *Let $\mathcal{E}_0$ denote a set of lines such that $\{i, j\} \in \mathcal{E}_0$ if and only if there exists a transmission line, either connected or disconnected, between bus i and bus j. An unobservable topology attack does not exist if and only if the set of buses with protected PMUs is a vertex cover of $(\mathcal{V}, \mathcal{E}_0)$.*

Theorem 7 implies that designing an optimal PMU-based protection is equivalent to finding a minimum vertex cover of the topology. Finding a minimum vertex cover has been a long-standing problem in graph theory, and it was shown to be NP-complete [18]. Nevertheless, there exist many polynomial time approximation algorithms that we can leverage [37]. One of the most popular approximation algorithms is a greedy one that chooses the vertex with the maximum degree first [37]. In particular, the greedy algorithm starts with $\mathcal{V}_1 = \emptyset$ and works as follows:

- Step 1. Find the vertex v in $\mathcal{G}$ with the maximum degree and add the vertex v to $\mathcal{V}_1$.
- Step 2. Remove the vertex v and the lines incident to v from $\mathcal{G}$. If $\mathcal{G}$ has no line, return $\mathcal{V}_1$ and terminate. Otherwise, go to step 1.

Table 2 from [21] shows the cost of the PMU-based protection strategies for the IEEE test systems, found by the above greedy approach.

The relation between feasibility of an unobservable state attack and feasibility of an unobservable topology attack was also studied in [21]. In particular, it is shown that any protection strategy that prevents unobservable topology attacks can also prevent unobservable state attacks.

Theorem 8 (Theorem 3.1, [21]). *If an unobservable topology attack does not exist, an unobservable state attack does not exist.*

[3] Given a graph $(\mathcal{V}, \mathcal{E})$, a subset $\mathcal{V}_1$ of $\mathcal{V}$ is a vertex cover of the graph if every edge in $\mathcal{E}$ is incident to at least one vertex in $\mathcal{V}_1$. The cotree of $\mathcal{T}$ in $(\mathcal{V}, \mathcal{E}_0)$ is the graph obtained by removing all the edges in $\mathcal{T}$ from $(\mathcal{V}, \mathcal{E}_0)$.

Table 2 Placement of protected PMUs for IEEE test systems [21]

	number of protected PMUs	$\left(\frac{\text{number of protected PMUs}}{\text{number of all buses}}\right)$
IEEE 14	8	57 %
IEEE 118	61	52 %
IEEE 300	140	47 %

Theorem 8 implies that any protection strategy against unobservable topology attacks is sufficient for protecting the grid against unobservable data attacks.

The experimental results in [27] showed that we need to place protected PMUs to about a third of buses if we want to prevent unobservable state attacks. Comparing it with the cost of PMU-based protection in Table 2, one can see that there is about 17% increase in the fraction of protected buses, which is the additional cost for further protection against topology attacks.

7 Summary and Open Problems

This chapter presented an adversary model of data attack on power system topology and state estimation, and attack mechanisms and security measures were reviewed. In particular, the chapter focused on unobservable attacks, which can pass the bad data detection at the control center. Feasibility conditions for unobservable attacks were presented, and cost-effective attack mechanisms were discussed. In addition, we reviewed data-driven attack mechanisms, which can be employed by an attacker with little or no knowledge of the grid topology and parameters.

We reviewed a variety of security measures that aim to disable unobservable attacks. Especially, protection strategies based on sensor data authentication were reviewed with the discussion of the associated cost. For the control center with limited security resource, we reviewed several approaches to allocating the resource in an optimal way.

While a great amount of effort has been made to understand data attacks, there are still open problems that need to be addressed. Note that most existing countermeasures have focused on preventing unobservable attacks. However, Kim *et al.* pointed out in [24] that such countermeasures are not sufficient. Even when an adversary cannot control enough sensors to launch an unobservable attack, it may be able to launch an *observable* attack which will be detected by the control center but still mislead the control center with a biased state estimate. For instance, the data framing attack in [24] injects bad data such that certain subset of normal sensors are framed as malfunctioning sensors while some adversary-controlled sensors appear to be normal. Such an attack renders state estimation affected by the adversary-controlled sensor data that remain undetected. It was shown in [24] that the data framing attack can successfully perturb the state estimate with a half of the adversary-controlled sensors that are required for an unobservable attack.

To truly eliminate the attack effect, the control center needs to be able to remove the bias in the state estimate introduced by the attack. Tajer *et al.* in [42] presented a Bayesian estimation framework to estimate the attack vector, but their works rely on the assumption that the prior distribution of the attack vector is known, which is not so practical. To our best knowledge, not much effort has been made in this direction.

References

1. Final Report on the August 14, 2003 Blackout in the United States and Canada: Causes and Recommendations. U.S.-Canada Power System Outage Task Force (2004)
2. Common Cybersecurity Vulnerabilities in Industrial Control Systems (2011)
3. Vulnerability Analysis of Energy Delivery Control Systems. INL/EXT-10-18381 (2011)
4. Abur, A., Expósito, A.G.: Power System State Estimation: Theory and Implementation. CRC (2000)
5. Bi, S., Zhang, Y.: Defending mechanisms against false-data injection attacks in the power system state estimation. In: 2011 IEEE GLOBECOM Workshops, Houston, TX, USA, pp. 1162–1167 (2011)
6. Bobba, R.B., Rogers, K.M., Wang, Q., Khurana, H., Nahrstedt, K., Overbye, T.J.: Detecting false data injection attacks on DC state estimation. In: First Workshop on Secure Control Systems, CPSWEEK 2010, Stockholm, Sweeden (2010)
7. Choi, D.H., Xie, L.: Ramp-induced data attacks on look-ahead dispatch in real-time power markets. IEEE Transactions on Smart Grid 4(3), 1235–1243 (2013)
8. Coleman, T.F., Pothen, A.: The Null Space Problem I. Complexity. SIAM J. Alg. Disc. Meth. 7(4), 527–537 (1986)
9. Cui, S., Han, Z., Kar, S., Kim, T., Poor, H., Tajer, A.: Coordinated data-injection attack and detection in the smart grid: A detailed look at enriching detection solutions. IEEE Signal Processing Magazine 29(5), 106–115 (2012)
10. Dán, G., Sandberg, H.: Stealth attacks and protection schemes for state estimators in power systems. In: Proc. IEEE 2010 SmartGridComm, Gaithersburg, MD, USA, pp. 214–219 (2010)
11. Esmalifalak, M., Nguyen, H., Zheng, R., Han, Z.: Stealth false data injection using independent component analysis in smart grid. In: IEEE International Conference on Smart Grid Communications, pp. 244–248 (2011)
12. Giani, A., Bitar, E., Garcia, M., McQueen, M., Khargonekar, P., Poolla, K.: Smart grid data integrity attacks. IEEE Transactions on Smart Grid 4(3), 1244–1253 (2013)
13. Handschin, E., Schweppe, F.C., Kohlas, J., Fiechter, A.: Bad data analysis for power system state estimation. IEEE Trans. Power Apparatus and Systems PAS-94(2), 329–337 (1975)
14. Huang, Y., Esmalifalak, M., Nguyen, H., Zheng, R., Han, Z., Li, H., Song, L.: Bad data injection in smart grid: attack and defense mechanisms. IEEE Communications Magazine 51(1), 27–33 (2013)
15. Huang, Y., Li, H., Campbell, K., Han, Z.: Defending false data injection attack on smart grid network using adaptive CUSUM test. In: 2011 45th Annual Conference on Information Sciences and Systems (CISS), pp. 1–6 (2011)
16. Hull, J., Khurana, H., Markham, T., Staggs, K.: Staying in control: Cybersecurity and the modern electric grid. IEEE Power and Energy Magazine 10(1), 41–48 (2012)
17. Jia, L., Kim, J., Thomas, R., Tong, L.: Impact of data quality on real-time locational marginal price. IEEE Transactions on Power Systems 29(2), 627–636 (2014)

18. Karp, R.M.: Reducibility Among Combinatorial Problems. In: Complexity of Computer Computations, pp. 85–103 (1972)
19. Khaitan, S., McCalley, J.: Cyber Physical System Approach for Design of Power Grids: A Survey. In: IEEE PES General Meeting, Vancouver, BC, pp. 1–5 (2013)
20. Khaitan, S., McCalley, J.: Design Techniques and Applications of Cyber Physical Systems: A Survey. IEEE Systems Journal, 1–16 (2014)
21. Kim, J., Tong, L.: On phasor measurement unit placement against state and topology attacks. In: IEEE International Conference on Smart Grid Communications (2013)
22. Kim, J., Tong, L.: On topology attack of a smart grid: undetectable attacks and countermeasures. IEEE Journal on Selected Areas in Communications 31(7), 1294–1305 (2013)
23. Kim, J., Tong, L., Thomas, R.J.: Data framing attack on state estimation with unknown network parameters. In: The 47th Asilomar Conference on Signals, Systems, and Computers, pp. 1388–1392 (2013)
24. Kim, J., Tong, L., Thomas, R.J.: Data Framing Attack on State Estimation. IEEE Journal on Selected Areas in Communications 32(7) (2014)
25. Kim, J., Tong, L., Thomas, R.J.: Dynamic Attacks on Power Systems Economic Dispatch. In: The 48th Asilomar Conference on Signals, Systems, and Computers, Pacific Grove, CA (2014)
26. Kim, J., Tong, L., Thomas, R.J.: Subspace Methods for Data Attack on State Estimation: A Data Driven Approach. ArXiv e-prints, arXiv:1310.7616 (2014)
27. Kim, T., Poor, H.: Strategic protection against data injection attacks on power grids. IEEE Transactions on Smart Grid 2(2), 326–333 (2011)
28. Kosut, O., Jia, L., Thomas, R.J., Tong, L.: Malicious data attacks on the smart grid. IEEE Transactions on Smart Grid 2(4), 645–658 (2011)
29. Krumpholz, G.R., Clements, K.A., Davis, P.W.: Power system observability: a practical algorithm using network topology. IEEE Trans. Power Apparatus and Systems 99(4), 1534–1542 (1980)
30. Liu, Y., Ning, P., Reiter, M.K.: False data injection attacks against state estimation in electric power grids. In: Proceedings of the 16th ACM Conference on Computer and Communications Security, pp. 21–32 (2009)
31. Monticelli, A.: State Estimation in Electric Power Systems: A Generalized Approach (Power Electronics and Power Systems). Springer US (1999)
32. Monticelli, A., Wu, F.F.: Network observability: Theory. IEEE Trans. Power Apparatus and Systems 104(5), 1042–1048 (1985)
33. Morrow, K., Heine, E., Rogers, K., Bobba, R., Overbye, T.: Topology perturbation for detecting malicious data injection. In: 2012 45th Hawaii International Conference on System Science (HICSS), pp. 2104–2113 (2012)
34. Ott, A.L.: Experience with PJM market operation, system design, and implementation. IEEE Trans. Power Systems 18(2), 528–534 (2003)
35. Paar, C., Pelzl, J.: Understanding Cryptography: A Textbook for Students and Practitioners. Springer (2010)
36. Page, E.S.: Continuous inspection schemes. Biometrika 41(1/2), 100–115 (1954)
37. Paschos, V.T.: A survey of approximately optimal solutions to some covering and packing problems. ACM Comput. Surv. 29(2), 171–209 (1997)
38. Rahman, M., Mohsenian-Rad, H.: False data injection attacks with incomplete information against smart power grids. In: IEEE Global Communications Conference, GLOBECOM (2012)
39. Sandberg, H., Teixeira, A., Johansson, K.H.: On security indices for state estimators in power networks. In: First Workshop on Secure Control Systems, CPSWEEK 2010, Stockholm, Sweeden (2010)

40. Scambray, J., McClure, S., Kurtz, G.: Hacking Exposed: Network Security Secrets and Solutions, 2nd edn. McGraw-Hill (2000)
41. Stoica, P., Nehorai, A.: MUSIC, maximum likelihood, and Cramer-Rao bound. IEEE Transactions on Acoustics, Speech and Signal Processing 37(5), 720–741 (1989)
42. Tajer, A., Kar, S., Poor, H., Cui, S.: Distributed joint cyber attack detection and state recovery in smart grids. In: 2011 IEEE International Conference on Smart Grid Communications (SmartGridComm), pp. 202–207 (2011)
43. Tillmann, A., Pfetsch, M.: The computational complexity of the restricted isometry property, the nullspace property, and related concepts in compressed sensing. IEEE Transactions on Information Theory 60(2), 1248–1259 (2014)
44. Vukovic, O., Sou, K.C., Dan, G., Sandberg, H.: Network-aware mitigation of data integrity attacks on power system state estimation. IEEE Journal on Selected Areas in Communications 30(6), 1108–1118 (2012)
45. Xie, L., Mo, Y., Sinopoli, B.: Integrity data attacks in power market operations. IEEE Transactions on Smart Grid 2(4), 659–666 (2011)

Author Index